The War against the Seals

By the same author

Britain and the Persian Gulf, 1894–1914

Britain, India, and the Arabs, 1914–1921

Mudros to Lausanne: Britain's Frontier in West Asia, 1918–1923

Hardinge of Penshurst: A Study in the Old Diplomacy

Master of Desolation: The Reminiscences of Captain Joseph J. Fuller

Alta California, 1840–1842: The Journal and Observations of William Dane Phelps, Master of the Ship "Alert"

The War
against the Seals

A History of
the North American
Seal Fishery

BRITON COOPER BUSCH

McGill-Queen's University Press
Kingston and Montreal

© McGill-Queen's University Press 1985
First paperback edition 1987
ISBN 0-7735-0578-4 (cloth)
ISBN 0-7735-0610-1 (paper)

Legal deposit second quarter 1985
Bibliothèque nationale du Québec

Printed in Canada

CANADIAN CATALOGUING IN PUBLICATION DATA

Busch, Briton Cooper
    The war against the seals

    Bibliography: p.
    Includes index.
    ISBN 0-7735-0578-4 (bound)   ISBN 0-7735-0610-1 (paper)

    1. Sealing – United States – History.   2. Sealing –
    Canada – History.   3. Sealing – History.   I. Title.

SH360.B87 1985   639'.29'0973   C85-098367-3

The lines from Kathleen Raine's "Eileann Chanaidh" are reprinted
with permission from *Collected Poems* (1981) published by George
Allen & Unwin.

for my father

NIVEN BUSCH

who also writes books

Contents

Illustrations

FIGURES

Tables

Preface

To conceive of this book was not difficult. While transcribing the hand-written memoirs, fascinating to me, of an American elephant sealer of the mid-nineteenth century (subsequently published in the Mystic Seaport Museum American Maritime Library series as *Master of Desolation*), I combed the shelves of that haven to maritime researchers, the G.W. Blunt White Library at Mystic, for a general history of the North American sealing industry. It was not there; it had never been written.

While occasional logs and journals exist, sealers themselves seldom wrote of their trade. Whalemen over the course of a long career would probably visit all the major whaling grounds, but sealers more commonly specialized, whether ice-bound Newfoundlanders stalking harp or hood seals off the Labrador, Aleut sealskinners on the Pribilof Islands in the Bering Sea, or "elephanters" on aptly named Desolation (Kerguelen) Island in the Indian Ocean; perhaps they saw no point in writing of what to them was commonplace. Moreover, since seals, at least fur seals, were a finite resource, fixed in space at some home rookery during their breeding season – however great their annual migration in the intervening months – sealers were a secretive tribe, reluctant to reveal sources of supply, especially in the later days when sealing might be illegal. Still, no part of the industry has been without its commentators and historians, and three aspects have received substantial attention: early nineteenth-century sub-Antarctic sealing, North Pacific pelagic sealing in the 1890s, and Newfoundland sealing since the 1960s. Biologists, diplomatists, ecologists, sociologists, economists: the industry touches so many aspects of society as to demand an impossible accumulation of expertise from the professional historian. Nevertheless, with all due apologies for interloping in fields not my own and for inevitable errors, I have assayed the task.

The origin of the hunt by man for any wild creature by land or by sea may be perceived only dimly, and that for seals is no exception. Where there were seals and man, the latter killed the former, for food, for fat to burn as fuel, for fur clothing or leather harness. Middens and shell mounds reveal to archaeologists and anthropologists what coastal peoples ate, and how they caught their prey. Small seal pup bones (needing careful work, being the most fragile), for example, testify to raids on rookeries, not harpooning at sea. But by "sealing" in this book is meant the commercial exploitation of seal herds by man, for the organized production of profit, whether that profit be in the fur, the skin, the fat, or – least common of all – the meat for consumption, human or animal. I have therefore concerned myself only incidentally with subsistence sealing, such as that of my Newfoundland "landsmen" informants, some of whom are really only interested in meat for their own freezers. The same is true where seals have been hunted simply as a nuisance, real or perceived, such as harbor seals and sea lions from British Columbia to California, regarded as rivals for fish to commercial fishermen.

My story, therefore, is first of the New England sealers of the late eighteenth and nineteenth centuries who hunted the world for fur seals, and in the hunting brought several species close to extermination before turning their attention to elephant seals for their oil. Second, I have been concerned with the Newfoundland sealing industry over the course of the nineteenth and twentieth centuries. Third, I have studied the exploitation of the Alaska or northern fur seal of the Pribilof Islands. Though some discussion of the Russian era of Bering Sea control is essential, my focus has been the post-1867 years following America's purchase of Alaska. Finally, I have explored the lesser-known but also significant sealing in the Californias, Alta and Baja, Australia and New Zealand – insofar as North Americans were concerned – and the Indian Ocean. Though Norwegians and Russians and Scots and Japanese and Argentinians and Australians and South Africans (and that is not a complete list) have all at one time or another "gone a-sealing" in the sense of this book, a history of sealing by the peoples of the world must await another day: as I hope to show, North Americans, whether from Canada or the United States, made enough impact on a worldwide scale to merit special attention.

A word needs to be said about humanity, and about morality. Some consideration is given to the first in this book. For example, it seems clear that the quickest and least painful way to kill an infant seal is a solid blow on the snout – it is the most "humane" way. Whether it is moral is a different issue, one which each reader must face in his or her own way; I have tried to avoid this aspect, at least until the epilogue. Sealing in the late twentieth century is an emotion-charged issue for many people, and with good reason – though justice, if that is the right word, is not all on one side. For myself, I have never killed a seal, and I very much doubt I ever will. Though a hunter of land game in the family enterprises of my youth, I have not as an adult hunted wild animals. I do eat meat – even lamb chops, if the price is reasonable – with little or no thought to the source of the repast.

The first confession will have brought some Newfoundlanders of my acquaintance to throw the book down in disgust, unread, with the cry that there is no earthly way that I could understand what sealing means to them. I have talked with others, in my experience even more fanatic in defense of their cause, who see little to distinguish me, as eater of the flesh of his fellow creatures, from the bloodthirsty pup-slaughtering sealers they so dearly wish to confront. But I have tried, as much as I can, to stand off and grapple with the question of sealing over past generations with the detachment expected of a professional historian.

I do not think anyone can study and write about some aspects of this industry with total dispassionateness; as the reader will discover, my own inclination is to pull for the seals. But to leave it at that would not be fair. In the nineteenth century, which is the focus of this work, sealing was vital to the lives of many people, and dominated the Aleuts of the Pribilofs. It had a similar role in parts of Newfoundland. The merchant wealth of more than one town was built on furs or fat, with St. John's, Victoria, and New London chief among them. For the historian who is describing that role, sealing ought to be as susceptible to impartial historical analysis as other industries such as mining or forest products – both of which are also subject to heated controversy among observers. In the preparation of this book it has been necessary to meet and talk freely with – and it is hoped, to understand – partisans of every position. I trust I have been able to give as much sympathy and understanding to the "swilers" of Newfoundland as to the Greenpeace cadre at their Vancouver headquarters who told me how wrong the sealers were.

Such discussions were the joys of creating this work (the sorrows will emerge in the course of the text). I take much pleasure in, and give warm thanks for, the new acquaintance of so many people in varied walks of life scattered about the world. Above all, I owe a special debt to the natural scientists who showed me or told me about marine mammals and their habits, whether on the site or writing from distant corners (and also explaining, from time to time, how wrong their colleagues were). Sealers may have been inarticulate, but biologists are not. The reader should not be deceived by the weight of citations from the natural sciences in my notes and bibliography. I am not a mammalogist, and I have relied upon their work extensively, though not always accepting their view of history. But this book would not have been possible without the help of those natural scientists; I hope they will not have cause to regret their contribution to a work which must really fall into the social sciences – and perhaps ultimately in the humanities.

Readers who are put off by lists of acknowledgments should press on at this point. I owe much to many people, doubtless too many to list, but I intend listing them anyway, with heartfelt apologies to those kind individuals who helped me but find themselves now depersonalized under some institutional title, and even more to those who are unaccountably omitted. First, the book would have not have been possible without the assistance of many Canadians from coast to coast. In Newfoundland, my thanks to the staffs of the Provincial Archives, the Centre for Newfoundland Studies and the Folklore Archive of the Memorial University of Newfoundland, the Maritime History Group (upon whom I imposed in the 1981 conference), Bobbie Robertson of the Newfoundland Historical Society, Cassie Brown, and Dennis Monroe. Ian McLaren of Halifax helped with the results of his seal research. In Ottawa, I benefited much from the holdings of the National Library and Public Archives, and the library of the Department of Fisheries and Oceans. My thanks also to T. Hsu of the Statistics Group, Economic Policy Branch, of the latter department. In Vancouver, I was assisted by the directors of Greenpeace who put their files at my disposal, as well as the staffs of the Vancouver Maritime Museum and the Library of the University of British Columbia. The extensive holdings of the Provincial Archives of British Columbia in Victoria were of great assistance, as were the suggestions from Barry Gough on their use. I must also thank in the same city R.R. Godden of the Victoria Maritime Museum, the University of Victoria library, and Jim Boutelier for his hospitality.

In the United States, I benefited much from discussions with participants in a conference on the sea in Alaska's past held in Anchorage in 1979. In Seattle, my thanks to Delphine Haley; the staff of the Marine Mammal Laboratory (especially Charles Fouler, Michael Tillman, Cliff Fiscus, Victor Scheffer, Anne York, Hiro Kajimura, Roger Gentry, and Karl Kenyon); Walter Kirkness of the Pribilof Island Program; the Federal Archives; the University of Washington Library Northwest Collection; the Coast Guard Library; the Washington Historical Society, and, in Portland, the Oregon Historical Society.

My several visits to California were much brightened by the hospitality of Michael and Carol Leach, Bea and Winston Miller, Art and Barbara Balinger, and Sue and Niven Busch. My thanks also to the Librarians of the Bancroft Library, the California Historical Society, the Santa Barbara and San Diego Historical Societies, the Huntington Library, the Monterey Public Library, Stanford University Library (Special Collections and Archives), the Federal Archives in San Bruno, the California Department of Fisheries Library in Long Beach, the National Maritime Museum (formerly San Francisco Maritime Museum); Burney J. Le Boeuf of the University of California, Santa Cruz; Jane Bailey and the Friends of the Sea Otter in Monterey; Adele Ogden; and my fellow participants in the conference held at San Jose State University in 1980 on West Coast Maritime History. Stella Rouse very kindly put her wealth of old Santa Barbara lore at my disposal.

In New England, I owe a special debt to the staff of Mystic Seaport Museum, particularly of the G.W. Blunt White Library. I cannot list all the people who helped at Mystic, but I owe a special debt to Bill Peterson, who over the years has watched for references to sealing for me in his own researches on the history of the New London area. I have benefited also from responses to papers delivered at the Southern New England Maritime history symposium held at Mystic, and an address which I delivered to the New London County Historical Society. The Kendall Whaling Museum in Sharon, Massachusetts (with special thanks to Carol Tobol), and the Old Dartmouth Historical Society in New Bedford provided valuable source material, as did the New Bedford Public Library. Dave Henderson of that city lent me his extensive knowledge of activities along the Baja California coast. The Federal Archives at Waltham, as well as the Harvard University Libraries (especially the Houghton and Baker Libraries) were invaluable. I owe warm thanks also to Sandra and Ken Martin, the latter especially for sharing a memorable voyage to Nantucket to benefit from the expertise of Edouard Stackpole.

Much contemporary information on marine mammals and American policy can only be found in Washington, D.C., and I would like to express my gratitude to the Marine Mammal Commission (especially Peter Major); the Office of Marine Mammals and Endangered Species (Georgia Cranmore); the Library of Congress Manuscripts Division; the Smithsonian Archives (William Diess); the National Archives (Renée Jusserand); G. Carleton Ray of the University of Virginia, and Clayton E. Ray of the Smithsonian Institution, who introduced me to the Marine Mammal Group and gave me free run of the Smithsonian's invaluable Kellogg Library. Craig van Note of Monitor, the conservation, environmental, and animal welfare consortium, gave me some excellent suggestions as well.

By sometimes lengthy correspondence, experts around the world have given me the benefit of their knowledge: Rhys Richards in New Zealand, J.S. Cumpston in Australia, Peter Best in South Africa, and in England, L. Harrison Matthews. Some of the more obscure printed sources used in this work were to be found only at the British Library. I owe a particular debt to the overseers of the Science Library (also part of the British Library), who – no doubt in desperation as my requests began to fill cart after cart – gave me freedom to roam among their periodical holdings.

I owe a different sort of debt to a number of Colgate University undergraduates who over the last few years labored on often thankless tasks connected with this book: Kirk Beckhorn, Catherine Campbell, Julia Digel, Fern Fryer, Hugh Jones, Leslie Keenan, Heather Korsvik, Kim Lake, Lynn Meyer, Gail Mulligan, Meg Timor, and Patti Takas. The staff of the Colgate University Library uttered no complaint at my inordinate use of interlibrary loan services. Lois Wilcox typed the manuscript with care and attention, while Rosalie Hiam maintained the illusion that I and not she was chairing the Department of History.

To all those cooperative people, my thanks indeed for their help, without which a book of this scope simply would not have been possible. My final and deepest appreciation, however, goes to those several experts who generously gave of their time to read and comment upon part or all of the manuscript in draft form: Burney Le Boeuf in California, David Sergeant in Montreal, and the team reading by the biologists and managers of the National Marine Mammal Laboratory in Seattle, whose always helpful but sometimes conflicting remarks did so much to reinforce my understanding of the difficulties in studying seals. They have saved me from numerous errors, large and small, but I take full responsibility for those that remain.

Hamilton, N.Y.   B.C.B.

PART ONE

Prologue

*And at this moment, with all the stars hidden
by clouds, as my walls and roofs were shaken by
the wind, as the sea roared hellishly below, the
ships cast loose from the quay and set forth on
their journey. One, perhaps, was bound for the
river Don, with passengers for the Ganges, the
Caucasus, the Indies, and the Eastern Ocean.
My heart bled for these unhappy men. And when
I could no longer follow the ships with my eyes,
moved and stirred I picked up my pen again,
exclaiming: "Oh, how dear to men is life, and
how little account they take of it!"*

Petrarch, quoted in
Morris Bishop, "Petrarch," in *Renaissance Profiles*,
ed. J.H. Plumb

CHAPTER ONE

Boston Men
and Stonington Sealers
1783–1812

*"Sea-dogs; for sea dogs is my sayin'.
They tell of seals getting scurse; but I say, it's all
in knowin' the business."*
Watson, the Stonington sealer, in
James Fenimore Cooper, The Sea Lions

The sea otter of the Northern Pacific is an amazing creature. Ever a delight to the observer, it is constantly on the move, diving, swimming, grooming, and above all feeding with what appears inexhaustible energy. So too is its hunger to feed that energy inexhaustible, for *Enhydra lutris* must eat a third of its body weight each day.[1] Otters select from among more than two dozen species of food resource from the near-shore kelp beds which are their home, but their preference is for those of high caloric value: sea urchins, abalone, crab. To get at its prey, an otter will crack mollusc shells with a stone, and many a fog-bound hunter was led to the otter he sought by the sound of stone on shell or the cries of gulls hungry for scraps.

And indeed otters have been hunted, for the animal's glory, and also its sorrow, is its loose, luxuriant pelt. William Sturgis, a famous trader of otter skins, left no doubt of his opinion in addressing a Boston audience in 1846: he said, according to a reporter, "that it would give him more pleasure to look at a splendid sea-otter skin, than to examine half the pictures that are stuck up for exhibition ... In fact, excepting a beautiful woman and a lovely infant, he regarded them as among the most attractive natural objects that can be placed before him."[2]

Otters are unique among God's creatures in that they inhabit frigid waters yet lack that layer of blubber beneath the skin which protects whale, seal, and walrus – indeed, every other sea-borne mammal. The otter's salvation lies instead in the incredible refinement of its pelt – some fifteen square feet in a full-grown adult. A vast number of tiny fur hairs (650,000 per square inch), lie flat on the animal and thus trap air pockets which keep the otter warm and buoyant. Any pollutant, such as petroleum, which interferes with that air layer means death from hypothermia. This fact explains both the otter's constant groom-ing and the difficulty of keeping the species in captivity (aside from their expensive appetites). Since condition of the pelt is all-important, the otter cannot moult in one sudden season as do some seals, but rather, like a human head of hair (and indeed, it is its anthropomorphic characteristics which tend to make the otter so attractive a creature to man), its pelt is replaced one hair at a time, and thus is always "prime." The result of this main fact of otter existence – a perpetually prime luxuriant coat – has much to do with the rise of commercial sealing, for it was as an adjunct of the northwest American luxury trade in furs to China that the first New England sealskin rush occurred.

The rise of American trade to China can conveniently, if arbitrarily, be dated to 1783, although individuals from the colonies had reached the Far East before then. But 1783 was the first year in which merchants of the newly independent United States of America were free to trade to the Orient; in the same year appeared the publication in Connecticut of John Ledyard's narrative of Cook's last Pacific expedition, in which Ledyard had participated as corporal of marines.[3]

Few paid any attention to Ledyard then, despite his account of Cook's accidental discovery of the immense value of otter pelts. First trading for them at then unnamed Vancouver Island in 1778 for winter clothing, Cook and his men were astonished in subsequent encounters with Rus-sian and Chinese merchants to learn of the enormous demand for these furs to ornament and line the clothing of North China's elite. With the publication of Cook's own account shortly thereafter (1785), enter-prising New Englanders suddenly saw the gleam of good profit, their desire for which was considerably enhanced by the troubled economic state of the new nation.

America at the time was badly in need of trade to replace the all-important West Indian connection, now officially closed by British regulation. Unofficial links were not adequate, and Atlantic coastal trade was no substitute, for much of it had been in commodities earmarked for transshipment away in the West Indian trade. The slave trade, too, was increasingly out of favor, as waves of independence and "rights of men" eddied onto those shores which traded in decidedly unfree indi-viduals. A general depression, intercolonial tariffs, and a desire for prestige for the new nation all contributed as well to the search for new means to economic development. At the same time, the British East India Company, which licensed trade beyond the Cape of Good Hope, and the South Seas Company, which included the Pacific Northwest in its domain, could no longer prohibit American trade to the Far East.[4]

The first recorded voyage, that of the New York vessel *Empress of China*, in 1784–5 by way of the Cape, found obstacles beyond Chinese restrictions on where and how trade was conducted. Above all, the Chinese were uninterested in much of what America had to offer in trade. Early cargoes were made up of commodities such as ginseng, an aromatic root of limited supply and demand used for medicinal purposes and Spanish silver dollars, for which demand was unlimited but supply dwindling. Clearly other trade items were required to exchange for the teas, silk, and porcelain so much in demand, and even the early New England vessels, their owners now alerted by Cook's *Journal*, shipped furs in the hope of finding a Chinese market. The sloop *Experiment*, sailing from New York in 1785, took a loss on squirrel skins, but the very mode of costume of northern China showed that the market must be there.[5] In fact, that same year the market was tapped by a British brig in the command of James Hanna which sailed from Macao to the northwest and back to China with a lucrative cargo of sea otter skins.[6]

The first Massachusetts ship to reach Canton, the *Grand Turk* of Salem (1786), carried mainly ginseng. The United States did not enter the otter trade until the joint voyage of the *Columbia Rediviva* and the *Lady Washington* under Captains Robert Grey and John Kendrick, and the New York brig *Eleanora* of Simon Metcalf, reached the northwest. These three vessels traded around Nootka Sound in 1788–9, and Grey's *Columbia*, returning from Canton to Boston in 1790, reported that otter skins were the answer – though because a number of tea-carrying vessels had beaten him home and saturated the market his own voyage was not notably successful. Others soon followed in the wake of these pioneers, and by the turn of the century a dozen or more vessels were active in the trade – winning some substantial fortunes for the participants, among whom was the above-mentioned William Sturgis, a foremast hand on his first visit of 1798–9, but by 1812 head of the leading mercantile firm of Bryant and Sturgis.[7]

British rivals attempted to compete, but the dual pressure of mercantilistic regulations and the long struggle with France soon eliminated their initial advantage. That Boston merchants like Sturgis were well placed socially and politically did not hurt either; John Quincy Adams (senator, 1803–8; first minister to Russia, 1809–12), for example, was close to many of the northwest merchants.[8] A tabulation of ships operating on the northwest coast by Raymond Rydell tells the story:[9]

	AMERICAN VESSELS	BRITISH VESSELS
1785–94	15	35
1795–1804	68	10
1805–14	43	3
1815–19	54	4
1820–5	53	1

The approximately 150 vessels making over 200 trips to the northwest coast meant enough men and gunpowder to insure displacement of both Spanish and British interests and the growth of a maritime presence of vast import in the later history of the area.[10]

Sealskins, too, were to be an important item in the China trade, but the American search for seals – in this case for oil – predates that for fur seal skins or even sea otters. Shortly after the Revolution began, several leading American whaling owners organized a fleet to cruise off the Brazil coast and then rendezvous in the Falkland Islands for the winter of 1775, where they would complete their cargoes with oil from elephant seals – the whole to be sold in English markets by Nantucket whaling entrepreneur Francis Rotch, who had arranged for British protection. The object of Rotch's plan was to protect the whaling fleet from the obvious dangers of war, but the result was probably the first planned expedition to the Falklands to exploit elephant seals. Despite a promising beginning, the imposition of a substantial British import duty on alien oil forced the American whaling/sealing owners to look elsewhere.[11]

Some New England merchants turned to otters, but Rotch, who had spent much of the Revolutionary War in self-imposed Falklands exile, stuck to sealing. In 1784, he fitted out his family's ship *United States* at Nantucket for elephant sealing in the Falklands under Capt. Benjamin Hussey. Although big for such work at 1,000 tons, the *United States* brought to New York in 1786 some 13,000 fur seal skins as well as 300 tons of elephant oil. The fur seal skins were sold and put aboard the *Eleanora* and taken the same year to Calcutta and on to Canton, thus originating the American fur seal trade with China.[12]

These skins were of the southern fur seal, *Arctocephalus australis*, known more familiarly to the early explorers as "sea bears" for their coats. Once they lived in vast numbers on the Falklands and the adjoining coasts – both eastern and western – of southernmost South America. Both fur seals and sea lions are of the family Otariidae (eared seals). This family differs physically from Phocidae (true seals, harbor seals, grey seals, etc.) mainly in the visible appearance of a small pointed ear, although other substantial differences exist in head shape, means of locomotion, and habitat. For sealers, the important distinction within the family of eared seals was the denser and softer underpelt which fur seals had and sea lions – called at first "sea wolves" and later "hair seals" – did not. The term "hair seal" unfortunately is most confusing, for while applied regularly to sea lions (in this case, the South American species, *Otaria byronia*), it is also used by sealers in reference to the harp seal of Newfoundland, a member of the Phocidae family and an important part of the history of sealing. Sealers distinguished between coarse, unmarketable "hair" pelts and softer "fur" skins, and only then between varieties of fur seal according to value of pelt, and, finally, between bulls or "wigs" (because of the curly hair on their heads?) and females or "clapmatches" (perhaps from the scraping sound of movement over rocks, akin to early flint on steel). Though over thirty species of seal, sea-lion, and walrus exist, fortunately for the nonscientific historian only about nine of fur seal, several of true seal, and the one ubiquitous walrus have been exploited commercially by North Americans.

Americans were by no means the first large-scale purveyors of seal-skins to Imperial China. That honor goes to the Russians, and involves a fur seal species of considerable importance to this story, *Callorhinus ursinus*, the northern fur seal.[13] Russia quite early on traded furs from Siberia to China, but fur seals were not part of the exchange until the *promyshlenniki*, the independent fur hunters, pushed outward from Kamchatka to the Aleutians following Vitus Bering, a Danish explorer in the service of the Russian crown. Bering met his death in 1741 on Bering Island – where his grave is still marked by a rusty chain and a few cannonballs[14] – which, with Cooper Island, forms the Commander Islands (Komandorski Ostrova), still in the late twentieth century a breeding-ground for the northern fur seal. Skins taken from these islands proved useful in Russia's China trade, but a fur seal skin worth three roubles ($1) or less in China was hardly to be equated with an otter skin worth up to fifty times that sum, and the Russians pushed on wherever otters were to be found.[15] Along the way, navigator Gerassim Privilof discovered the islands subsequently named for him – and the Pribilofs were the most important breeding-ground of all for fur seals in North Pacific waters.

The Pribilofs were uninhabited at the time of their discovery in 1786, but the fur-hunters soon transferred enough Aleuts from their home islands to the south to these new-found rookeries to exploit the resource. Meanwhile the chase for otters had led the Russians southward along the Alaskan coast, and in the process the populations of both otters and Aleuts were decimated, the latter often in the forced hunt of the former. Unlike coastal tribes, the Aleuts had no interior, no forest preserves, to which to retreat in defense. Otters were similarly vulnerable, and in the half-century following 1740, roughly a quarter-million were taken from these waters by Russians and their competitors. No completely accurate figure can be calculated given the considerable wastage from poorly prepared skins, the many otters killed but unrecovered, and smuggled skins, all of which went unrecorded. Those which were recorded were worth in total at least some $50 million; by comparison, the approximately 2.5 million fur seals skins taken in the same era were worth only $1 each.

The whole process of seal and otter hunting was disorganized and wasteful, so in 1799 St. Petersburg gave the Russian-American Company a twenty-year monopoly on the area's trade and resources. Tours of inspection were inaugurated and larger plans of development outlined which, following Russia's "passion for distance,"[16] included not only the northwest coast but also California and Hawaii. Alas for such designs, the Boston men had already largely taken control of the trade. Nevertheless, the Russians developed their administrative center of Sitka and in 1812 established their outpost at Fort Ross, some sixty-five miles north of San Francisco, which for a quarter of a century served as a base from which to hunt sea otters. Even then, as will be seen, the Russians could not provide the necessary sea power at such distances.

Meanwhile, Russia operated a profitable, if secondary, trade in seal-skins.[17] The fur seal's pelt, though far less luxuriant than that of the sea otter or many land fur-bearing mammals, is nevertheless suitable for fur clothing, once properly prepared. In that phrase lies the problem, for the pelt is composed not only of the soft hairs lying close to the skin but also of a layer of stiffer outer guard hairs which keep the soft underfur from matting flat on the body and must be separated from it. An economical way to do this seems not to have existed much before 1750 and is generally regarded as a Chinese invention. Fur seal pelts need not be used directly for clothing, however; the fur itself was sometimes taken off the actual skin, particularly where the skin was badly preserved (above all if sun-dried rather than wet-salted) and the fur was thus likely to come off in any case. Loose fur could be compressed for the preparation of felt, a common and popular Chinese clothing material.

Sealskins were reaching Chinese markets from Russian sources at least by the end of the eighteenth century (there was little indigenous Russian market for Bering Sea sealskins while yet there were otters to take). It was thus no surprise that American traders soon thought of fur seal skins as an item of trade. There was little point in trading for skins from the Russians, and less in attempting to visit the Pribilofs, whose existence, let alone location, was scarcely known. Sealskins were available after all for the taking on the outward journey to the northwest, on both sides of the South Atlantic. Though the fur seals of South West Africa (*Arctocephalus pusillus*) differ slightly from *A. australis* of

South America, as usual sealers cared only if there was a distinct price differential.

At least as early as 1610, African seals were taken by the Dutch for oil and hides.[18] On the opposite coast, Uruguayan seals which frequent the Lobos and other small island groups in those waters had been cleaned off by the time of the first protective legislation in 1820.[19] Developments on the South Atlantic islands inhabited by *A. australis* are somewhat easier to document, at least where American vessels are concerned. In 1790–1, for example, the ship *Industry* of Philadelphia took 5,000 skins from Tristan da Cunha.[20] The Falklands by that time had been long exploited, one more reason for European states to compete for these strategic islands, "one of the most frequently discovered, named and forgotten groups in all the seas," as Hugh Mill remarked.[21] Many might agree with Samuel Johnson, writing in 1771, on the absurdity of warring with Spain (or Argentina) over "the empty sound of an ancient title to a Magellanic rock, an island thrown aside from human use, stormy in winter, barren in summer, an island which not even the southern savages have dignified with habitation, where a garrison must be kept in a state that contemplates with envy the exiles of Siberia, of which the expense will be perpetual and the use only occasional, a nest of smugglers in peace, in a war a refuge of future buccaneers."[22] But when Johnson wrote, the "seal rush," like the extensive trade around the Horn in general, had yet to come, and while both lasted, the Falklands had considerable value.

The first sealers at the Falklands were not totally effective in cleaning off what is, after all, an extensive group of islands, for as the first big cargoes were taken richer harvests awaited the sealers at still unexploited rocks along the South American coast. From the Falklands it was no great leap to Tierra del Fuego where British sealers were active by the late 1770s, and soon were clashing with Spanish patrol vessels from Montevideo. Statenland (Isla do los Estados) off Tierra del Fuego's tip had a substantial herd available to those willing to risk the rocks and surf – as was the master of the 250-ton brigantine *Hancock* of Boston in the 1790–1 season. The 400-ton *Butterworth* was probably following an established pattern when in the next year she left a party of sealers on the island to be retrieved with their catch of skins on the return voyage.[2]

Even whalers landed men to augment their catch with skins (on the other hand, sealers were not normally equipped to render whale blubber). For whaler, sealer, or merchantman, the addition of a partial cargo of 10–15,000 skins was an excellent form of insurance, particularly for China-bound merchants when otters became hard to find and alternatives, such as sandalwood or bêche-de-mer (sea slug), grew scarce in the early nineteenth century.

The easiest way to collect fur seals was to take them from the islands off western South America, for there were substantial rookeries on the Juan Fernández Islands – Más Afuera and the larger Más a Tierra, often known, confusingly, as Juan Fernández; since 1966 the islands are named Alejandro Selkirk and Robinson Crusoe, respectively, to honor Defoe's novel and its nonfictional model.[24] Seals also were to be found on the small and barren rocks of San Felix and San Ambrosio to the north. Here the take was not of the South America fur seal, *A. australis*, but two different species, *A. philippii* of the Juan Fernández Islands, and *A. townsendi* of Guadalupe Island and the Baja coast further north. Scientists, with few samples to work with, long disagreed whether these were separate species or subspecies, but all acknowledge the reason for large seal herds: their ability to feed on the rich upwellings from the Humboldt Current, including the same fish stocks which fed the gigantic bird flocks which in turn made the guano islands of this coast so famous.[25]

On the islands of Mocha and St. Mary's (Santa Maria), closer to the coast, large sea lion ("hair seal") rookeries were available, as well as nearby safe anchorages from which to engage in contraband trade to the coast, though the local reception varied unpredictably from warm welcome to seizure and arrest, depending upon local and international conditions. These islands – at least the ones further away from the coast – were conveniently located on the direct route to the northwest, despite the wide westward swing required to reach the Sandwich (Hawaiian) Islands where sealers, whalers, and traders replenished stores and often, particularly in the early days, wintered, thereby avoiding both the prevailing northerly winds down the western North American coast and hostile Spanish coastal authorities.

As with the otter trade of the northwest, the first large sealskin cargoes from the several islands were British, but the Americans were close behind and soon dominated the trade. In 1793, for example, Capt. Josiah Roberts's Boston ship *Jefferson* took 13,000 skins from St. Ambrose while the vessel lay "off and on," stopping only when the seals left the island[26] (pups are born in early summer – January, in this hemisphere, though the species has been little studied by mammalogists; the herd would be off the islands as a rule over the next few months). The biggest harvests were made at Más Afuera, whose vast seal population was generally undisturbed by Spanish interlopers as yet. At first the seals and sea lions were "so thick on the shore, that we are forced to drive them away, before we could land, being so numerous, that it is scarce credible to those, who have not seen them," wrote Woods Rogers, who in 1709 rescued Alexander Selkirk ("Robinson Crusoe") from the island.[27]

It was not easy to land upon Más Afuera, but the sight of vast seal herds was too great a temptation to be long resisted. Edmund Fanning, whose account remains the best record of early American sealing in these waters, visited Más Afuera in 1797 as captain of the 100-ton *Betsey*. The first boat sent to shore met only disaster.

When within one or two hundred yards of the desired point, the boat was struck by an over-sized breaker, and capsized, the bow striking against a rock under water, and breaking her into two parts in the first blow; the men, by swimming and diving, and managing the best way they could, were at length enabled by God's Providence, to reach the land, every man maimed or bruised in a greater or less degree, by the violence with which the surf had thrown them against the rocks, losing every article they had except those on their bodies; as for the boat, little of it was found, for literally speaking, it was stove into more than a hundred pieces.[28]

All in the day's work, as the sealers were to find around the world, for if fur seals managed to survive the slaughter which followed, it was due to their love of crashing surf and windward rocks. Try as sealers would to exterminate every last adult and pup for its skin on the assumption – accurate enough – that otherwise another sealer would perform the act, some of each species survived, though only barely enough in some cases. At Más Afuera, a small volcanic island some seven by four miles, 500 miles off the coast of Chile, and endowed with plenty of wood and fresh water (and, after they were left by sealers and whalers, goats), the early rush was staggering. The small New York vessel *Eliza*, probably the first American visitor, managed in 1792–3 to take off 38,000 skins which brought $16,000 in Canton – a record low price.[29] Though sealers like Fanning were not about to hold back where British predecessors had revealed the potential for turning a nice profit, it appears that no further American effort was made at Más Afuera for the next four or five years.

Amasa Delano, who participated in the early commercial sealing here, estimated that for seven years from the first American sealing visits ("about the year 1797"), more than three million sealskins in total had been taken from Más Afuera by ten to twenty ships a year to the Canton market.[30] In 1805, the Spanish authorities closed the islands to American sealers, though few fur seals remained. Delano's estimate, which has long been taken at face value, seems too high for reasons which will be explained. Nevertheless, every seal that could be taken, was. "Indeed," wrote Fanning,

so anxious were the officers and men to make sure of filling the ship, that even after the hold was stowed so as not to have room for any more, then the cabin, and finally the forecastle, were filled, leaving just space enough for the accommodation of the ship's company; and yet there was remaining in stacks on shore, more than four thousand skins; with these, a boatswain and boat's company were left, to take charge of, and add to, until a vessel from our owners should call for them. These dry skins, after being stowed on board ship for a few days, in tiers, will settle very much. This was the case with our own; in the course of a couple of weeks we were enabled to clear the men's abode in the forecastle, and in like manner a portion of the cabin ... Thus, in the space of ten weeks, by perseverance and industry, was our little ship completely laden to cross the Pacific to Canton, for a market ... At the time of leaving Massafuero, there was, according to our computation, between five and seven hundred thousand fur seals there.[31]

Delano adds further details:

I have carried more than one hundred thousand myself, and have been at the place when there were the people of fourteen ships, or vessels, on the island at one time, killing seals. The method practiced to take them was, to get between them and the water, and make a lane of men, two abreast, forming three or four couples, and then drive the seal through this lane; each man furnished with a club, between five and six feet long; and as they passed, he knocked down such of them as he chose; which are commonly the half grown, or what are called young seals.[32]

The seals were then skinned, with a thrust to the heart first to insure death. Some men, he reported, could skin sixty seals an hour, a very fast rate indeed, but since some of the fat and meat were left on the skin, less care was required than might have been the case. The skin was later stretched for "beaming" or separating the skin from fat and meat, and then pegged out on the ground for two or three days of fair weather, and then stacked; provided the skins were kept dry, Delano advised, they would keep for two years. So many skins were pegged out along the Patagonian coast that it came to be called "New Haven Green." Though one fine day was enough for drying, in the area of the Horn several weeks of constant turning might be necessary before the pelts were dry enough. The most disheartening aspect of the process, at least at Más Afuera, was the total lack of harbors: every skin had to be carried, sometimes long distances, to the few landing places for ship's boats, and then laboriously rowed several miles out to waiting vessels. Wastage, inevitably, was high, even with a skilled crew. Delano is very forceful on the need for the latter:

In voyages for seals, you must have men who understand the business, and not raw hands, who will certainly make it a losing enterprise. Out of twenty, which should be the least number for a crew, the captain and six others at least ought to be able to teach the rest their business with skill. Such a set of men will do more and better than twice the number of those who are untaught. Let every man depend on his share of the seals for the voyage. In no other way will the men do well ... The shares to prime seamen, or sealers, should be one per cent or a hundreth part of the voyage ... The money is to be divided after the expense of the boats for carrying the skins to Canton is deducted – no other expenses are to come from the skins. The prerequisite of the captain should be ten per cent on all that can be realized from the cargo in the return of the ship.[33]

The financial arrangements could be complex. On the ship *Enterprise* of Providence, Rhode Island, her master Sylvester Simmons left Ephriem Stubbs, probably one of his mates, on Más Afuera with a crew of thirteen men for eight months of sealing in the spring of 1800.[34] Simmons thought it wise to draw up a contract with Stubbs on the spot, in which the latter would receive "every Fortieth skin that may be taken by the said Ships Crew." The fear, obviously, was that Stubbs might deal on the side:

And the said Simmons agrees to advance the said Stubbs in China or any other Port where the Skins may be sold, one fourth part of the Skins which may be due him and a Note for the remaining three fourths payable in Thirty days after the Ships Arrival in Providence. And it is further agreed that the said Stubbs is to have a privilege of one Ton in the Ship from China to America. And the said Stubbs further agrees to Forfeit his Skins Goods Chattles and one Thousand Pounds Lawfull Money if he trades buys or sells to the Amount of one Seal Skin during the time he is in the Sealing Business. Signed sealed and delivered ...

Provisions landed with Stubbs were agreed to be adequate for the eight months, but, in a separate document, the possible delay of the *Enterprise*'s return was foreseen, and Stubbs was authorized "to draw bills on the Said Simmons or the other Owners of the Ship to purchase Provisions of Ships that may be at the said island or on the same." If the ship did not return in a year, Stubbs was to buy food to last another six months, and after the year was out, he would receive 5.5 per cent of any extra skins taken (the crewmen would get 1/24th).

No advice was given to Stubbs on what to do if the *Enterprise* did not return at all. But in a very similar undated document of instructions for Lovett Mellen (or Mullen) from Simmons – probably issued in the following season – the omission was corrected: "If [after two years] from the date hereof, Then Providing the said Simmons should not be heard of nor no prospects of the Owners in America sending a ship out to take the said Mellen and Skins of[f]. Then the whole of the Skins obtained by the above mentioned Fourteen Men, to fall to the charge of said Mullen to be Convey'd to Canton and the proceeds to the Owners in America to the best advantage." Mellen's skins were in fact taken to Canton by another vessel, at a freight of 25 per cent (paid in skins) of her 13,502 pelts. Subtracting damaged skins, Simmons was left with 10,023, which fetched $6,679 on the Canton market in early 1802, via the well-known agency of ex-Bostonian Sullivan Dorr, who acted as agent in China for many of those sealing vessels which sailed without their own supercargoes.

Simmons received his pay in bales of silk, less $119 worth of provisions. By December of 1802, he had doubled this value of goods to $13,300 of silk and another $1,230 of tea, presumably combining the return from his 1801 voyage with earlier credit or late-arriving skins in order to complete his homeward cargo – in many ways a typical voyage. The skins had been sold for 67 cents each, and Mellen, assuming his unspecified share to be the same as his predecessor Stubbs, had 250 skins worth $168 with which to find his homeward ton of goods: not really enough of a capital base to buy silk or tea of that weight, meaning that Mellen would require additional capital of his own, or to combine with that of others in the crew, to make the best use of his opportunity.

Since the average crewman received one skin in every hundred, he would be considerably less well off than Mellen. But when the whole of the voyage, including sale of China goods, was considered, the average return for able seamen, as computed by James Kirker,[35] was in the neighborhood of $120–240 each, enough, particularly if he had with him funds from an earlier voyage or friends at home willing to trust him to invest for them, to bring him nearer what was then the ultimate goal of most seamen: a farm. Few sealers stayed long in this rather grim and dangerous business, which explains the trouble so many captains had in recruiting experienced crews, even in traditional sealing ports such as Stonington and New Haven.

The real profits, clearly, were to be made by owners, master, and supercargo, whose function was to transact business along the way, above all at Canton. The owners normally took half the skins, the officers another quarter, leaving the same for the rest of the ship's company. If a crewman received $200, his master might make ten times that. Joel Root of Stonington shows what was possible for an enterprising supercargo. Root collected a cargo of fur and hair skins from St. Mary's, Más Afuera, and St. Ambrose for the New Haven ship *Huron* in 1803–4, and when he arrived home in 1806, had amassed a considerable fortune. In the process he spent a year on Más Afuera buying skins from the "alone men," or "lopers," men sealing on the island without any connection to a particular vessel hoping thus to amass more profit.[36]

Root's 9,000 skins, purchased for 25 to 50 cents each, brought 95 cents each at Canton. China goods bought with the proceeds (and another $7,000 worth of credit) at Canton and sold at Hamburg added another 75 per cent profit. Root then wisely invested at St. Petersburg in a mixed cargo for the United States which eventually returned him a total profit of $20,000. But the days of such quick gain from this island were fast dwindling. As it was, Root upon arrival had been "not a little surprised to find on the island more men than seals." By 1806, both seals and sealers were gone from Más Afuera, save for one alone man, John Wright, who stayed on with two dogs, Rover and Bully. In 1815, when Captain Barnard visited the island, he found neither Wright, nor dogs, nor seals.[37]

But even while seals were still plentiful, no sealing voyage was a guaranteed success. The brig *Rebecca*, which departed New York in mid-December 1797, is an example.[38] Uriel Coffin, her master, took her first to the Cape Verde Islands off the West African coast, as was common throughout the days of sealing, to obtain cheap salt from the salt pans of the island of Sal, and fresh provisions. By May 1798 she was at anchor at the Falklands, to spend two months collecting seals for skins and geese for food, departing in late June – mid-winter – with roughly 500 of each, bound around the Horn for Mocha.

For the next three months, while the vessel followed an irregular schedule between Más Afuera, Mocha, and St. Mary's, the men took what seals they could find. At Mocha, a boat's crew lived ashore, at first under their upturned boat, "suffering considerably in that time from the Inclemency of the weather," as the log-keeper, second mate Elihu Bunker, himself a part of the shore gang, put it. Eventually they built a crude shanty adjoining the "seal yard" into which they herded seals for convenient killing (though too many seals were likely to demolish their rough fence). By the end of the season, only 7,000 seals had been taken, hardly a full cargo, and Coffin determined to "winter" over, that is, spend the southern summer on the coast of Chile, only to be alarmed by news of "a disturbance between France, Spain, and America." Contracting with other equally nervous sealing masters to convey their cargoes home, the *Rebecca* returned to New York with 14,000 hair skins (sea lions, for leather), and 13,000 furs. This cargo, much of which did not belong to the *Rebecca*, was still deemed better "than to Risque another Season for the Sake of 10,000 more, as our Sails and Rigging would Consequently be in bad Order at that period." In mid-May the brig was back in New York, but her hold hardly contained the riches of China to show for her 22-month voyage.

By the turn of the century, such indifferent voyages were likely to be the rule, and success the result only of exceptional exertion. The Salem (later Nantucket) ship *Minerva*, leaving New England at the end of 1802, was exploring the Chile coast six months later, but found that "Seals was scarce" and when found, it was only "on Places whare it was Impasable to take them."[39] In July the *Minerva* found the ship *Dispatch*, another sealer, in the harbour of Coquimbo, where she had run in for provisions, eighteen months out and "but 10,000 of Skins on board." St Ambrose produced no seals, but far too many of another of nature's creatures: "It is remarkable for flys as Sure as you open your mouth you will Catch it full of flies at the time when I was Striving to keep my mouth & Nose clear of them So as the Breath I happened to take of My hat. To My Perplexity when Put it on Again I had at Least one Pint of Flies in the Crown of It."[40]

Altogether a horrid place, it seemed to Reuben Jones, the *Minerva*'s master, and he hurried south to Más Afuera, where he landed two boats with crews and provisions for six months, despite the fact that no seals could be found on the island. Remnants of the once great population still turned up, however, and when the *Minerva* returned in January of 1804, Jones's "head sealer" had collected 6,378 skins, more than any other gang which had spent the same season on the island – but this was where once every crew could collect a full load in a few weeks.

The *Minerva*, by perseverance, was able to return home with a full cargo of 30,000 skins in April 1805, but only at much cost in hardship, including considerable scurvy and one death (from a boat upset in the surf). The ship's log, though incomplete in some respects, makes clear not only the physical dangers facing sealing crews but also the "Wandering Dutchman" atmosphere aboard the vessel, condemned for months to tramp about, rarely permitted by nervous Spanish officials to remain long in one spot: "thus these diabollical Sons of Vulcan treats us. Whilst we like the Wandering Jew in A State Find the Boistrous Main our Mildest Retreat & only Asylum."[41] Mutiny – thankfully only on other vessels – theft of the cargo by the crew, encounters with whalers: there was indeed adventure to be experienced, and even nature to be observed with wonder, for example off Cape Horn: "It is Singular that Such a Fridgid & Dreary Coast Should Abount with Sea anamels. Such as Olavertrases Pinguins Seals Sea Eliphants & Co. all Skiping and Playing as if overjoyd and thankful for the Particular Plenty Bistowd on them by the Creator."[42]

Despite such admiration, the fur seals were soon gone from the more accessible islands. Within a year or two of the *Minerva*'s arrival home – again, without voyaging to China – sealers would find very little profit to be gained at Más Afuera. Already, however, seals were being taken from areas considerably to the north. The northwest voyages of the "Boston Men" were mainly for otter, but when in their search the vessels worked south into California waters, fur seals were again available for the taking, though in considerably smaller numbers than in the southern hemisphere. Here the Russians were already active, but their resources were limited, and the Russian–American Company was further handicapped by its high administrative overhead and Chinese regulations which required that Russian goods, including furs, enter China only at the land crossing point of Kiakhta, 300 miles south of Irkutsk on the Siberian-Mongolian border. On the other hand, the otter chase was best pursued by experienced Aleut hunters in their *baidarkas* or skin boats (and no seamen, however knowledgeable, was capable of managing a delicate *baidarka* without training from youth).[43]

The stage was thus set for a marriage of convenience: the Russians needed American vessels to carry supplies and transport the *baidarkas* at long distance, and the Americans needed the Aleuts and their skin boats.[44] The first cooperative venture seems to have been arranged in 1803 by Capt. Joseph O'Cain, who had several times visited Russian outposts in the northwest trade, but in this year came as master of his own swift and newly built 280-ton *O'Cain*. O'Cain made a contract with Governor Alexander A. Baranov at Kodiak to carry hunters to California and share the proceeds. When O'Cain returned with a valuable cargo in four months, Baranov used his share to buy guns and ammunition from the *O'Cain*'s hold with which to attack the decidedly unfriendly Sitka (Tlingit) Indians, who had wiped out the Russian settlement at Sitka in 1802 (the Tlingit's guns had also come from American traders).

In 1805–7 the successful experiment was repeated, this time with former mate Jonathan Winship, Jr, as master of the *O'Cain* with his own brother Nathan Winship as mate. The vessel transversed the California coast with its hunters, camping ashore at night, with grapeshot-loaded field pieces to keep any hostile Indians at bay, but trading cheap goods for valuable otter pelts wherever possible. At the end of 1806 they were also taking fur seal skins at Cerros (more generally known today as Cedros) and Todos Santos Islands off the Baja California coast, according to William Dane Phelps, who summarized the logs of the *O'Cain* in his unpublished manuscript, "Solid Men of Boston in the Northwest." Phelps was a reliable observer with a serious interest in the history of this coast; his remarks in any case are confirmed by reports from (Baja) Mission San Francisco Borja of eighty canoes, each with two Indians, in Bahia Sebastian Viscaino in 1806.[45] Winship's fur sealing on the Baja California coast may well not have been the first, but it is the first of which there seems to be record. The Anglo-Russian local cooperation lasted only a few years until on the one hand America was otherwise involved in the War of 1812, and, on the other, the Russians became more concerned that their control was slipping. After 1808 and above all after the establishment of Fort Ross four years later, they were attempting to collect otters on their own along this coast.

The Russians and Americans may have noticed – but most probably cared little – that the fur seal taken off California was not the northern fur seal from the Pribilofs, but a different species, now known as the Guadalupe fur seal, *Arctocephalus philippi*, and in particular the sub-race or separate species (depending upon which mammalogist is consulted), *A. townsendi*, once found in substantial numbers from the Farallon Islands off San Francisco Bay well down the Baja coast and on Guadalupe Island. *A. townsendi* was one of four seal species which were taken commercially along this coast, and all will be considered in due course. The Guadalupe population, along with that of *A. galapagoensis* at the Galápagos Islands further south, was small to begin with, and was soon reduced to near-extinction by the unrecorded sealers who first exploited these islands, helped along by Winship and his successors, even though the Russians, at least, paid only half as much for *A. townsendi* or *A. galapagoensis* pelts as for the thicker and silkier northern fur seal pelts. The Galápagos, however, offered the additional attraction of the galápago, or giant tortoise, which could be kept alive in captivity for long periods and was an important fresh food staple. The *Dromo* of Boston, for example, the first sealer known to make directly for Baja California, in 1808 stopped off for a supply large enough to last three months.[46]

It is not always easy to pinpoint the prey, not least because modern scientists are in disagreement on taxonomic details. Of particular interest in this regard are the Farallon Islands – "Los Farallones de los Frayles," the headlands of the Friars, seven rocky islets which are barely visible from San Francisco's coastal heights in clear weather. These rocks were home to both Guadalupe and northern fur seals. They were not exploited by the Russian-American voyages until Nathan Winship's return visit in 1810 in the *Albatross*, when Winship reported that sealers from two other American vessels were already there. Winship's men, principally "Kanackers," or Hawaiian Islanders, took at least 30,000 fur seal pelts that year and another 53,000 the next.[47]

The logs Phelps studied show that at least five Boston vessels, the *O'Cain, Mercury, Albatross, Isabella,* and *Charon* had men sealing at the Farallons in 1810–12, though no longer in partnership with Russians. Since at least 30,000 seals were killed between August and December (in 1810), when the northern fur seal is at sea on its migratory paths, it would appear that the bulk of this take was *A. townsendi*. The total for the *Albatross* alone was 73,402 skins, together with some 631 otters from the mainland, which, with a few assorted fox, beaver, mink, and land otter, brought over $157,000 on the Canton market. Phelps estimated that in three years 150,000 seals were taken from the Farallons, nothing like the huge catches at Más Afuera, but equally ruinous to this much smaller base population. Phelps's calculations are supported by a somewhat unlikely source. The remains of shell mounds left by California Indians on islands along the coast show *townsendi* to have furnished a high percentage of aboriginal sea mammal catches, thus implying that they lived here in considerable numbers.[48]

Though again the rush had reduced the population, the Russians between the establishment of Fort Ross in 1812 and its abandonment and sale in 1841 continued to hunt seal and otter where they found them. Probably they took fewer than 20,000 skins off the Farallons in that time, but men were often stationed there to take both sea lions for their meat (to feed the Aleut hunters) and fur seals. The experience, as related by Zakahar Chichinov, an employee of the Russian-American Company, could be grimly disastrous.[49] Chichinov, who worked as a sort of clerk at Fort Ross, where his father was a carpenter, was sent on such a meat-gathering expedition to the rocks in 1819. The crew, including Chichinov and his father, numbered nine men; not until six months after they had begun clubbing and salting sea lions (they had no firearms) did a vessel appear to take off the meat and such furs as they had taken – but the men were left to winter on the rocks with little in the way of shelter or supplies, and only brackish water from pools to drink. The fur seals they killed could not always be preserved owing to the lack of salt. It was now that real disaster set in, as Chichinov recalled more than half a century later: "All the next winter we passed there in great misery and when Spring came the men [only seven were left alive] were too weak to kill sea lions and all we could do was to crawl around the cliffs and gather some eggs and suck them raw. On the first day of June of that year [1820] my father lost his balance while trying to reach out for an egg and fell into the water and as he was too weak to swim the short distance to the shore he was drowned. His body was not washed ashore on the island and I never saw it again."

A few weeks later help arrived, and Chichinov ended his career as superintendent of the company's Sitka shipyard; but the hardships and magnitude of these expeditions deserve some reflection. American seafaring tradition involving Cape Horn tends more to remember the gold-hungry landlubber's single passage to California than to recall the more dangerous calling of the sealers, by their very profession bound to approach the most dangerous of rocky shores. "An American has been known," wrote a French observer of the 1820s, "to leave a detachment of his crew at the Falkland Islands to double Cape Horn, ascend to the north, leave a second detachment on the rocks before San Francisco, in California, 2,500 leagues from the other, then repass the Cape with some men, collect his detachments on both coasts and purchase in China with the produce of their fishery, a cargo for the United States."[50]

Save for the Farallons, this could be a description of the voyage of the 270-ton ship *Amethyst*, Capt. Seth Smith, which left Boston in September 1806, bound for the California coast, with three years' provisions aboard for her crew of fifty-two officers and men.[51] In late November, she discovered – to her master's satisfaction if not to that of subsequent geographers – Gough Island in the South Atlantic, on which she at once left a party of sealers and then made for Guadalupe Island. By the time of her return to New England in 1811 – this time from Canton – the *Amethyst* had bid fair to clean off the fur seals of both islands and helped in the depredation of Cerros, San Benito, Natividad, and the other seal haunts along the Baja California coast, all the while keeping a weather eye cocked for interfering Spanish authorities.

Of these particular "ports of call," Guadalupe was the worst. There was no fresh water on the island, and it was commonly brought by a longboat decked over and rigged as a shallop with two masts. Men sickened and died, or deserted into the interior never to be seen again. The survivors had by necessity to make themselves comfortable, as Lewis Collidge, keeper of the *Amethyst*'s journal, makes clear:

My hut was compos'd of rock except the top which was of sea elephant skins, neatly sew'd together when green. The door was a large flat stone, which I could remove at pleasure My furniture consisted of a stone table about two feet square, which projected from the side of the rock, which made the inclosure on this table stood a lamp, a tin pot, two large muscle shells for spoons and a jug of "Honest Water" cold as ice. my seats was two points of the back bone of a whale, my bed was plac'd in a small vacancy at the back part, principally of skins, a blanket and a Sandwich Island Matt. In the aperture's of the rock were plac'd Skins of a great variety, and from the roof hung the intire skin of a Dog fish, stuff'd. This always show'd which way the wind blew as exact as the compass. A few volums of Shakespear on the wall. So much for my Hermitage.

The *Amethyst* and another sealer, the *Triumph* of New Haven (with whom the *Amethyst*'s men had vast celebration on the fourth of July), took at least 35,000 seals off Guadalupe. As with the Farallons, the back was broken of the home herd of the Guadalupe fur seal before 1812.

While sealers were working the California coast, their fellows had already entered other waters – some already known to merchantmen, such as the distant Indian Ocean, others desolate, treacherous, and unknown, such as those south of the Falklands and Cape Horn. Though the precise discovery date is known for some islands in the Antarctic Ocean, the northern boundary of which is commonly regarded as the "Antarctic Convergence," the meeting of colder polar waters with the warmer, saltier, and less dense oceans under which they sink, no one can be sure when the first sealers, disappointed in small pickings from the Falklands, or Statenland, searched for yet unhunted beaches. Only the island of South Georgia, however, appears to have been sealed before 1800.

South Georgia, some 1,600 miles due east of Staten Island, is fairly sizable, 100 by 30 miles at its widest and longest points, with mountains rising to 9,000 feet.[52] The environment is not overly attractive, with winters which produce temperatures of −15° C. and 6–10 feet of snow. The island averages some 300 days of precipitation a year of which well over 200 include frost and snow. The principal danger, however, aside from surf, is the fierce and sudden gusts of wind which pour down the fiords in which any visiting vessel must anchor, "striking her first on one bow, then on the other, causing her to sheer and roll as much as though she was in a gale at sea," recorded Edmund Fanning, notwithstanding the fact that his vessel, the *Aspasie*, was moored with three anchors ahead and two astern, and her yards and topmasts had all been sent down to give the wind less purchase.[53] Thomas Smith, in the 1816–17 season, saw a small shallop driven ashore by waves and one man carried off by the surf to be smashed to bits, literally, among the rocks: "Of the remains of Johnson, one thigh bone and one arm were the only parts of the body that could be found."[54] This, it should be remembered, was in the austral summer sealing season, which lasted only from the arrival of the seals in November to pup until their departure in February – so short a season that skins could not be dried and they were of necessity salted first and later dried on the Patagonian or Chilean coasts.

Cook visited, and described, South Georgia in his second circumnavigation, the account of which was published in 1777; British sealers were here at least by the time the *Ann* of London visited in 1791.[55] Despite claims for some earlier visits the first clearly documented American sealers were the brig *Nancy*, Daniel Greene, master, and the Brigantine *Polly*, under the command of Roswell Woodward, in 1792–3.[56] Of average size for sealers (respectively 143 and 95 tons), they were owned in part or in full by Elijah Austin, an enterprising New Haven merchant. Little is known of the actual voyages, save that while the *Polly* returned directly to the United States, the *Nancy* went on to China and eventually paid customs of over $10,000 on a China cargo in New York. Greene was predictably reticent about the sealing aspect of his voyage: his original destination was given only as "fishing." The American snow *Sally*'s master, Captain Farmer, could hardly be so circumspect: his ship was wrecked on South Georgia in August 1797.[57]

When Fanning arrived on his first visit for the 1800–1 season, he found, as he had expected, traces of other sealers and a British vessel already at work. By the time he departed with 57,000 skins gained by virtue of his early arrival and skillful use of two shallops to communicate with his gangs ashore, seventeen more vessels had arrived to share in the hunt, taking for themselves, according to Fanning, another 55,000 skins.[58] James Weddell, himself a sealer and explorer of considerable experience, noted in 1825 that "the number of skins brought from off Georgia cannot be estimated at fewer than 1,200,000."[59] Many vessels went on to other rookeries; Fanning found five more vessels at St. Mary's where he went to replenish wood and water and to dry skins, and where he learned, "There was upwards of thirty sail of American sealing vessels on this coast, whose cargoes were destined for the China market."[60]

Once again, South Georgian seals were soon exhausted, and the sealers set out for new sources of supply. These voyages would now lead to the Indian Ocean, but as seals disappeared in numbers large enough to justify the investment in long voyages to China, sealing vessels returned far more often directly to their home ports than to Canton. Sealing required oversized crews, profitable if they could be left on productive islands (especially if they could live off local resources), costly if they had to be carried for the entire voyage. Some sealers stayed on in southern waters to produce oil from elephant seals: a valuable enough commodity, but not to be compared to the profit from furs.

Twenty years after the rush began, the procedure had changed. Now sealing vessels brought their cargoes, from whatever source, home for sale at auction, and middleman purchasers found a final market – increasingly in Europe, where a London furrier's 1796 discovery of a process like that of the Chinese to remove the guard hairs opened up new use for the sealskins, if and when available. Few sealers left American ports, however, after 1805 – none at all in the three years 1808–10, according to the research of Kenneth Bertrand, and only one each in 1811 and 1812.[61] In the latter year the crew of the brig *Nanina* of New York, Charles H. Barnard, master, included at least four unemployed former sealing masters. After 1815 voyages picked up a bit but were likely to be for elephant oil and to be organized by the few merchants determined to keep up their interest in sealing, such as James Byers of New York, for whom Fanning sailed on several voyages.

Stonington entrepreneurs also did not consider the trade to be at an end. Fanning's son, William A. Fanning, who had been a junior partner in Byers and Company of New York, joined with other Stonington men, including Capt. James P. Sheffield, to purchase the brig *Frederick* (1818) and construct the 130-ton brig *Hersilia* (1819. The latter was sent forth with Sheffield (himself a former employee of Byers) in command in search of the promised land of unexploited fur seal islands, the totally mythical "Aurora Islands," which at best might be taken to mean the very dangerous Shag Rocks, 115 miles northwest of South Georgia.[62]

Though Sheffield found no new rookery, while in the Falklands he learned of the discovery of the South Shetland Islands by William Smith in the British merchant ship *Williams* of Blyth. Smith had seen the islands when rounding the Horn from Montevideo to Valparáiso in 1819 then returned to make a more detailed investigation the same year.[53] When the *Hersilia* reached home in 1820 with the news (thus confirming other reports received from South America), the sealers were again off and running to this new "gold strike" 400 miles south of Cape Horn. James Byers, who would have been one of the first to hear, urged the United States government to survey and name – and thus to claim – the islands, but despite the interest of President Monroe, who inquired about the practicability of sending a frigate, the British had first claim. When the discovery was made public, cries were not unnaturally heard that the sealers had known of the islands for some time but had kept their discovery secret. But as Bertrand has argued, Byers might well have acted before 1820 had this been true.

In any case, in 1820–1 thirty American vessels were at work among these cold and snow-covered rocks. Dr. James Eights, who was here a decade later, gives some idea of the impression left by the South Shetlands:

Although many of the scenes about these islands are highly exciting, the effect produced on the mind, by their general aspect, is cold and cheerless to an unusual degree, for on their lonely shores the voice of man is seldom heard; the only indication of his ever having trod the soil, is the solitary grave of some poor seaman near the beach, and the only wood that any where meets the eye, are the staves that mark its dimensions; no sound for years disturb the silence of the scene, save the wild screech of the sea-birds as they wing their way in search of their accustomed food – the incessant chattering of the congregated penguins – the rude blasts, tearing among the icy hills – the sullen roar of the waves, tumbling and dashing among the shores, or the heavy explosions of the large masses of snow falling into the waves beneath, to form the vast icebergs which every where drift through the southern ocean.[64]

Much attention has deservedly been given to the exciting adventures of these intrepid voyagers, particularly in the search by historians for the first man to sight the continent of Antarctica.[65] There are several candidates for this distinction, notably the young Stonington sealer, explorer, and later clipper master, Nathaniel B. Palmer. Minute study has been given to those logs that remain of vessels such as the *Hero*, *Huntress*, and *Huron*, but we must be concerned with seals – for example, with the results of the *Hersilia*'s first South Shetland visit, from which she returned in May 1820 with 8,868 skins (she had too many trade goods in her hold and not enough salt to take more) which sold for $22,146. Not surprisingly, she sailed again two months later for a return visit.

This time, for the 1820–1 season, the *Hersilia* had more than enough company. The Fanning interests sent five vessels, under the lead of another famous sealer, Benjamin Pendleton, in the brig *Frederick*. Another Stonington group sent four more ships; it must indeed have been a hectic spring for this small Connecticut town as owners and masters scurried about to find the capital, provision the vessels, and select their crews. William A. Fanning's five ships, for example, were owned by at least thirty-two people (Edmund Fanning was not a direct partner, but surely took an interest in such goings on). The meaning of success is shown by Palmer's quick rise: away to sea at the age of fourteen (a big lad, he grew to six feet, and remained active into his seventies), he was second mate of the *Hersilia* in 1819, and now, at the tender age of twenty, was named master of the small sloop *Hero* (44 tons, five men, including Palmer), in addition to a holding share in the ownership of the *Hero* and the schooner *Express*. Fanning's vessels returned in the spring of 1821, save for the *Hersilia*; sent into the Pacific to gather hair seal at St. Mary's, she had been seized by Chilean nationalists, with a cargo of some 21,000 skins.

Their competitors also accumulated fur pelts, though how many is not clear. The New Haven ship *Huron* combined with the Nantucket schooner *Huntress* to bring back 2,500; the second Stonington group took 20,000 or more on their own. Byers from New York sent five vessels, but they arrived late, were beset by storms, and only two reached home again, with small cargoes. Three Boston vessels, including the *O'Cain* (now owned by Abiel Winship), three from Salem, five from Nantucket, two from New Bedford: at least thirty American vessels and two dozen from Britain competed for the seals of the scattered Shetlands, and sometimes physically. Several cases are recorded, in logs and journals, of this or that crew being driven off by earlier-arriving gangs, and at one point a bloody confrontation between American and British sealers was narrowly averted (though battle lines were not always drawn along national frontiers). Meanwhile, a very limited resource was soon exhausted. The total numbers were not large when compared to those estimated by Fanning and Delano for other islands:

if each of the more than fifty vessels took 5,000 skins (a few took more, but most took less, and several did not return at all), the total would amount to 250,000 skins: quite enough effectively to destroy, or at least substantially deplete, this herd.

The next season, 1821–2, the results were clear. Fanning and Pendleton combined to send another fleet of six ships with eighty-five men – part of a total effort little less than the previous year's – but so unsuccessful was this visit in gathering skins that the fleet had to turn to "hair seals" from South American islands to complete their cargoes, and Palmer, now in command of the 80-ton sloop *James Monroe*, spent most of his time exploring for new seal islands. In the process, he reached (and may have discovered) the South Orkneys, 200 miles east of the South Shetlands, but they yielded up no seals.[66] Seals actually have been taken from this group from time to time, but so icebound are the Orkneys as to be frequently unreachable until mid-January, a month after mid-summer. The next year, the British sealer-explorer James Weddell also found no seals there but saw enough positive signs to return in 1823. James Morrell (whose reputation suffers from his exaggerated claims for the importance of his own efforts, and his possible borrowings from others – notably Weddell – in his later published account) in 1823 reached the South Sandwich islands, but with the same discouraging results. The main haunts of the fur seal in these waters – the South Shetlands and South Georgia – were not to be duplicated in this quadrant: sealers would have to look in quite different directions now, particularly the Indian Ocean.

The Cape of Good Hope had long before been rounded by Europeans, and had Indian Ocean waters been as well supplied with seal islands as Cape Horn, no doubt there would have been sealers in plenty. But aside from Madagascar and the islands in the more temperate seas to the north such as the Maldives and the Seychelles, the islands of the Indian Ocean are few and far between, and in the north very seldom visited by seals. Still, it is no surprise that sealers soon turned east, after the exploitation of Tristan da Cunha, the Falklands, and perhaps Bouvet (or Bouvetoya, as its Norwegian owners designate it). The latter was first discovered in 1738 by Jean-Baptiste-Charles Bouvet de Lozier (or Lozier-Bouvet), a captain in the service of the French East India Company which still in the age of Louis XV sponsored exploration. Since Bouvet misplaced the island by six to eight degrees of longitude, it was difficult for others to find later on the basis of his account.[67]

Seals were to be found, once the Cape was passed, on the Prince Edward Islands, some 1,300 miles southeast of Capetown. Discovered in 1772 by Marc-Joseph Marion, Sieur du Fresne, who called them the "Astral Islands," they were rediscovered four years later by Cook. Their history is often overlooked; so far to the south that they are seldom considered part of the Indian Ocean's history, they are equally too far from Antarctica to receive much attention from historians of that continent's discovery.[68] Cook gave them their present designations, preserving Marion's name in the name of the larger of the two islands which compose the group and calling the other – with the same sort of confusion found at Juan Fernández – Prince Edward.

Both are small, Marion some eleven by seven miles, rising to over 4,000 feet, and Prince Edward with a circumference of sixteen miles. Twelve miles apart, they are separated by a deep channel, and both are normally surrounded by kelp. The approach is dangerous, and in the 1870s observers from the oceanography vessel HMS *Challenger* left good advice: "The prevalence of fog and mist not only renders the islands difficult to see, but also prevents the position of the vessel approaching them being ascertained with certainty, so that a course cannot be shaped to make them, or pass at a given distance on either side of them with any great degree of confidence, it is therefore advisable to avoid their neighbourhood unless some considerable object is to be gained by visiting them."[69] Seals, of course, were such an object. The date of the first sealers is, as usual, unclear, but Fanning considered the islands well known and worked by 1802, by which time the sealers had regular shore establishments to facilitate their trade.[70]

The Crozets, another 600 miles further east, form a group of six islands and some outlying rocks, with just the same dangers. Julien Crozet was Marion's second-in-command; the islands were discovered and claimed for France (and a bottle with a note to that effect buried on Possession Island) in the same 1772 voyage. Once again, the first sealers are uncertain, but Edmund Fanning made the claim for his brother Henry, master of the New York ship *Catherine*, who landed a gang on Possession in 1802 or 1803; the usual rush soon followed.[71] Charles Goodridge, who spent two years on Possession as a castaway sealer in 1820 and left a detailed account of his experience, still could see the many detailed wooden tallies of skins taken from the early rush, showing that high if uncertain numbers had been harvested.[72]

There is some debate over the species. Although clearly it was the Antarctic fur seal, *Arctocephalus gazella*, some biologists believe two subspecies are involved, *A. tropicalis gazella*, the Kerguelen fur seal, which lives principally south of the convergence, and *A. tropicalis tropicala*, the fur seal of Amsterdam and St. Paul which stays on the north, or warm, side. The difference is mainly in head and bone size and – as will be seen, perhaps a significant factor – lactation period. Both species occur at Marion Island and possibly interbreed to produce a hybrid. As always, sealers cared very little and took every animal in sight.[73]

Once the first herds were cleaned off, the islands were visited only sporadically through the century. Sir James Ross, for example, in 1840 found a gang of eleven men on Possession, one of whom had been there for three years. The islands are ruggedly volcanic, with high dark cliffs, and weather which "may be described generally as bleak, boisterous, and foggy," as the *Challenger*'s men put it. But there were freshwater ponds, fur and elephant seals, and wild pigs on Hog Island (released there by the sealers, who soon found that piglets grow to be dangerous semi-wild hogs, scarcely palatable in any case since they fed mainly on penguins and bitter Kerguelen cabbage, *Pringlea antiscorbutica*).[74]

Not for another 800 miles to the southeast is land encountered again. The substantial island of Kerguelen was discovered in 1772 by the Breton nobleman Yves-Joseph de Kerguelen-Tremarec and visited in 1776 by Cook who called the island "Desolation" from its bleak appearance. Kerguelen is fairly big – roughly eighty by twenty miles – with a long coastline caused by a myriad of fiords, bays, and inlets. The prevalence of bare rock, mountains, and glaciers gives it a very forbidding aspect, particularly on the exposed western coast where the winds appear never to cease. Nevertheless, the existence of several useful harbors, together with "cabbage," penguins, birds, seals, and the fact that, since it lies north of the convergence, the temperature seldom falls below freezing, not only made life possible but explain why Kerguelen became in the mid-nineteenth century the principal sealing base in the Indian Ocean.[75]

The first American sealer was one of two candidates. The first is Capt. Simon Metcalf, already encountered carrying the first sealskin cargo to Canton in the New York brig *Eleanora* in 1788.[76] In 1792, sailing out of Mauritius, Metcalf visited Kerguelen (always called "Desolation" by Americans), though he found few fur seals at the time. Two Nantucket whalers, the *Asia* (Elijah Coffin) and the *Alliance* (Bartlett Coffin), with a schooner, the *Hunter* (purchased as a shallop at Mauritius) collectively provide the alternate candidate, for they were at Kerguelen's Christmas Harbor by mid-December of the same year, 1792. They departed the following March, leaving Bartlett Coffin's grave – the first of several Americans to die on the island. British whalemen and sealers seem not to have exploited the island's resources until the East Indiamen *Hillsborough*, on its return from a voyage to Botany Bay with convicts, spent eight months there in 1800.

In the next decade British – but few American – sealers were at the island in considerable numbers. Seven vessels, for example, visited in 1805 taking not only fur and elephant seals, but also the right whales which frequented the island's bays. After a general slump in 1809 and the temporary flooding of the London market with skins from Australia, the industry declined, to be revived in mid-century by Americans. But the fur seals were gone by the mid-1820s, despite occasional visits for elephant seals and whales. Interestingly, Heard Island, 260 miles to the southeast of Kerguelen and also rich in seals, was not discovered until the visit of Erasmus Rogers in 1847.

The list of Indian-Antarctic Ocean seal islands is completed by Amsterdam and St. Paul, 1,000 miles northeast of Kerguelen. First discovered by the Portuguese in the sixteenth century, they were explored by the Dutch seaman Vlaming in 1696 and described by several subsequent visitors, including Cook; since they lie further north than Prince Edward, the Crozets, or Kerguelen, they were nearer the standard courses for sailing vessels in the pre-great circle era, hence their earlier discovery and exploitation. They were commonly visited by fishermen from the Mascarene Islands to the north, and two unsuccessful attempts to establish colonies were made from Réunion at the end of the nineteenth century.[77]

Not surprisingly, both islands were sealed at an early date, though seals were plentiful when the British officers Lt. George Mortimer and Sir George Staunton visited in 1789 and 1793 respectively. Staunton was given a thorough tour of a fully regularized sealing operation by the "chief sealer" for the *Emily*, an Anglo-French vessel operating from Mauritius under American flag. The identity of the first American to arrive is as usual uncertain – perhaps it was Metcalf again, who certainly knew of sealers fitting out for these islands in Réunion, or perhaps the *Warren* of Newport which passed by bound for Calcutta in 1791 and returned to seal the next year.[78] What is clear is that despite their inaccessibility (only St. Paul has a safe anchorage for boats, and neither has a harbor for substantial vessels), these small volcanic rocks were soon denuded of their seals: by the middle of the first decade of the nineteenth century, there was little point in expecting big cargoes there.

From the Indian Ocean, the logical next step was Australia, whose southern coast and islands were to provide a further rich harvest. When the port of Sydney (Port Jackson) was established in 1788, much of the south coast was unknown (early visitors swung wide to the south of Tasmania); it remained so until almost the end of the century. The Bass Strait was discovered only in 1797, although sealers soon worked its islands. The important rookery on Kangaroo Island lay 200 miles to the north of regular shipping routes and was discovered in the normal sense only in 1802 by Matthew Flinders. Biologists as usual are uncertain about species; unquestionably the Australian fur seal, *Arctocephalus doriferus*, inhabited Kangaroo and extended its range into the Bass Straits, though some believe that a separate Tasmanian species, *A. tasmanicus*, was the main prey here.

The first sealer to call at Sydney (not necessarily the first to work these rookeries) was the snow *Fairy* of Boston in 1793, arriving from

St Paul (where she had found five men who had been left there two years earlier to take skins and had in turn left her own gang).[79] The Rhode Island brig *Mercury*, at Sydney in 1794, may have been a sealer; the ship *Otter* of Boston (Capt. Ebenezer Dorr), bound in 1796 from Amsterdam to the Pacific Northwest (where it was to be wrecked) certainly was.[80] There is no evidence, however, that any of these vessels did more than provision in Australian waters in their hurried transit to and from Indian Ocean sealing grounds to Nootka and thence Canton.

Already, though, the British Enderby whaling interests had entered into the sealing business through their *Britannia* of Falmouth, which arrived in Sydney in mid-1792 under the command of William Raven. Raven made not for nearby Australian islands but for New Zealand, where Cook had again blazed a trail for sealers, remarking in 1773 that seals were to be found in great numbers at Dusky Sound. Raven left a gang of a dozen men there which in ten months took 45,000 seals; their hut was the first European dwelling in New Zealand. The voyage was not repeated for some years; Raven operated under East India Company license, and company restrictions for some years barred Sydney merchants from exploiting New Zealand resources. Early in the next century, the firm of Kable and Underwood of Sydney established a permanent station at Dusky Sound.[81]

While Raven was at New Zealand, sealing had already begun in the direction of Bass Strait. The pioneering entrepreneur seems to have been ex-Royal Navy gunner Charles Bishop.[82] After a successful journey to the northwest for otters sold – along with his original vessel – in China, Bishop bought the brig *Nautilus* in Indonesia and reached Sydney in 1792. He was looking for seals reported on the "South Cape of New Holland," still thought at the time to be modern Tasmania, and took 3,000 to 4,000 gallons of elephant seal oil and some 5,200 skins from the Cape Barren Islands northeast of Tasmania. Bishop was soon followed by a number of vessels, large and small; it is hard to understand how the Bass Strait avoided formal "discovery" and description for another five years. The seals were everywhere. As Joseph Banks described them in 1806, "The beach is encumber'd with their quantities and those who visit their haunts have less trouble in killing them than the servants of the victualling office have who kill hogs in a pen with mallets."[83]

Sealskins and seal oil were most useful to the infant colony, since they were among the very few commodities which could be obtained locally and yet found a market overseas for cash with which to purchase other needed goods. Furthermore, the investment for sealing was small – a longboat and crew would do – and expertise was quickly mastered to skin and peg out the skins (they were commonly dried first for the

China market, not wet salted). Whaling, on the other hand, required substantially bigger vessels and was liable to run afoul of the powerful Enderby interests.

At least by the turn of the century, the Furneaux group (discovered in 1773 and named for a colleague of Cook's), Kent Group, and King Island were all being sealed along with the other rocks in this strait.[84] Substantial cargoes were returned and shipped abroad, and local Sydney merchants could well rise to eminence on this trade. But the role of sealing was really temporary. It did not, after all, encourage industrial development aside from shipbuilding – and there it was secondary, for the 40–90 ton vessels so useful for sealing found their standard employment carrying timber and wheat in all but the summer sealing season. On the other hand, agriculture suffered by the diversion of manpower to sealing, to say nothing of the escape it offered to convicts unhappy with their lot. Organized sealing soon cleaned off the major rookeries, but decidedly disorganized gangs of convicts lingered on for at least another generation. There was little profit in any case to be earned from Sydney merchants who paid low shares or "lays" (1/80th was common), and that pay often made in skins which the merchant bought at artificially low prices or more commonly set off against supplies charged at equally inflated prices.

Kangaroo Island, with its 120 miles of rugged coastline, was a particularly famous home for the "sea-rats." It had seals and other animals to live on and salt to pan and trade for the few necessary outside commodities, rum above all, which these sealers – men and their native women – needed. For many decades, colonial authorities could not control these piratical dens. They tried numerous expedients, such as restrictions upon shipping or new settlements to control seals and other resources and to keep the convicts divided. Van Dieman's Land (Tasmania) was first settled in 1803 in precisely this fashion.[85]

While American sealers were not part of this disreputable era of Kangaroo Island's history, they certainly did share in the extensive sealing of the first few years, as did vessels from Britain and India. There may have been other ships from New England in the Bass Strait, but none is recorded – if the lucrative trade from 1792 onwards in whiskey and rum to a thirsty colony be excluded – until the arrival of the Fanning-owned brig *Union* at Kangaroo Island in 1802. Isaac Pendleton of Stonington, her master, apparently had missed the Crozets and required a considerable haul of skins to complete his cargo. He wintered on Kangaroo, built a small schooner, the *Independence*, for a tender, and then, with 5,000 sealskins from Kangaroo and King Islands, arrived in Sydney in 1803. Here he entered into a business arrangement for sealing and trading with a local notable, ex-convict (transported for theft) Simeon Lord, one of the most important, and crafty, of the early

Australian sealing entrepreneurs. The Pendleton-Lord contract eventually produced 60,000 skins from the Antipodes Islands, but in the process Pendleton had been murdered at Tonga and the *Union*, "a ship of great venture," wrecked at Fiji in the hunt for sandalwood (to which many disappointed sealers had turned in the last few years before the War of 1812).[86]

Knowing of Australian seal resources was one thing, claiming them another. Nathaniel Cogswell, who shared in a venture organized at Ile de France (Mauritius) in the 90-ton schooner *L'Entreprise*, found this to his cost. Like the *Union*, Cogswell's vessel missed her Indian Ocean island targets, in this case St. Paul and Amsterdam, and, though denied formal permission to seal by the Sydney authorities, went to work in the Bass Strait, only to be wrecked on the Two Sisters rocks in 1802. Isaac Percival, in the *Charles* of Boston, clashed with local sealers over the spoils at the Kent Group in 1803; similarly, Amasa Delano in the ship *Perseverance* with her tender schooner *Pilgrim* found the following year that the determined hostility of local established sealers was the main obstacle at whatever rock he paused. By 1804, unfortunately for Delano, the height of the Bass Strait sealing industry had passed, the resource over-exploited. Sydney sealers made their motives very clear: "Their ideas were, that no foreigner had any right to that privilege near the colony." The tension was as great as that at the South Shetlands. "These men," reported Delano,

practised many impositions, such as stealing from me, enticing my men to run away, conspiring to steal my boats, and to cut my vessels adrift. They would sometimes go on to an island, where my people were waiting for an opportunity to take the seals that were about it; and if not able to take them themselves do something to frighten them away. They would say and do all in their power to irritate and vex my people, in order to cause them to do something that was reprehensible ... I kept clear of an open rupture for some months; during which time some of my people left me, as they said, 'because they would not be tied down to such close orders as to be obliged to put up with any insults from such villains.'[87]

Delano managed to obtain a partial cargo in the strait, but few Americans followed in his wake. Not only were the seals dwindling and the many rivals hostile, but also colonial legislation increasingly limited American activity. Governor King in 1804 issued a proclamation barring British sealers from serving on American vessels, prohibiting the construction of any vessel over 14 feet in keel (thus eliminating, he hoped, the essential tenders), and ordering that vessels calling at Sydney could clear outward only for their original destination, a regulation which would not stop sealing, but which would prevent a sealing vessel from returning safely to Sydney with its cargo.[88] Not only sealing aroused such response; authorities had little love for the American habit of making off with escaped convicts, an enterprise which sounds humane but was commonly initiated for the men's skills – skills which were just as useful in the colony. In their defense, it should be added that the sealers' own desertion rate in any civilized port (and some not so civilized) was high.

On the whole, it may be said that of the sixty-odd American vessels which visited Australia before 1812, the twenty which were sealers made a rather special contribution: "American sealing in the Bass Straits and whaling in Western Australia waters undoubtedly engendered proprietary feelings about local resources on the part of Australian settlers, and thus contributed to a feeling of colonial, if not national, identity."[89] The Americans were closed out of British imperial commerce by the War of 1812 and did not return to Australian waters until Sydney was again opened to American shipping in the 1830s.

By that point, the industry had changed substantially. By 1805–6 at least 100,000 skins had been taken from the strait, and the herd was large enough only to interest local sealers in small gangs.[90] Early prices for skins in London had been a high 15/- to 25/- for a prime skin, and the first direct Australian shipment to London in 1803 seemed promising: the hat industry above all had a use for the fur, "seal wool," though it preferred salted skins to dried. But direct shipment was direct competition, and the East India Company did not view it with favor. A glut of skins meanwhile brought the price down to five shillings by 1808. When low prices were coupled with high war-imposed insurance charges, it is not hard to see why an extensive colonial carrying trade in skins did not develop. American sealers, however, were still interested in this part of the world if decent cargoes were available – and they were, but now from the outlying islands of New Zealand.[91]

Most such islands had been discovered before the turn of the century, for example, Bounty Island, discovered by Captain Bligh in 1788.[92] Bounty, 400 miles east of the nearest land, had no vegetation and no fresh water, but this was insufficient obstacle to sealers, who left only such traces of their passing as a hut thatched with bird wings. The Snares, another group sixty-five miles southeast of Stewart Island, were discovered in 1791 by Broughton and Vancouver, and they too were soon visited by sealers. One sealing vessel left a gang of four men on the Snares in 1810, after the island had already been thoroughly sealed off: no vessel came to rescue them until 1817! The Chatham Islands were discovered by Broughton in the same year as the Snares, and the Antipodes, or "Pentantipodes" as they were then called, in 1800 by Captain Whitehouse.

In the case of the Antipodes, a considerable haul was made by the Pendleton-Lord interest, which left a gang here in 1804. The 60,000 skins taken off these islands eventually formed part of the cargo to China of the Nantucket ship *Favorite* (after the demise of the *Union*): 87,080 skins crammed into a vessel of 245 tons must have been a record cargo, though she paid off her crew on the basis only of 32,000 skins in what appears to have been a rather sharp deal for the owners.[93] So thoroughly were the Antipodes sealed that it is not certain whether the seals were *Arctocephalus fosteri*, the New Zealand seal which has become reestablished there from New Zealand in the twentieth century, or another stock: all that can be said is that Antipodes pelts commanded particularly high prices and were known as "upland seals."[94]

The Auckland Islands were discovered by whaling master Abraham Bristol in the *Ocean* in 1806; the next year three gangs were combing these small rocks. In 1810 Campbell Island was discovered by Frederick Haselburgh, master of the brig *Perseverance*, in the course of sealing the Auckland Islands for Robert Campbell and Company of Sydney. In 1809–10, perhaps earlier, the various bays and rocks of Foveaux Strait were explored and hunted. None of these sources provided vast harvests, but all were sealed for at least a few years and revisited periodically.

Macquarie Island, however, was another story, for it was to yield rich harvests of skins and elephant oil after its discovery in 1810, also by the *Perseverance*, though the news was first spread about by the *Aurora*, a New York brig on charter to Simeon Lord for Campbell Island sealing. Named for Lachlan Macquarie, then governor of New South Wales, the island is 600 miles from either New Zealand or Tasmania, and 400 from the nearest island. Though an oblong of twenty by two miles wide, it has no really safe harbor, and only sealers who were landed and retrieved by substantial vessels were likely to be secure. On the other hand, Macquarie is on the warm side of the convergence, and conditions there, though as wet as other sealing locales, were not as much a hardship as at South Georgia, for example.[95]

As described by the Russian explorer Bellingshausen, who visited the island in 1820 and, surprisingly, has left the most complete account, sealers here had not only elephant seal for food (at least the flippers) and oil, and bird and penguin eggs, but also another sort of cabbage (*Stilbocarpa polaris*) which grew wild over the island and served as an effective antiscorbutic.[96] Sealers have been blamed for adding to the species here in a most undesirable way, as noted in 1815 by the *Sydney Gazette* while bemoaning the dim prospects for any recovery of the sealing industry:

[Macquarie wildlife was] totally obliterated by the ravages committed on the

younger seal by innumerable wild dogs bred from those unthinkingly left on
the island by the first gangs employed upon it. The birds which were formerly
numerous and were found capable of subsisting a number of men without any
other provision, have also disappeared from the same cause. Their nests, which
are mostly in inaccessible situations, have been dispoiled of their young, and
the older birds themselves surprised and devoured by these canine rovers, which
as they multiply must every day diminish the value of one of the most productive places our sealers were ever stationed at.[97]

Strangely, the only other reference to dogs was by Bellingshausen,
and that from hearsay evidence only. But there can be no question
about the seal harvests, estimated by the same newspaper to have been
100,000 in the first season, but only 5,000 or 6,000 in 1818. J.E.
Cumpston, an authority on the history of these islands, estimates over
120,000 fur seals were taken by the end of 1812, with an additional
350 tons of elephant seal oil.[98] Ian Kerr puts the total at 140,000 for
1810–13 on Macquarie and Campbell islands.[99] Mary Gillham, a biologist who spent some time on Macquarie, concludes the take was 180,000
in three years, adding that a decade after discovery there were no more
fur seals on the island, though it would continue for some time to be
sealed for elephant oil.[100]

Macquarie had given a new, but last, lease of life to the Australian-
New Zealand seal rush. But the second decade of the nineteenth century
was one of few seals and less sealing, certainly by Americans, though
Australian colonists remained active. When sealing revived somewhat
in the 1820s, the main prey was likely to be elephant seals everywhere
in the southern hemisphere. The habits and utility of the elephant seal
will be discussed in due course, for it was the base of an industry of a
sort different from fur sealing, and one moreover which reached its
prime only in the middle decades of the nineteenth century in a general
era of peace and industrial demand.

By the coming of the War of 1812 (and in some areas the early
1820s), the first great seal rush was ended, each of its many claims
having been worked out until the returns of this form of "soft gold,"
as the Russians were wont to call the sea otters (perhaps "soft silver"
is a fair equivalent) rewarded only small-scale enterprise, like the solitary
prospector with donkey and gold pan wandering the deserted environs
of once-thriving boom towns. As with mineral gold, the attention of
large-scale capital would return only if and when the seals did. But the
effects of that first rush were large.

For the northeast American ports which sent out sealers, there was
money to be made as well as risk to encounter. From Salem to New
York, many owners and masters had gone "a-sealing," but above all
it was the Connecticut shore which had specialized in this industry,

and for the Fannings, Pendletons, and Palmers of Stonington, the preeminent sealing town, it was the making of their fortunes and of the fame of their home port. For America as a whole, however, fur sealing rather merges into the entire enterprise of the northwest trade, since for long sealskins were only a supplement to the otter trade, itself only a means of access to China – and the real profits were always in the China commerce. But skins, otter and seal, did bring America into that commerce, and opened up both the Pacific and the Orient to American enterprise, maritime and commercial, as well as adding the thrill and fame of discovery of new lands.

It has been argued by Thomas Philbrick that, until the 1840s, the main American frontier was maritime, not landward, and America's interests and energies, literary as well as economic, pointed in that direction.[101] It was no accident that two of the leading figures in early nineteenth-century American fiction both used a setting of sealers for important works: Edgar Allen Poe in his long story, *The Narrative of Arthur Gordon Pym of Nantucket* (Pym is rescued from dire danger by an English sealer, and visits Kerguelen briefly), and James Fenimore Cooper in his important novel, *The Sea Lions*, a quotation from which begins this chapter. Cooper never went sealing (nor did Poe), but he knew the sealers of Long Island Sound with whom he had associated thirty years before the publication of the novel. Both works have been discussed from varied literary and psychological perspectives, but the simple fact remains that the setting of sealing and sealers was deliberately chosen by both authors as natural and appealing to a prospective audience.

The significance of the fur trade for the northwest coast, in which Americans now took such an interest, really falls outside the bounds of this work, whether as evidenced in the establishment and pursuit of territorial claims and commercial interests, or in the reduction of the local population (through such gifts of civilization as smallpox, venereal disease, and whiskey), or even the permanent alteration in the total world-view of those same people, as Calvin Martin has argued in his important work, *Keepers of the Game*.[102] The sealers who went there, however, paused along their course in South American ports – Valparaíso, Concepción, Lima, Coquimbo – and along the California coast long enough to leave an impact. Whether through their tales of American independence, or the more practical contribution of firearms, however marginal the influence of sealers (which has no doubt been exaggerated at times), they contributed to the growth of a Spanish-American independence movement.[103] For Australia, finally, sealing was important in the early development of the economy of the colony, and here too Americans had a role to play.

But the greatest impact was upon the fur seal population of the southern hemisphere. On island after island, coast after coast, the seals had been destroyed to the last available pup, on the supposition that if sealer Tom did not kill every seal in sight, sealer Dick or sealer Harry would not be so squeamish. No fur seal population has yet recovered to the levels once shown by fur seal catches to have existed; indeed, some species may have been exterminated or at least pushed to the very edge of extinction; and those which have survived have sometimes been altered, occasionally radically, in life history or habitat.

Still, the question remains, how many? Any figure is but a guess, but at least we may estimate the general level. To recapitulate:

On Más Afuera (Delano's estimate, which seems too high but may include St. Ambrose, St. Mary, and St. Felix)[104]	3,000,000
Galápagos, Guadalupe, and the Baja California coast (another guess, perhaps too low)	150,000
Farallons (Phelps's estimate)	150,000
South Shetlands (another guess)	250,000
South Georgia (Weddell's estimate)	1,200,000
Macquarie, New Zealand Islands 1810–12 (Kerr's estimate)	150,000
Bass Strait and the New Zealand Islands, 1800–10	150,000
Tristan, Bouvet, and Indian Ocean Islands (another very conservative guess)	150,000
Total:	5,200,000

While this calculation may be on the high side for South Georgia or various New Zealand or Baja California islands, it ignores long-existing sealing on the Uruguayan and West African coasts, says nothing of hair seals, discounts elephant seals taken only for oil, and completely overlooks sealing in the northern hemisphere aside from the Pacific Ocean (since these furs were not bound for China, as will be seen in the next chapter).

By no means all of these skins found their way to China. James Kirker has estimated that 2,500,000 fur skins were unloaded from American vessels in China in the period 1792–1812.[105] This is a far cry from 5,200,000 but, if 1,500,000 taken by British and French vessels are subtracted, the total is reduced to 3,700,000, which is much closer to Kirker's figure. If we then adjust to compensate for overestimates – particularly by Weddell – as well as for skins sent to Europe and not to China, and skins lost by shipwreck or other disaster (a not uncommon experience), Kirker's figure seems reasonable.

Two and a half (let us say three) million skins would have earned something under $3 million for their owners. Prices ranged from 42 cents to $1.12, but on average fur skins yielded about 90 cents each.

That $2.5–3 million was a considerable sum with which to invest in China goods for further profit in its own right. The ultimate price was paid by the fur seals, and for some species the struggle for survival did not end in 1812. The future of sealing, however, lay in much more organized commercial ventures in the northern hemisphere, in the chilly waters of Newfoundland and the foggy Bering Sea.

PART TWO

The Swilers
of Newfoundland

Me rope upon me shoulder
Me gaff all in me hand
Both day an' night 'tis my delight
To kill swiles in Newfoundland
sealers' song: G. A. England,
Vikings of the Ice

CHAPTER TWO

The Schooners of the Outports, 1790–1863

Then here's to Capt. Farquar
Likewise his gallant crew
May you be spared for many years
The "whitecoats" for to slew.

Michael E. Condon, *The*
Fisheries and Resources of Newfoundland

As long as there has been man in Newfoundland, there has been sealing. The first European discoverers found the native peoples making good use of seals; seals influenced many later northeast Newfoundlanders in the choice of settlement; sealing had much to do with the subsequent concentration of wealth and power in the chief Newfoundland city of St. John's. Fishing as a profession in Newfoundland waters could never be called easy, but so remarkable was the seal "fishery" (a natural and useful term, not least because, quite logically, if seals be fish they are not – for Catholic purposes – meat) that over time it has played and still plays a considerable role in the folklore and indeed the psychological make-up of Newfoundlanders.

It is sealing that plays the role, not the seals. The old north European folk tradition respecting the spirit and power of seals, with charms of seal paws, or music to soothe the great grey *selchie* (Shetland Islands dialect for "seal") and calm the distress of his weird cry or "song," or the legend of woman-seal or seal-woman so dear to Irish and Scottish story-tellers, is largely meaningless to the Newfoundlander, though in fact it may be part of his own national heritage.[1] The difference is that in Newfoundland, man does not live in permanent juxtaposition with seals, but must go to sea to hunt them at certain periods of the year for commercial purposes. Perhaps the physical difficulty of the hunt leaves little room for mystery – though ceremonial dabbing with blood from the first seal has not been unknown. On the whole, while the Newfoundland vocabulary may have twenty words for seals of varying ages and conditions, and a hundred more for the several aspects of the industry, the Newfoundlander's songs are of great captains and grand catches and equally great disasters – to men and ships, not seals.

Only a minority of modern Newfoundlanders participate in the seal hunt, but most of the island's inhabitants have strong views on the subject, largely in defense of tradition against the interference of unknowing outsiders in the affairs of their island. A great many have read, and will urge the visitor to read, Cassie Brown's well-told tale of *Death on the Ice*.[2] By no means an apologia for sealing, it is rather the gruesome story of the spring of 1914, when 257 men lost their lives on the ice through greed, stupidity, misunderstanding, and sheer bad luck. The victims were the men, and their presence in the island's consciousness was only strengthened by the publication of this book in 1972. It is a sombre tradition to which *Death on the Ice* speaks, but it is a tradition fully reinforced for popular consumption in the outside world by the equally sombre etchings of David Blackwood, the photographs of John de Visser, or the words of Farley Mowat, though the perceptive student will also want to read Patrick O'Flaherty's critical study, *The Rock Observed*.[3] The disaster of 1914, after all, was only the worst of many, just as the focusing of world attention upon Newfoundland in the mid-1960s, as a result of the killing of harp seal pups, was only one more development in a complex story which already had generated considerable controversy.

The object of the hunt, and the controversy, is principally but not exclusively the harp seal, *Phoca groenlandica*.[4] A migratory animal, the harp seal each winter swims south from its summer haunts in the Davis Strait, retreating before the oncoming ice. Riding the Labrador current, the herd of seals closes the Labrador coast, near enough to be taken by native harpoon or net – hence the ancient onset of sealing. Into the early nineteenth century, the population was perhaps four million, though estimates have reached as high as eight or ten million. Though the migration would take three or four weeks overall, when the denser groups passed a given point on the coast, they must have presented an extraordinary spectacle. In 1984 the world harp seal population was around three million, and recent studies by Canadian government scientists put the Newfoundland herd at roughly two million.

At the Strait of Belle Isle the herd splits, or, more precisely, one substock traditionally moves through the strait to the Gulf of St. Lawrence (the "Gulf" herd). The majority (the "Front" herd) moves along the

eastern Newfoundland coast as far as Bonavista Bay. It was long thought
that they continued on to winter feeding on the Grand Bank, but scientific study has never confirmed this belief. Gulf and Front stocks do
intermingle, particularly the young seals, but they nevertheless seem
to have separate identities, at least from a chemical standpoint. "Bio-
cide" residues such as PCB and DDT are higher in Gulf seal carcasses,
not surprising given the greater agricultural and industrial concentration
in the St. Lawrence area.[5] Biologically there is little to distinguish Gulf
seal from Front seal, and why some go one way and some another is
one of the unexplained mysteries about harp seals: as with many aquatic
creatures utilized by man, seals are still little known in some respects,
despite all the scientific inquiry which has been directed towards them.

The Strait of Belle Isle is passed in January. Through the winter, the
seals feed in open water; by late February, breeding-time, the females
are in top condition. Though surprisingly few seals are actually observed
during this period, the Gulf herd feeds mainly in the estuary of the St.
Lawrence, and the Front herd primarily but not exclusively in White
Bay. In late February, the breeders haul up to whelp on new ice which
is found both in the Gulf (usually north and west of the Magdalen
Islands), and off southern Labrador, drifting south with the current to
Newfoundland.

Precise location of the whelping seals depends upon ice and weather
conditions: the ice is constantly on the move. Harp seals are gregarious
and tend to concentrate in one or two main "patches," three to ten
miles on a side, meaning that the area of a patch may vary from ten to
a hundred square miles (but more probably is long and narrow in form:
sixty to eighty square miles). The "main patch" may be over 300 miles
from St. John's or, more rarely, may be so close as to be worked from
land. In 1880, the spring of the "Walrus," a vessel of that name loaded
twice in sight of land.[6] The potential area to be searched by sealers is
thus some 250,000 square miles, so it is not surprising that the seals
might be missed during a short season when fog and low visibility are
common.

Breeding ice is pack ice, no more than a foot or two thick at most.
Depending upon sea conditions, it may be in huge floes or smaller pans,
but it must be thin enough for the harp seal to cut breathing holes.
Whelping of the entire herd – some 250,000 to 400,000 pups annually
– takes place very quickly at the end of February and the first week of
March. So regular is this event that the sealing season since the late
nineteenth century is set by legislation at precise dates, though the actual
opening date has varied according to the type of vessels used and the
degree of conservation attempted.

The pup when born is the "whitecoat" of popular concern, small
(18-20 pounds), appealing (its sharp claws are in fact to be respected),

and largely immobile. At birth the pup has a coat of white – actually transparent – foetal hair or *lanugo*, which lasts ten days or so and then is shed in patches over a few days, during which the whitecoat becomes a "raggedy jacket" and hence is less valuable for the fur trade. The clear coat serves the purpose of absorption of solar heat during nursing, yet somehow the seal's outer coat is cooler than the ice with which it is in contact, for it does not melt the ice in any way.[7]

The nursing period is very short, usually nine days. Harp seal milk is amazingly rich, containing 25-40 per cent fat (cow's milk in comparison normally is about 5 per cent fat). In this time, therefore, the pup balloons to 90, perhaps 100 pounds, over half of which is fat. Now the mother deserts the pup, which soon takes to the water itself, at this point having shed all traces of its whitecoat, as a "beater" with a black-spotted light grey coat. The entire process from birth to independence is a quick three weeks (less than two as a whitecoat), no doubt as a necessary adaptation to the short life of the thin ice floes. At times bad storms, especially those which produce "rafting" of one floe upon another, may bring high mortality among newborn pups and even whelping females.

In late March the females ovulate and mate with the males who have been hovering about the fringes of the patch. Development of the fertilized egg is delayed, and actual pregnancy lasts about eleven and a half months until the following spring whelping season. Breeding does not take place until age four or five (density dependent, the breeding age has fallen from over six years on average in 1952 to the current rate; since the introduction of sealing quotas in 1971, the population is thought to have increased, but the increase does not yet show in average reproductive age).[8]

Immature seals ("bedlamers" from the French "bête de la mer") arrive later in the patch area. Adult males may be five or six feet long and weigh 200-300 pounds. Both sexes live to the age of twenty, perhaps thirty; seal ages up to twenty or more may be read by section studies of canine teeth. The harp-shaped outline on the backs of mature animals appears only gradually as a dark pattern on a silver grey background at the age of eight or nine, later – sometimes never – in females. Bedlamers and adults alike leave the breeding grounds for more northerly ice to moult their hair coat, again in dense aggregations. When the moult is over, and the ice breaks up, the herd swims north to the arctic summering areas in the open water and among the ice floes of the eastern Canadian arctic or along Greenland's west coast. Gulf seals follow their own pattern during this entire process and do not leave the Gulf until late April after the ice has melted. In both herds, the newborn seals move south with the drifting ice into April and May and then swim north somewhat later than their elders.

Throughout this cycle the harp seals have been accompanied at a distance on their ocean side by another hair seal species of the same family (Phocidae), the hood seal, *Cystophora cristata*. Hood or bladdernose seals, so-called because adult males have a bladder which when inflated in excitement forms a sort of hood, prefer rougher, heavier ice. They are also less gregarious and thus less densely packed (and harder to hunt), remain in family groups, and are considerably bigger (an old "dog" hood may weigh 900 pounds). Since unlike harp seals hoods are quite willing to defend themselves and their young, it is understandable that this considerably smaller stock has been less exploited by men. Hood pups are more active from birth and do not go through a whitecoat stage, though the early coat, an attractive dark blue over silver grey (the pup is called a "blueback"), is most desirable to the fur industry. Hood seals off Newfoundland are an offshoot of the main hood concentration near Jan Mayen Island, and they have seldom amounted to more than 7 per cent of the total catch,[9] for which reason they have been of less concern to conservationists, just as hood seal hunting has attracted little attention from sealing opponents.

The scientific literature on both species is substantial, and though it cannot be said that all questions are answered, still much is known. Feeding habits, for example, from the standpoint of time and place, are not all that clear, nor are they a particular concern of this study. But what seals eat is important to fishermen, who regularly claim that there is a direct and clearcut relationship: more seals mean fewer fish for man. The probability is really the opposite, i.e., fewer fish mean fewer seals, since in the food chain process, seldom do higher elements exterminate the lower. On the contrary, each level's population is dependent on the next down the scale. Be that as it may, seals do eat fish.

The general rule in regard to feeding habits is twofold: first, most seals feed on a variety of prey, beginning in youth with all-important small crustaceans, and second, the bigger the seal, both in average size and chronological age, the deeper the dive and the larger the prey. Harp seals, which can dive at least to 100 fathoms, primarily eat capelin (small, schooling smelt-like fish), polar cod, cod, shrimps and other crustaceans, and bottom flatfish (halibut, plaice) varieties from Canada's east coast. It is the capelin, in particular, which brings the seals near the coast on their southward migration and into the reach of the hunters. Hoods eat some capelin and larger, deeper varieties of fish and squid.

The earliest seal hunters needed only to await the arrival of the seals. The concern of this study, however, is not the subsistence hunting of seals by natives or settlers along the coast but the commercial utilization of their fat, meat, and skins, and for that reason pre-European exploitation and modern sealing by landsmen to fill their family freezer chests need not detain us. It should be emphasized that the first Europeans to harvest the waters of Newfoundland took whatever was valuable and near at hand, and this meant not only the famous cod fish (the real meaning of "fish" in Newfoundland) but walrus and seals as well. Biscainers, Normans, Jersey Islanders, West Country Britons – if seals were convenient, all took them for their meat and skins, though cod might be their main goal. Jacques Cartier reported on Indian sealing in the mid-sixteenth century, but it was not until 1593, if reports be accurate, that the first vessel sailed from Britain (the *Marygold*, 70 tons, of Falmouth) specifically outfitted for seals, with butchers and coopers aboard. When she reached Newfoundland waters (presumably the Gulf), she found a St. Malo vessel already " 'almost full freighted with morses [walrus]' " – for walrus, now virtually extinct in the St. Lawrence/Newfoundland area, then provided an even easier haul than seals.[10]

But the *Marygold* was not soon followed. Though sealskins were always useful for winter clothing, the preliminary preparations and subsequent methodology necessary for cod and for seals were quite different, and the former was the more valuable cargo to Europe from North American waters. After all, oil could then better be provided by the Greenland whalers – at least as long as whale and walrus stocks survived. Seal meat was of more interest to settled residents, and the population of Newfoundland was slow to develop. For this there were several reasons, including the rival pull of New England, the complication of international conflicts (Newfoundland was not given up by France until 1713), and above all the opposition of British merchants outfitting for Newfoundland waters annually from West Country English ports to the competition of rival "planters" in the colony – with much talk along the way of keeping alive a "nursery for sailors" to man the fleet.

Nevertheless, despite such obstacles, settlement did grow, especially after the Treaty of Utrecht ceded Newfoundland to Britain. Fishing crews were left over the winter to take furs on land and seals in the spring (far more men were needed to fish than to sail fishing vessels, and carrying home such surplus manpower was simply an expensive drain upon supplies). By the end of the eighteenth century, the first British population base in the Avalon Peninsula was expanding northward. In 1763, Labrador and the Newfoundland coast from Bonavista northward to Belle Isle were placed under Newfoundland control; by 1770, Fogo, Greenspond, Bonavista, and several Trinity Bay settlements were well established, though the total population in 1765 was still only some 15,000.[11] This expansion, as so often the case, was encouraged by several factors, including the French retreat and the overpopulation of Avalon,[12] but the availability of seals as a major food resource in a slack fishing season without question played a considerable role. Fishing for cod was in some ways an agricultural enterprise, meaning that the more the fishermen ate the more they developed the industry (whereas other trades involved more direct competition with labor for food production), and sealing had a similar function.[13]

The early sealing of these settlements was principally by nets in an art perfected on the Labrador coast by seasonal European fishermen – Bretons especially.[14] On that coast (and Quebec's) netting was a practicable proposition from November to Christmas. The basic method was well evolved at least by the 1750s, probably earlier. A substantial wide-mesh net (8–10 inches was standard, but 15 inches possible) was placed in a "tickle" or narrow gap between an island and the shore or between two islands. The nets might be 50–70, or even 100, fathoms long and 10–14 feet deep. If the seals in their chase of capelin, and now driven by noisy men in boats, avoided entangling and strangling themselves in the net (seals will not jump over, though perfectly capable physically of doing so), they found themselves caught or "pounded" when a second (or more) submerged net, which was left slack until pulled taught with a windlass, was raised behind them.

Not every sealer had the advantage of a "tickle," and the seal "frame" was also common. Working on the same principle, a net was firmly anchored by "killicks" (wooden anchors weighted with stones) parallel to the shore. Poles fixed to the ends of the net served as marking buoys. A series of "stopper" nets ran perpendicular from this net to the shore, to be raised at the proper moment. The fixed net could also be run out from the land to an anchor offshore, with the stopper net running back at an angle forming a convenient trap for penning and catching seals. The whole might be carried away by heavy weather or ice – or whales – or the seals might swing too far to sea, but on the whole it was a lucrative enterprise. By the early nineteenth century, a netting station could be a substantial investment, with nine or ten men, a small house, and two or three dozen nets; in the 1840s, there were an estimated 1,700 nets in Bonavista Bay alone.[15]

Quite logically, if the seals did not come to the shore to be netted, the men could go to the seals, first in small open boats or ice-skiffs. Seal nets really did not have to be fixed to shore or bottom, and "shoal" nets could be used in the open water where bigger seals might be taken. The "foot" of the net would be weighted with grapnels or stones. Similarly, seals could be shot; since seals have a tendency to sink when killed this was a wasteful process best done on the ice (a wounded seal, of course, might slip off the ice and disappear). Shooting seals was more difficult than it sounds when the weapon was an 8-foot muzzle-loader (each marksman had a subordinate or "dog" to carry his powder horn and bullets) and the target a moving seal's head in the water.

Going to the ice always had an element of risk and necessitated the development of vessels more substantial than small open dories. Shallops, perhaps 30 feet long, decked fore and aft but open in the waist, provided sufficient size and security to venture further from shore and remain for a day or two if conditions warranted. Such vessels, in fairly wide use in the eighteenth century, were hardly the end of the process, particularly since Newfoundland sealers had before their eyes the example of larger Scottish whalers (30-50 tons) which occasionally took seals, at least by the 1790s. By 1800 the Newfoundland schooner fleet was under construction and a large-scale seal fishery in the making; by 1806, according to the bible of sealing statistics carefully collected by L.G. Chafe (*Chafe's Sealing Book*), the shallops were gone.[16] John Bland, in a letter of 1802, said sailing from large boats "has not been general longer than nine years,"[17] but now employed between two and three thousand men. The heaviest commercial sealing, however, had to await the further French retreat and new demand created by the later Napoleonic Wars.[18] By 1815, the typical vessel was 30-40 tons, with a crew of eight to ten, working perhaps up to 100 miles offshore in the spring, the crew probably having spent the winter trapping for furs, netting seals, or cutting logs.

Fish, seals, forest resources: all interlocked in a rapid expansion of Newfoundland's economy, a development moreover which was not physically dependent upon any one port city (finances, as will be seen, were another matter). George Cartwright, who for many years worked the Labrador for seals and salmon, typically based his activities alternately on Fogo and England direct.[19] By 1825 there were sixty to seventy vessels sailing from St. John's but another 200 from various Trinity Bay ports. The forty years 1820–60 were the height of the seal fishery, peaking in 1857 with nearly 400 vessels and some 13,000 men engaged.[20] But by that year the best seal harvest under sail had already occurred. In 1804, when the fishery was in full swing, the catch was 84,000; but the traditional high year was 1831, with a catch of 687,000 harp seals. The figure is Chafe's; Shannon Ryan's recalculation from Colonial Office records reduces it to 601,742, leaving higher yields (all over 600,000) for the years 1840, 1843, and 1844 – the maximum being 685,530.[21]

The actual number is less important than the level sustained: figure 1 attempts to modulate discrepancies in the data by providing five-year averages. The high levels of harvest clearly were passed by mid-century. Put another way, the harp seal herd could not sustain losses of half a million seals each year, despite the increase in technology which accompanied the introduction of steamers. If those averages are multiplied by a factor of five to produce annual totals and compiled for the entire period 1800–1915, the total take was at a minimum the enormous

The Schooners of the Outports

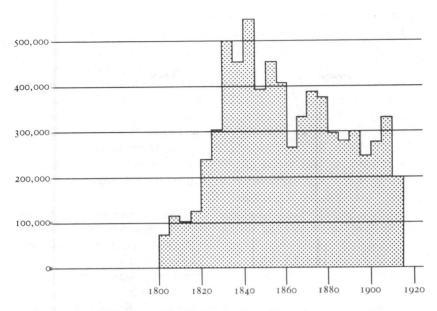

FIGURE I · Newfoundland Seal Catch, Five-Year Averages, 1800–1915

Sources: Shannon Ryan, "The Newfoundland Cod Fishery in the Nineteenth Century" (MA thesis, Memorial University of Newfoundland, 1971); *Chafe's Sealing Book*; J.S. Colman, "The Present State of the Newfoundland Seal Fishery," *Journal of Animal Ecology* 6 (1937): 145–59.

figure of some 35 million seals, the "greatest, most protracted mass slaughter ever inflicted upon any wild mammal species," as Fred Bruemmer, well-known harp seal defender, has put it.[22] But to reach 35 million the sealers needed only to average 300,000 a year (note that we are using 115 years in this calculation), and although frequently below this average, the annual catch often rose above it, and the figure takes no account of seals killed but not recovered. The loss of older seals was particularly high. "In my opinion," wrote Abram Kean, one of the most famous of old-time sealing masters, "for every [adult] seal saved and brought to port 20 would be sunk."[23] Kean could only have referred, however, to adult seals shot in the water, and such seals were never a very high percentage of the total catch.

Not only seals were lost: approximately 400 vessels and 1,000 men were destroyed between 1800 and 1865, though few statistics exist on pre-steam losses.[24] Given the number of ships and men engaged in this dangerous profession the figure is surprisingly low, since nearly that many vessels were engaged in each of the peak years – and the dangers of ice and weather were ever-present. It was just those difficult conditions which prevented the hunters from totally destroying the herd, though it is clear that the take was beginning to fall off by mid-century.

The sealers, however, had little leisure to share the long-term view. Their triumphs and tragedies were measured in "springs": the "Bona-vista Bay" spring of 1843, when most of the seals were taken close inshore, some even on land, in that bay; the "spring of the growlers" of 1844 when many heavy pans of sunken ice caused extensive damage among the fleet; the spring of the great fire of St. John's (1846); the "spring of the Wadhams" (1852), when the seals were found near those desolate rocks and forty vessels were driven to their destruction by rafting ice pushed by a north-northeast gale; the "first Green Bay Spring" of 1862, when the seals were taken in that bay so close to land and so numerous that " 'the women and dogs made ten pounds a man'."[25]

Before mid-century was reached, the success of the schooner fishery had already brought some substantial changes, beginning with the in-troduction of larger vessels. Originally few schooners were over 30 tons, since it was believed that anything over 60 tons was too heavy for the work and too difficult to manage in constricted outport waters. But in 1819 the first 100-ton vessel (the 104-ton brigantine *Four Broth-ers* of Brigus, still at work in the 1870s) was built, and by 1830 the prejudice was gone.[26] Several 200-ton vessels were built before the coming of steam, the largest probably the *Thomas Ridley* of Carbonear, built in 1852 – 260 tons gross, or 190 net. Extra size and strength helped, as did the taller masts which facilitated spotting the seals and catching the wind in the ice, where sailing could be difficult and quick-ness was essential to catch sudden-opening leads between floes. For this reason many schooner-rigged vessels of 60–80 tons used for various fishing chores were turned into "beaver hats" for the seal fishery, with one large temporary foremast topsail rigged to help through the ice and to back the vessel when jammed – larger "jackass brigs" had topsails on all masts, though the variety of local rigs could blur the distinction.[27]

Vessels in the 100–150 ton range had a decided advantage working the Front. Since they carried more sealers per ton of vessel, they were also more economical to operate. By the 1830s the tendency was to build brigs, and somewhat later (1850s) brigantines, though there is some disagreement about the chronology. These vessels could be used during the rest of the year in the carrying trade, taking cod to Europe. Good years with high catches were followed by a rash of construction, for a major haul could cover losses of several years, and a few good years in succession could make an investor rich. By the 1860s, when Newfoundland was hit by depression, the carrying trade was saturated, and any ships built for sealing were once again schooners. But by then few were needed, and a fleet of nearly 400 vessels in 1857 dwindled to less than half that in 1866.[28] The social damage, so to speak, had been done.

In the 1850s, seal products comprised 24 per cent of Newfoundland's

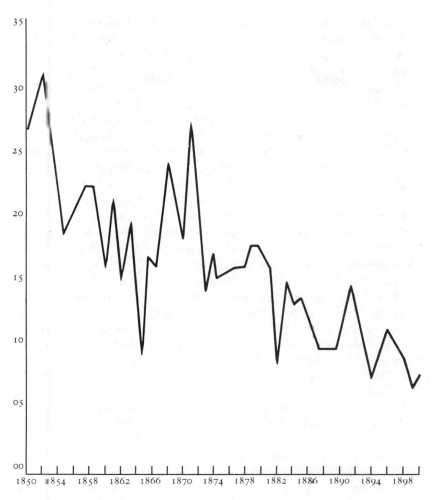

FIGURE 2 · Seal Products as a Percentage of Newfoundland Exports,
1850–1900

Source: Shannon Ryan, "The Newfoundland Cod Fishery in the Nineteenth Century" (MA thesis, Memorial University of Newfoundland, 1971).

exports.[29] Unfortunately for the fishermen-sealers of the outports, the profit from this activity was drifting out of their hands. The shift to larger vessels had often reduced former share-holding participants to wage workers controlled by a most deleterious truck system, discussed below. The difficulty was that while outport families might acquire the resources to build or share in a shallop or small schooner, anything bigger represented too much capital, not only because such vessels were bigger and made longer trips (requiring investment for stores), but also because their life expectancy was shorter: twelve years for a schooner, but only six and a half for a brig and four for a brigantine.[30]

As vessels grew, in other words, ownership became more concentrated, at first in the hands of the one or two successful families in each settlement. Even these found the larger brigs beyond their resources, and in a steady progression – contrary to common belief, a progression which did not begin only with steamers – ownership became concentrated in a few centers, above all St. John's, whose merchants already occupied a monopolistic position in supplying the island's population with essential imported commodities. The "Water Street merchants" – in Newfoundland, not a term of endearment – saw solid prospects in larger sealing vessels which also served the Labrador fishery and the carrying trade. By the 1860s, when steamers arrived, the mercantile elite of St. John's and Conception Bay (Brigus, Harbour Grace, Carbonear) held 90 per cent of the nonschooner fleet.[31] (From 1840 to 1899 the total Newfoundland registry was 3,895 schooners, 630 brigantines, 190 brigs, and 59 auxiliary schooners.)[32]

Full centralization of control had to await steam, but the concentration of profit and savings was already pronounced. Until steam took over totally in the 1870s, however, schooners still gave employment to outport men, not only as fishermen but also, as a necessary adjunct, as carpenters, blacksmiths, and shipwrights needed for the upkeep of the fleet. The schooner represented the days of outport financial independence relative to what was to come. Smaller vessels had meant more dispersal of the shipbuilding industry (44 per cent of those 3,895 schooners were Newfoundland-built). But already ownership had narrowed: in the same 1840–99 era, 3.4 per cent of owners held 35 per cent of registries, and 44 per cent of tonnage registered in St. John's; according to the calculations of Eric Sager a mere seventy people held 31 per cent of all tonnage, sail or steam, registered in Newfoundland in that sixty-year period.[33]

Not only did the concentration of ownership speed the growth of a financial elite; the truck system went a long way to create, at the other end of the social spectrum, a dependent labor class ("servile" would be a correct term in the economic sense, but it is not a word which really fits Newfoundlanders). Simply put, in this "lingering remnant of feudalism," the merchant owned the means of production in the form of the larger vessel. (How important that ship was can be realized by recalling that, unlike most fields of investment, the basic resource – fish – and the territory – the open ocean waters – were always available to public access.) The merchant hired men for their labor, in exchange for which he provided not cash but enough goods to keep man and family to the next season, much as a serf was fed for his farm labor.[34]

The fishermen repaid the merchant with his catch of fish, but the

merchant controlled the grading or "culling" of fish, and hence the
price, and the price of essential food and commodities was as high as
one might expect from monopolistic control – the company store, in
a sense. This was particularly true when the newer, larger brig became
the only supply vessel reaching a particular outport.[35] In such a system
there is little need to encourage productivity: the profit lies in the difference between the value of the catch sold once the merchant gets it
to market and the value of the supplies advanced, not on the fisherman's
debt ledger but on the merchant's cost account. Profits, as a result,
were likely to be put into increased production only so long as more
units of production – ships – could be expected to be profitable; otherwise the money went to general trade, overseas securities, and the
like, for there were as yet no comparable secure landward investment
opportunities.

The system was not totally without merit. It was some six months
from the catching of fish (the only stable commodity) through its salting
and storing to its final sale, and the fisherman needed goods on credit
during much of the year. The merchant, often the only one in a small
outport, needed some cushion against a very uncertain fishery, and the
severity of the system might be modified by the close personal relations
expected in a small isolated community.[36] But a bad year meant substantially increased debt in a system from which there was little hope
of escape. Most fishermen did not try, but rather took refuge in a fiercely
egalitarian society based, in the outports, upon an isolated if not solitary
life, with little room for initiative in a state of an economic dependency
which "enslaves its victims and makes them love their chains."[37]

A further element was introduced in the mid- and later nineteenth
century, for with the growth of the island's population the merchants
could no longer carry the burden of supporting its inhabitants, obliging
the government to step in and provide relief in particularly hard times.
The merchants, increasingly remote in St. John's (and thus operating
normally through the smaller outport middlemen), provided credit for
gear, salt, and provisions: the "credit" system differed from the earlier
"truck" system in that the merchant was providing credit, not the actual
goods. A significant Newfoundland bank failure in 1894 added yet
another step in the pyramid as Canadian banks became principal sup-
pliers of Newfoundland financing, but at the bottom of the pyramid
little changed. For the fisherman, though his technical relationship to
the larger merchant might resemble that of sharecropper more than of
serf, the economic essentials were the same in practice.

It's this way. Ye see there's lots of poor fisherfolk all 'long this coast and islands
that never sees a dollar from one year end to another.

'Fore they goes to cod fishin' in summer the merchants give 'em grub to keep their families all summer while they're away. Fishin' goes on til October, and by December they've got nothing, so has to go to the merchants again to get "tick" in provisions to last 'em through the winter. Then to pay this off they hev to go to the seals in the spring or they won't get no more credit, as the merchants also own the seal vessels. Only the captains make any money at the seals and they're good fellers as a rule, but if they lose a vessel or let their men "break out," as they do at times, they're soon as poor as the rest o'us. It 'ud make your heart sore to see the way lots o' these islanders come aboard the sealin' vessels in the spring – wi' pinched, half-starved faces, and hardly 'nough clothes to stand a summer breeze.[38]

To rise out of this situation took unusual drive and probably physical prowess as well. The career of Sir Michael Cashin, who pushed and brawled his way from Cape Broyle fisherman to Newfoundland minister of finance and customs, is a case in point.[39] Few had Cashin's courage to turn to another line of work (in his case, the challenging job of salvaging wrecks, then on to supplying fishermen as a lesser merchant and fishing entrepreneur). Cashin had grit, good connections, and capable fists; few fisherfolk could follow his example. Fish could be sold to an independent buyer, but then where would next year's credit come from? Fishermen lived in dread of being "cut off" for bad catches, or independent behavior (including political views), or above all selling fish outside the system. Hostility and bitterness and retreatism were a natural result. Sealing, though linked to this credit system, offered at least some amelioration of its effects, since income from sealing, while less certain, might also provide badly needed cash, the lack of which was certainly the reason for the credit system in the first place.

At first sealers were paid wages, but soon a division of proceeds was established which remained common on seal vessels under sail, allotting one-half the proceeds to the crew and one-half to the owners, while the master (if not himself the owner), received a set royalty of four to six pence a seal, or 1/– to 1/3 a hundredweight or quintal (two terms for the same British weight of 112 pounds) of "sculps," the sealskins with attached layer of fat.[40] Owner's costs were not deducted before distribution, a shareout which seems most generous when compared to the American whaling industry, but it should be remembered that the larger merchant-owners set the buying price of sculps which was very different from the selling price of hides and oil.

A mid-century brig, for example, with an average-sized crew of forty-five men, might bring in 5,000 seals.[41] Buying prices varied from year to year, and depended upon whether the load was old seals or whitecoats, but a buying price of £1 a quintal would produce roughly £2,500 for the voyage (assuming two sculps a quintal, a bit high for whitecoats but low for adult seals). The men's share of £1,250 would give each man nearly £28. This calculation is oversimplified but a reasonable expectation for a man going to the ice. Twenty or thirty pounds was a goodly sum of money with substantial buying power, particularly when the sealer might have no other cash income during the year. Put another way, income from sealing was not simply supplemental income but rather a basic cash nexus around which other aspects of life could be organized – social events such as weddings, or improving living standards however marginally. The chance always existed that the vessel would fail to find the main patch and return empty, but equally the possibility existed of making a quick catch and returning for a second, with double the reward for a few weeks' work. Sealers might supplement their income as well by selling flippers to St. John's consumers – as George Batten of Baseneed (Conception Bay) did, investing 35 cents with a friend for a used pork barrel to hold them, then selling them for $1.80 a dozen (shortly after World War I) – enough for an evening of St. John's entertainment.[42] Here again, the first vessels into port commanded the highest prices for the new season's flippers for pie or stew, but the market was soon saturated.

Seldom in Newfoundland life is windfall reward achieved without cost. Sealing was dangerous work; a man who did not choose to go on the hunt might – as in the account quoted above – be forced to go for reasons beyond his control. Nor could all those who wished to go find places or "berths," certainly not on the bigger and more successful vessels. Berths might be sold, and stowaways were commonplace – it was general practice to check each vessel carefully before sailing. Owners, with limited supply to meet considerable demand, found they could charge the men a fee, "berth money," merely for the privilege of stepping aboard.

Whitecoats were taken by the entire crew on the ice, but if older seals were to be shot the procedure was to launch smaller boats or punts from the larger vessel, and here experience and marksmanship were important. In a four-man boat, bow gunners were at a premium, and so to a lesser extent was the gunner in the stern. The two middle men – "batsmen," for they did little but row and carry a gaff or "bat" when on the ice – came cheaply, all of which explains why berth money charges were inversely related to ability. The earliest fees seem to have been about £1 (20/–) for a batsman (greenhand), 10/– for aft gunners, and free berths to bow gunners. By 1842, when catches were reaching all-time highs, berth charges had risen to 60/–, 40/–, and 20/– respectively – even, in some accounts, 70/–, 50/–, and 40/–. The fees appear to have varied from area to area and not surprisingly were likely to be higher in St. John's.[43]

In March of 1842, in a rare display of solidarity, the sealers forced the St. John's owners to reduce the charges to 40/–, 30/–, and 0. The men, according to one participant's recollection recorded many years later, "assembled at the head of King's Road, and with Bradley the fiddler, a piper, a drummer, marched through the town, visiting all the wharves, and searching the ships for those not in sympathy with them; such men had to fly for their lives ... The town was small then, and the strikers made a big showing. Besides rum was plentiful in those times, and it was not wise for the merchants to hold out too long. I pay more for water now than I did for rum in the old days. You could get it at O'Mara's then, the primest Jamaica, for three shillings a gallon."[44]

A demonstration of the same magnitude occurred in 1845 (there seems to have been minor trouble in 1843, but accounts are skimpy). A large sealers' meeting at Brigus forced a reduction of the charge to 10/– a man, with all gunners receiving free berths. The hero of the day, all accounts agree, was Capt. Henry Supple, an educated fisherman-captain, who seems also to have had an active role in the 1842 strike. Supple assumed the role of spokesman for the men, who rewarded him by drawing him in a sleigh from Brigus to Harbour Main on the road to St. John's. Such events were most unusual, and when they did occur the men found need for a spokesman from a social level closer to that of the owners. Only one more strike occurred in the nineteenth-century sealing trade, when in 1860 a berth money strike at Brigus was broken by one captain whose men backed him in sailing for the ice. In 1902, as will be seen in the next chapter, the sealers would act one last time.

Berth money was not the only charge to the crew, for any clothing, tobacco, or equipment needed by the sealer had to be obtained on credit. For this advance or "crop" the sealer's share of profit was reduced at the end of the voyage, and as might be expected the sealer was not given bargain prices for the gear he purchased. With the coming of steam and the greater regularization of the process, it was common for the sealer to be charged $12 for $8 or $9 worth of goods (at best), a considerable markup for a few weeks' worth of credit (Newfoundland adopted the dollar as currency in 1866, hence the shift in denominations). Nor was this the only burden; sealers might be expected to work without pay (though normally they were fed) to prepare the ship, cutting firewood and the like, perhaps for weeks before sailing, or at least to bring aboard a specified amount of firewood as a sort of berth money in kind. Steamship owners would add further refinements. The sealer was unlikely, as a result, to realize his full share, and thus for more than one reason "lived on hope."[45] That hope was not shared by all, as a mid-century verse testifies:

Come down with the killock
And out with the line;
Of fish about here, boys,
There is good sign.

Tis early in April
What matters for that;
We are not like the fools
Who are hunting for fat.

Hunting for fat, boys,
They never may get;
And losing the fish
They could get in their net.

A bird in the hand
Is worth two in the bush;
And off to the ice
Go fools with a rush.[46]

But such cynicism was the exception, not the rule. It is going too far to characterize outport life as "a peasant culture which produced little that has lasted, apart from a canon of derivative folk songs,"[47] but in those songs can indeed be found the more popular mythology of sealing, as in the quotation that begins this chapter. For there was more than mere hope of financial reward in going to the ice. The significant change in routine, the excitement which passed from man to man, the danger – all made of sealing a rite of passage into manhood for many and called for explanations from those who could or would not go, like the Spanish Pamplonan who would not run with the bulls, or the lioness Masai. The population grew but the number of sealers declined, and the more this was the case, and the more esoteric the experience, the greater the social credit to the successful sealer. Even the most ardent modern opponents of sealing, such as Robert Hunter, Greenpeace activist and historian, have been persuaded of the social significance of sealing: "Swiling lay at the heart of their sense of identity, for it had always been their central myth that it was on the ice, in the darkness and cold, that manhood itself was earned," an interpretation which only echoed what the Newfoundland House of Assembly had reported in 1913: "There are few young men in the colony who have not been on the 'ice' and an expedition is looked on as a test of manhood."[48]

As with other rites of passage, this too was swathed in ceremony: the flags and bands and songs and sirens and cannon blasts which accompanied the annual sailing of the fleet; the bands playing as (when necessary) the combined crews cut a channel through the ice of the narrows guarding St. John's harbor and towed the vessels out (perhaps 4,000 men working together); the appearance of Father Neptune once the Strait of Belle Isle was crossed much as in the crossing of the Equator.[49] This latter initiation was required even of those who had sealed from the land; that it was "going to the ice" which required the ritual, helps to explain why, though many went to the ice several times, even for many years, many men also went only once or twice.[50]

The masters, no matter how hard-hearted, were folk-heroes, larger than life – men such as Sam Blandford or Arthur Jackman, and most famous of all, Abram Kean, rewarded with the OBE for bringing back his millionth seal pelt. No wonder that the sealers celebrated in St. John's on their return, or expended their share in new finery with which to impress their outport homes.

Greenspond is a pretty place,
So is Pinchards Island,
Me Ma will have a new silk dress,
When me Da comes home from swilin'.[51]

And indeed, success at the sealery meant increased standing at home, more credit from the local supplier, perhaps greater security in the cod fishery, and a berth in the next sealing season for himself or a relative. Unconsciously exaggerating the hardships, elaborating the special technical language and mythology, all contributed to the mystique. Sealing was a way of mastering the environment, exemplified in common words deliberately managerial and exploitive as the sealer sailed in search of a good "harvest" or a "bumper crop."

But sailing for the ice was only the beginning.[52] Now the sealer had to help sail the vessel, though the ships were greatly overmanned for the simple purpose of sailing from point to point. The claim is sometimes made that sealers preferred steam since they were not expected to work steamers; they did indeed prefer steam, but for reasons unconnected with leisure time – on a steamer, a man was liable to spend many hours shifting coal from hold to bunker. The real work came when the patch was found and the ship was "burned down in the fat." Once over the side with his gear, the sealer might have miles of rough ice to traverse, hard enough going but harder still returning with seal skins in tow. It was exhausting: "Oh my God, the day, we couldn't get a day long enough, we couldn't get a day long enough out to the ice, you get in among fat and you couldn't get days long enough."[53] Indeed, the sealers commonly left before light and returned after dark.

Sealing gear varied from man to man, but tradition and experience insured that each was outfitted much alike with canvas or cotton jacket (there was little wool for homespun save a cap here and there), sealskin trousers and boots (or perhaps rawhide for the latter), "ninny bag" on his shoulder with some hard biscuits, a piece of pork or cooked seal, perhaps raisins and raw oatmeal; a canteen with water or molasses (for water might be available from pools on the ice) mixed with "Radway's Ready Relief," a patent medicine whose basic ingredient was alcohol (strong drink as such was forbidden on most vessels); goggles to help prevent snow blindness, perhaps with wool wrapped around the rims to prevent freezing to the skin; a sculping knife (homemade or bought with crop money, perhaps a "Greenheart" after the maker's name), with steel for sharpening; a line for towing sealskins; and, above all, the sealers' gaff.

The gaff, a multi-purpose staff of 6-8 feet with a hook at one end, was a most useful instrument, designed to help in "copying," jumping from pan to pan, something that Newfoundlanders were especially good at since many practised as youngsters, and masters preferred nimble wiry men for just this attribute.[54] The gaff was of great help in pulling from the water a man who had misjudged his leap, or whose ice pan had suddenly broken under him. But above all the gaff was used for killing seals, since a blow to the nose from the flat side would kill a whitecoat immediately (some sealers claimed that a kick in the nose was adequate). Hunting larger seals required a gun and ammunition as well, and the sealer might be required to carry a longer pole to flag piles of skins which had to be gathered (a practice more common after steamers were introduced, as will be explained). The sealer needed to be hardy and in condition for this work, but even the best of conditioning would suffice little for the greatest dangers – not an icy bath, unpleasant as that might be, but sudden storms, or loss of direction, or being forced to remain on the ice overnight – all possibilities only too real.

Once having found and killed his seal, the sealer had to remove the sculp. Experts could skin a whitecoat in half a minute, beginning with a long slice from chin to tail down the belly. The sealer separated the carcass, now abandoned on the ice except for choice edible parts such as flippers, heart, and kidneys (which might be laced in the sealer's belt); the ice would melt eventually, and what the birds left of the seal would always find a spot in the ocean's food chain when the ice melted. A knowledgeable man could tell by the sound of his knife as much as the feel when he was cutting properly and not nicking meat, bone, or – what was worse – the skin.

Were seals skinned alive, before they were really dead? It is hard to believe that some of the 35 million were not, though the sealer was best advised to have his seal truly dead before skinning, for scratches from sharp claws were quite dangerous. In the haste and fever of killing, obviously some animals were merely stunned before skinning began, and it was this aspect of the hunt which as early as the 1840s excited the horrified criticism of observers and would be the focal point of twentieth-century protests. J.B. Jukes in 1842 may well have been the first to write in this vein, but certainly he was not the last (Jukes went to the ice in 1840). "When piled in a heap together the young seals looked like so many lambs, and when occasionally, from out of the bloody and dirty mass of carcasses, one poor wretch still alive would lift up its face and begin to flounder about, I could stand it no longer; and arming myself with a handspike I proceeded to knock on the head and put out of their misery all in whom I saw signs of life." In this case, the sealers were loading whitecoats into punts, and subsequently on board the vessel (a brigantine), before sculping. The image did not leave Jukes, and he makes more than one reference to the fact: "the vision of one poor wretch writhing its snow white wooly body with its head bathed in blood, through which it was vainly endeavouring to see and breathe, really haunted my dreams." He well understood the point, however: "Still there was a bustle and excitement in the scene that did not permit the fancy to dwell on the disagreeable ... besides, every pelt was worth a dollar!"[55]

Only four years later, Philip Toque made much the same sort of comment, his argument being that sealing was a nursery for moral evils: "It has a tendency to harden the heart and render it insensible to the finer feelings of human nature. It is a constant scene of bloodshed and slaughter. Here you behold a heap of seals which have only received a slight dart from the gaff, writhing, and crimsoning the ice with their blood – rolling from side to side in dying agonies. There you see another lot, while the last spark of life is not extinguished, being stripped of their skin and fat; their startings and heavings making the unpracticed hand shrink with horror to touch them."[56]

Toque, like most commentators on sealing, was not himself a sealer. John Fitzgerald of South River, Conception Bay, was, going first to the ice before World War I, and in his eighties he recalled the same aspect of the hunt:

I often took the pelt off a seal and the carcus [carcass] crawled right after, Ya, the fact is you just knocked the seal out. He wouldn't be killed. Any little tap like that (he taps on the table) on the nose – one of them whitecoats now you are talking about, a young seal and he's knocked out. Well you'd haul the knife right down through him right high, you know. Of yes, I often sculped a seal and the carcus ran off – all over the ice.[57]

This view naturally did not go unchallenged. As Wilfred T. Grenfell, famous Labrador missionary-doctor, who went to the ice in 1896, put it

Now the killing of young seals has been frequently described as brutal and brutalising, and the seal hunters depicted as inconceivable savages, and this not only by shrieking faddists or afternoon tea drinkers. But, to my mind, the work is not nearly so brutalising as the ordinary killing of sheep, pigs, or oxen, driven terror-stricken to the shambles, which are already reeking with their fellows' blood. Here the animal is too young to feel fear, and evinces no signs of it; no animal is wounded and left to die in vain. [58]

Always, however, the sealer preferred the animal dead (though even then there could be involuntary movement) to facilitate skinning and to avoid being scratched. A scratch or other open wound (including chafing sores; sealers seldom if ever used gloves) made the sealer liable to an occupational hazard known as "seal-finger" or "spekk-finger" (Norwegian for "blubber finger"). [59] Seal-finger was a serious infection which was the product of contact between such a wound and bacteria from a scar or wound on the seal's skin – thus a greater danger when dealing with older seals than with whitecoats – and usually occurred on the second and third fingers which the sealer used to tow the sculp short distances or to separate sculp from carcass. The infection was treated by various means, all ineffective until the coming of penicillin and even more aureomycin, before which the infection might very well spread to the bone and force amputation of one or more fingers. It was danger enough to give the sealer cause for caution – when he thought about it.

Once removed from the carcass, the sculp was laced together with one or two others on a four-fathom line, carefully worked with an eye on one end and a point on the other. Each sculp was placed skin down, overlapping the next, for it might be a long tow, and three sculps could easily weigh 150 pounds; a 5,000-seal catch would thus weigh 250,000 pounds, making some 1,000 barrels of fat, each barrel of which would eventually become 22 gallons of oil (22,000 gallons, or 4.5 per seal). Later with steamers such care was less necessary, for the sculp would be towed a shorter distance to a designated "pan" to await the steamer's return. The method of collection would become more elaborate, but the basic process of killing and sculping remained unchanged until the mid-twentieth century.

Preparation of sealskins and oil, on the other hand, underwent several technological changes in the nineteenth century. [60] The procedure by the end of the Napoleonic Wars was to cut the fat from the skins on landing, at which point the two separate commodities, skins and fat, began quite different routes to market. Separating the two was a much more delicate process than dividing sculp from carcass and was done by professional sealskinners, nearly all professional butchers or coopers (the butchers naturally took to the work; for the coopers, also skilled craftsmen, whose work was seasonal, skinning provided supplemental income). Such men with their 18-inch knives could take the fat off 300 skins in a nine-hour day. Records were kept on this too, and the top speed seems to have been some 600 in nine hours, under a skin in a minute; a crew of two dozen men from Job's, a famous sealing establishment, processed 67,000 sculps in nine days, which works out to just under two minutes each, for which they might be paid up to $15-20 a day, at a time when wharf wages were at best 30 cents an hour, and often less. Skinners were decidedly not sealers but skilled craftsmen, making ten times the pay of sealers, and they saw to it that the trade was not easily entered. The number of apprentices was strictly limited and the apprenticeship long: it might be six or seven years before a man could "step behind the table" (the apprentices tended the skinners and distributed skins by barrow after weighing them on scales). Twentieth-century skinning machines would, in due course, be the downfall of the sealskinner elite, but in their time they played a key role.

The fat itself was cut in small pieces (later also subsequently ground up), and left in vats of 15-20 tons' capacity to melt by the heat of the sun and the fat's own decomposition. Rods a few inches from the inside of the vat provided space for oil and water (rain) to be drawn off by spigots at various levels. A big vat or "crib" could be 20-30 feet square, and 20-25 feet high, built over a larger "pan" of water; when the oil dripped into the pan, it rose to the top of the water, leaving blood, bits of skin, and other residue in the water. Credit for developing the early method for drawing off the best "seale oyle" traditionally is given to Harbour Grace entrepreneurs but every sealing port of any size had its seal oil vats, though perhaps not such elaborate cribs, which seem to have developed somewhat later.

The first, best "virgin" or "white" oil, known in the oil trade as "pale seal oil" (and sold by the pound, not the gallon; a ten-pound cask held about nine gallons), took some two months to self-cure, at which point it would be drawn off into hogsheads, which would still need to be "trimmed" of water for some time. But such "cold drawn" oil, that is oil produced without artificial heat, rendered only 50–70 per cent of the oil, depending upon weather conditions and the state of the fat (whether whitecoat or mature), even after the fat had been turned periodically and shifted among the sections of "pounds" of the crib. The residue remained to be boiled out. The further the process went, the darker the oil became, and this "boiled seal oil" commanded a lower price. It was sold at first by the ton (256 imperial gallons) according to quality (in the twentieth century, by acid content: the older the oil, the higher the acidity) and then by the hundredweight. It was used principally for confectionery, but also for the illumination of mines and for various grades of lubricating oil.

But boiling was an unpopular process, producing a distinctive odor. The raw material, after all, was in an advanced stage of decomposition. According to a hostile critic of 1852, life in St. John's in the summer, when the process reached a conclusion, was unbearable. Though the sealing plants had moved across the harbor to the south side following the disastrous fire of 1846, "there still remain sufficient vestiges of the seal trade to cause a summer residence in the Town of St. John's to be anything but desirable. Even the country, for several miles around St. John's, affords no protection from these horrible stenches."[61]

The solution was to render the fat into oil by subjecting it to steam pressure. Daniel Archibald, the bemoaner of the "horrible stenches," developed such a process, which he claimed could turn fat to oil in only twelve hours. Steam-produced oil made redundant any distinction between cold and boiled oil, the resultant product being simply "seal oil," used for the same lubrication and illumination purposes as whale oil.[62] Though producers and residents alike found the change beneficial, not all consumers were pleased; steam-produced oil burnt more evenly and with less smell, but it also produced more smoke, a fact which miners in particular found displeasing.

Finally, the skins underwent a complex process of their own.[63] The procedure – again in the early nineteenth century – was to stretch the skins, once the fat had been removed by the skinners, and dry-salt them, then set them aside for several weeks of curing. At that point they were shipped away for more complex stages in the preparation of leather. Harp seal leather, because of its attractive grain, was used primarily for the luxury trade. Sealskin on the whole does not wear well in constant abrasive use; it is not successful in shoes, except for the upper leathers of fancy models, though larger and coarser hides were used for luggage and furniture covering.

Once at the tanner, the skins were cleaned again of residual flesh, and then placed in lime pits to loose the roots of the hairs. After perhaps a month of increasingly stronger solutions and frequent turnings, the skins were ready for the hair to be "beamed" off. Some of the best quality hair was used in a form of plastering, but most was simply consigned to rotting heaps for fertilizer or simply waste. The leather, on the other hand, was split – by the twentieth century, a process done by a clever machine – with the good outer grained leather reserved for tanning and eventual consumer products and the inside for glue or more fertilizer. The value of the finished product was its attractive grain and durability-to-weight ratio. It was much favored for gloves, pocket-books, wallets, card cases, and belts: by 1905, a good skin might well sell for $3 in the United States.

The preparation of leather had reached a stable level by the twentieth century, including steam preparation of fat. But steam did much more for Newfoundland than affect fat preparation: steam vessels changed the entire nature of the Newfoundland sealing industry in the last third of the nineteenth century.

CHAPTER THREE

The Wooden Walls
of St. John's, 1863–1916

"*I had always been a high liner, never coming
back without the seals, always with the best ship,
and with the best crew, and it was my proud
boast that I had never lost a man.*"

Capt. Bob Bartlett, *Sails over Ice*

For the introduction of the full-rigged auxiliary steam ship, Newfoundland had Scotland to thank. The first vessels so equipped, the ss *Polynia*
and *Camperdown*, were sent from Dundee to try their luck on Newfoundland seals in 1862.[1] Dundee was just emerging as the leader
among its rivals (Hull, in particular) through the evolution of the "wooden
wall," a ship-rigged vessel with a small auxiliary steam engine; competitors had gone too soon to steel vessels, when it was not yet realized
that steel was more fragile than wood in contact with ice unless hull
shapes were substantially altered. Unfortunately for Dundee, this supremacy was gained when the Greenland whale and Jan Mayen seal
fisheries were already overexploited and in decline, particularly under
the new (1850s) pressure of Norwegian competition. In 1850 forty
British and foreign vessels took 400,000 harp seals from the Jan Mayen
herd (never part of Newfoundland sealing), a pace which could scarcely
last: within a decade, the herd's depletion was quite clear. By the late
1850s, however, the first steam tenders were at use in active whaling,
and steam seemed to promise new rewards in fresh fishing areas –
Newfoundland waters, for example, where Scottish whalers had long
been active but had not yet entered the seal hunt. A further inducement

was the increasing demand for oil as British textile mills turned to wider use of jute, which required much oil for processing.

The year 1862 was a bad one for sealing, and the Scots had little luck. The *Polynia*, for all her 9-knot steaming ability (probably an overestimate even for calm waters) became locked in the ice in Trinity Bay and carried as far as the settlement of Heart's Ease, whose population trekked over the ice to the vessel, no doubt to regale the crew with stories of the horrors of Newfoundland sealing, for Scottish competition was not needed. The *Polynia*, with her propeller blades broken by ice, limped into St. John's for repairs. Though unlucky with seals, the vessel was yet able to rescue a number of sealing crews from the bad ice that year, and her mobility gave cause for reflection to the principal Water Street merchants. Scottish vessels would return in the 1870s and keep coming until the end of the century, but these later arrivals were not begrudged since they sailed with Newfoundland masters and crews who knew the waters and the techniques and at least processed their oil in Newfoundland. (The Dundee fleet dwindled along with the Greenland whales; neither, for practical purposes, outlived World War I, but the seals had had the effect of continuing the Dundee whale fishery longer than otherwise might have been the case.) It should also be noted that their local "Newfoundland" rivals in actuality had their head offices in Great Britain.

The potential of steam was noted by several St. John's entrepreneurs, beginning with Walter Grieve, himself an immigrant from Scotland around 1820 (his firm remained based officially in Greenock).[2] For the 1863 season, Grieve purchased the ss *Wolf* (30 h.p. engine for a 210-ton vessel), while Baine, Johnston and Company, established in St. John's at the turn of the century, bought the slightly smaller *Bloodhound*. By the 1860s Baine, Johnston was controlled (through inheritance) by the Grieve family; Grieve was a partner until he withdrew to found his own business in 1868. His namesake, Walter Baine Grieve (1850–1921), was for many years the head of Baine, Johnston, and in this capacity will be met again.

Wooden steam/sail vessels proved the means to further exploitation of the harp seal herd for several decades, though the fishery had passed its prime. The *Wolf* returned in 1863 with 1,340 sculps and the *Bloodhound* with 3,000, reasonably promising beginnings even with the added cost of coal. Steam gave a clear advantage in maneuvering among the floes – above all, a steamer could go astern – and thus a much better chance at a full cargo. A number of St. John's firms were drawn into the contest, including the three which were to dominate the sealing industry along with Baine, Johnston and Walter Grieve. Job Brothers and Company, established in 1730, bought the *Nimrod* (later used by polar explorer Shackleton) in 1867. Harvey and Company, one of the oldest firms (1763) and one of the few large mercantile firms in New-foundland which was not the branch of a British company (though it had extensive Bermuda connections), purchased its first wooden wall, the *Panther*, in the same year. Finally, Bowring Brothers, long active in sail sealing, was rather slower in turning to steam, though the *Eagle* (1871) was only the first of what was to be the largest fleet owned by any single firm. Although Benjamin Bowring had set up shop as a watchmaker in St. John's in 1811 and had his own wharf by 1823 the firm was still headquartered in Liverpool.[3]

More than a dozen other concerns owned wooden walls at one time or another, but very few had more than one, and fewer still were owned outside St. John's. By the mid-1870s, vessels sailed from Catalina, Bay Roberts, Greenspond, Trinity, Pool's Island, and Fogo, but until the 1890s half at least still sailed from St. John's, and nearly the whole fleet was outfitted there. Only Harbour Grace tried to compete in a major way, but the figures tell the story. In 1870, fifty-three vessels (2,825 men) sailed from that port, but in 1880 only seventeen (1,515 men), and in 1890 three small steamers (600 men). Munn and Company, the principal firm in Harbour Grace, was a steady rival of the bigger St. John's firms, and Munn family leaders commonly spoke for Harbour Grace in the Assembly. Munn was the only non-St. John's company to own one of the larger steamers for sealing (ss *Commodore*), but it could not continue to compete in the twentieth century.[4]

Of fifty-eight wooden walls which participated in the seal fishery from 1863 until the last survivors decayed away after World War I, thirty-five were operated by the five major St. John's firms, reduced to four when Walter Grieve retired to Scotland in 1879. Of these thirty-five, the majority (twenty-six) were purchased in the period 1866–74, after which it was a question of replacement rather than expansion. The vessels were bought abroad, since Newfoundland had no facilities for their construction, and they were seldom new. The initial rush was explained by high prices for cod and seals just when the output of the latter increased, with correspondingly low prices for used Scottish steamers.[5] With some famous exceptions, the life expectancy of these ships was not high. Only Baine, Johnston, Job, and Bowring could keep more than a single steam vessel at the ice, and only Bowring's was able consistently to send five or six ships.

These establishments, the elite of Water Street, pursued multiple interests. Bowring's, eventually the most famous, operated five or six sealers, several coastal trading vessels, and several more foreign-going cargo ships (for which purpose sealing vessels could also be used). Bowring vessels carried salt cod and seal oil and skins to Europe, the West Indies, and Brazil, returning with anything from salt (an important item for cod and sealskin processing) to molasses, not to mention man-

ufactured goods, all to be sold at a considerable markup to lesser merchants who added their own profit factor. Cashin at Cape Broyle operated
in just such a manner, and in addition obtained credit from Bowring's
to expand his own business. Bowring's allotted him a fixed number of
berths on the sealing vessels to maintain the local influence, economic
and political, useful to him in his Ferryland district as a candidate for
the House of Assembly.[6] Bowring interests were thus very widespread,
and included directorates or interests in cordage, cooperage, foundry,
paintmaking, banking, and mining firms; without maintaining a single
outport branch, the firm yet covered the island.

The result, therefore, of the coming of the wooden walls was the
continuation of the process begun with the large sailing vessels: the
concentration of wealth in the hands of Water Street owners. Schooners
still sailed, but the substantial catches were now made by steam vessels,
and there was little profit in building small wooden sailing schooners.
Outport economies, based as they were upon sealing and shipbuilding
(and maintenance) in addition to cod fishing, suffered everywhere. The
short-term effect was particularly serious in Bonavista and Notre Dame
bays, which were now cut off from the hunt except for those men who
were prepared to walk south for a berth, carrying or pulling their gear
by sled – in February conditions – the many weary miles to St. John's.
Northern men were good sealers, however, and steamers soon sailed
from northern ports with locally recruited crews (a development which
not unintentionally brought the point of departure closer to the probable
location of seals). The construction of the Newfoundland Railway,
begun in 1881 and carried on through Harbour Grace (1884), and on
to Gambo (1891), the railhead for Northern Bonavista Bay (Bonavista
itself was reached only in 1911), facilitated the travel of sealers, particularly when steel vessels once more concentrated sealing at St. John's.
The railway also had the countervailing effect, however, of undercutting outports distant from any railhead, such as Twillingate.[7]

But the vessels were owned in St. John's, and it was to that city that
the profits gravitated. While sealers might still make that hoped-for
cash income, the middle element of artisans and lesser merchants of the
outports lost ground or were destroyed. "The only consolation we can
lay to our hearts is that steam was inevitable; it was sure to come,
sooner or later, the pity of it is that it did not come later. Politics and
steam have done more than any other cause to ruin the middle class,
the well-to-dealers that once abounded in the outports."[8] D.W. Prowse,
who wrote these words in 1895, was observing Newfoundland in the
midst of a depressing financial crisis, but his conclusion has not been
discarded by modern historians, such as Ian McDonald, writing eighty-
five years later: "Lacking either the capital or the entrepreneurial skills
to respond to the new challenge, Conception Bay and the northeast

The Wooden Walls of St. John's

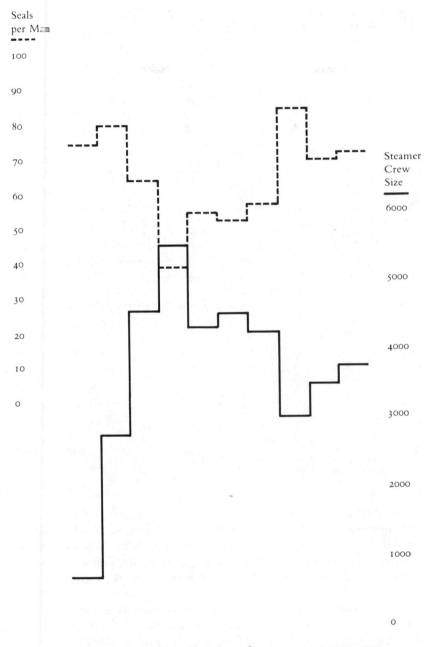

FIGURE 3 · Seals per Man and Steamer Crew Size, 1865–1915

Source: Chafe's Sealing Book (1923).

TABLE I
Newfoundland Steam Sealing, 1863–1913

Year	Number of Vessels (Second Trips)		Men	Seals	Year	Number of Vessels (Second Trips)		Men	Seals
1863	2		210	4,340	1888	15	(1)	2,907	195,191
1864	3		311	1,059	1889	15	(2)	3,083	252,887
1865	3	(2)	323	19,086	1890	15	(1)	3,309	165,052
1866	4	(2)	460	28,058	1891	15	(4)	3,352	280,928
1867	8	(5)	1,026	128,050	1892	15	(5)	3,838	303,196
1868	7	(4)	823	57,800	1893	18		4,092	109,304
1869	8	(7)	913	76,620	1894	17		3,809	132,792
1870	10	(6)	1,050	102,310	1895	16	(2)	3,784	234,993
1871	12	(10)	1,431	178,769	1896	21		4,506	187,517
1872	16	(7)	2,348	76,291	1897	20	(2)	4,943	126,628
1873	17	(17)	3,198	262,261	1898	18		3,810	241,708
1874	19	(14)	3,489	175,378	1899	18		3,503	268,881
1875	19	(13)	3,357	252,880	1900	19		3,760	353,276
1876	17	(13)	2,963	193,990	1901	19		3,836	345,055
1877	19	(19)	4,112	258,146	1902	20		3,798	274,219
1878	22	(16)	4,491	249,716	1903	22		3,378	317,560
1879	22	(13)	4,127	305,929	1904	21		3,328	284,473
1880	24	(20)	4,894	124,968	1905	22		3,532	177,100
1881	22	(17)	4,745	281,949	1906	25		4,061	341,836
1882	21	(17)	4,344	137,864	1907	22		3,513	245,051
1883	20	(16)	4,625	247,230	1908	18		3,141	213,863
1884	17	(6)	4,078	178,198	1909	20		3,580	260,302
1885	18	(5)	7,466	174,681	1910	19		3,364	333,349
1886	16	(12)	3,322	169,890	1911	18		3,973	304,594
1887	16	(9)	3,323	146,204	1912	23		4,179	175,430
					1913	19		3,609	272,965

Source: Chafe's Sealing Book (1923).

coast by 1884 saw the once proud schooner fleet reduced to fifty vessels ...
The number of sealers fell to 6,000, sealing skippers and schooner
owners were bankrupted, outport seal-oil refineries shut down their
vats, and the artisan class found itself reduced to the ranks of the fishermen. The shock to outport interests was brutal and the social upheaval
great."[9]

It was bad enough that a sealing work-force of 13,000 men in 1857
had been cut to under 6,000 twenty-five years later; jobs were also lost,
or substantially altered, in the Labrador cod fishery, for the schooner
fleet was also the Labrador fishing fleet, accounting for 25–30 per cent
of Newfoundland's cod exports. But the schooners were also ideally
suited, economically, for carrying salt cod to foreign markets. The

steamers, with their bigger overhead and larger cargoes (requiring urban markets), were not. "Newfoundland's marketing troubles began only after the steamship replaced the sailing ship at the seal fishery."[10] The export of cod became concentrated in St. John's as well; thousands of men formerly tied to the schooners now became independent small boat fishermen who outfitted in St. John's and had their craft freighted back and forth to the Labrador by "city" steamers. More of the Labrador catch was now sent to Europe direct by steamer or to St. John's for reshipment. The outports had no steamers for the European trade, which was now similarly in the city's hands, and those few but important outports which had carried on a significant foreign trade of their own now subsided into total dependence.

Still, there was naturally work to be done on the steamers in St. John's.[11] The hulls had to be sheathed with especially tough woods – greenheart from British Guiana, ironwood from Australia. Protective iron plates were added to the bows and hardwood slip boards built to protect the vulnerable rudder and propeller. (This vulnerability in part explains why the wooden walls remained full-rigged ships despite the expense entailed, though owners might well balance the cost of rigging against that of coal in the former's favor.) Reinforcing balks of timber were set within the bows and laterally at waterline and deck levels. Timber "pounds" or sections were boarded off in the hold, for a heavy load of sculps had to be kept from moving as much as possible. In addition to affecting the vessel's trim, movement generated heat and thus hastened decomposition; a full ship locked in the ice could see its entire cargo start to "run" as the fat melted. The decks were protected with a layer of boarding, without which the "chisels" or "frosters" attached to sealers' boots for traction on the ice would knife the deck to splinters. Coal, provisions, explosives (for blowing a way through the ice – in at least one case, that of the *Viking* in 1931, with the effect of blowing the entire vessel to bits), all had to be stowed aboard.

But the vessels were not built in St. John's, nor was subsequent maintenance all it might have been. What mattered was that these vessels, which had often started life as a Dundee or Norway whaler, got to the ice, and not that they remained in pristine condition. Used steamers, like sealers, were in oversupply; it must often have seemed as if whaleboats had been replaced by sealing punts but little else visibly altered. In the early twentieth century, when steel vessels began to be introduced, the wooden walls were even more likely to have no other purpose than sealing (a few were used for whaling, exploration, or transporting immigrants), and thus to be doomed to lie up in St. John's for all but a few weeks each spring. To their owners, general upkeep was an unwanted expense, an attitude which helps explain why so many vessels were lost in the hunt (forty-one of the first fifty, 1863–1900);

fortunately, the crews of lost ships were generally able to escape over
the ice. Of the fifty-eight vessels joining the fleet between 1863 and
1924, only six survived into the 1930s; but one, Bowring's lucky ss
Ranger, made sixty-nine trips to the ice between 1872 and 1941 under
nineteen masters, returning with a total of just under 950,000 sculps.

Such statistics were the delight of Newfoundland in the heyday of
the wooden walls, when bets were made on the first ship home or the
total catch for the year. Those who did not go sealing, after all, always
knew someone who did. Pride was taken in the greatest number of
steamers going to the ice (twenty-seven in 1880), or number of men
(5,815, counting second trips, in 1881). The smallest (ss *Ariel*, 79 tons)
was contrasted with the biggest (ss *Newfoundland*, 568 tons). Poor catches
– the greatest humiliation being to miss the seals altogether and return
"clean" – were compared with record hauls: the *Neptune's* biggest single
catch (some 42,000 seals in 1888) or the *Commodore's* bigger haul per
ton of 31,314 in 1872 (the *Commodore* was 290 tons to the *Neptune's*
465). The *Commodore's* record catch weighed in at 665 tons, and Capt.
Azariah Munden could not take it all on board, despite throwing coal
reserves over the side, but had to tow long lines of sculps into Conception Bay as he made for Harbour Grace, so heavily loaded "that the
crew could easily wash their hands over her side."[12] The biggest catch
for one year for one vessel (ss *Proteus*, in 1875, made two trips to the
ice for a total of 44,377); the quickest trip (the *Southern Cross* sailed at
8 A.M. on 9 March 1901 and returned to Harbour Grace at 6:30 on the
20th with 26,563 sculps); the biggest share per man – a figure of even
more commanding interest to sealers (the *Retriever* of Harbour Grace
brought back 23,400 seals in 1866, and each man received a share of
$303); the famous explorers, such as Scott's *Terra Nova* or – most
famous of all in this regard – the *Bear*, a vessel which has had at least
two biographies[13] – all these were at the heart of contemporary Newfoundland concern and later folklore.

And steamers provided fertile ground for the feelings of awe and
wonder which give rise to mythology, not simply from the many
disasters to vessels. The experience of going to the ice on a wooden
wall – an experience which, it must be repeated, was known first-hand
to a smaller and declining share of the population – differed in kind and
degree from sailing on a schooner. Most obvious was the increased size
of crews. Steamers nearly always carried over 100 men, commonly
200, and sometimes over 300, all the better to kill, load, and unload
the most seals in the least time and perhaps make a second trip to the
ice. A fair average would be 150 men; any ship over 300 tons was
likely to have close to 270 men, the maximum crew size by law as of
1898: both masters and sealers disliked overcrowding, which usually

did not increase the catch but did reduce each man's share (Abram Kean tended to blame the politicians for the overcrowding).[14]

Steamer owners put so many men on each sealer that a berth ticket was little more than permission to go aboard (though to owners, the money paid by the sealers was not an insignificant item) and certainly no guarantee of an actual berth. Berths had to be shared, and sealers provided their own bedding, which might be nothing more than hay from southern hills outside St. John's kept in a box or bag along with a man's clothing and food. But even the berths were likely to be turned into storage space for sculps if the trip went well, forcing most men to sleep when and where they could: the "ballroom" or fo'c'sle could only house a small percentage of such big crews. With minimal change of clothing, little opportunity to wash, and a sham of a toilet (one "outhouse" on the bow, another on the stern), the sealers soon resembled the demons of hell to those few outside observers who experienced a sealing voyage and wrote up their impressions.[15]

Many sealers had little experience with large vessels, and might be seasick in a swell; sealing, after all, was less "seafaring" than a landsman's hunt carried out on the ice. The resulting atmosphere of a dark, dimly lit hold, fogged by smoke from strong tobacco, and the reek of grime of all sorts is hard to conjure up in the mind, but witnesses found it all too believable:

Crowded like pigs on a sealing steamer, they cultivate a positive affection for dirt and regard it as a kind of honorable badge of their adventurous calling. During a voyage of several weeks they never take off their clothes, even to sleep. The oil from seal blubber fairly drips from their garments, dirt, soot, and tar adhere to their faces in steadily thickening strata, and when they finally enter port to strut the streets in unwashed glory they are incarnate emblems of filth and odor. A night in St. John's after the arrival of two or three lucky seal crews means bedlam for the city. Honest burghers fly the streets and look well to the doors and shutters o'nights.[16]

The "afterguard" of captain, second hand (who elsewhere would have been called "first mate"), the doctor (if any), "Marconi," the radio operator (again, if any), and the chief engineer lived separately.

Food was limited in quantity and quality, essentially hard bread or flour, salt pork, molasses, and tea, which the men cooked themselves (though not individually) in "cabooses," or stoves bricked in on deck. Duff, flour and water mixed with raisins and boiled in a bag, was notorious, hard enough in half an hour to make a serviceable football, according to the tall tales. Yet to some men these rations (with what might be filched from the afterguard's more copious and varied foodstuffs) was one of the appeals of the voyage: "It must be partly these glorious doughs [duffs], this unlimited molasses, flour, butter, pork, tea, dried fish, seal meat, etc., which lures so many to strive for berths. For, sure enough, many have been on uncommonly short allowance all winter, and some have hardly known what a good square meal means at all," wrote Dr. Grenfell of the men with whom he went sealing in the 1896 season. [17]

Vessels had tanks for only a few days' water for such a crowd, relying on refilling from fresh water ice pinnacles which form in the rafted ice fields. The ice made drinkable water when melted, but it was brackish and seldom slaked one's thirst even when turned into the strong brew known as "pinnacle tea." The ice had to be dragged across decks whose conditions is best left to the imagination and the sealer was likely to find more than tea in his mug. One reason why seal meat was abandoned except for organs and flippers for immediate consumption was that in such unsanitary conditions it would soon spoil.

Not only were living conditions aboard the steamers affected by the presence of big crews; for the same reason, more elaborate organization and tactics were necessary in the hunt itself. [18] The crew was divided into three or four watches of forty to fifty men each under an experienced sealer known as a "masterwatch," himself often perfectly capable of being master of a vessel were more commands available – many men who commanded their own ship in the cod fishing season went to the ice in lesser positions in the spring. [19] The other subordinate officers had to be capable – they included the second hand, "barrelman," who had the responsibility of spotting seals from the crow's nest, "sternman," who warned of ice nearing the propellers when going astern, "scunner," who from the foretop watched for leads in the ice, "tallyman," who counted the sculps coming aboard, and "wheelman" at the helm – but the masterwatch had the greatest role to play on the ice.

Leading his men in the designated direction for seals, he pointed subsections, "ice parties," this or that way, for so many men could kill a dense patch of 20,000 in a single day if the entire crew was well placed. Meanwhile the vessel moved on to leave other watches, which might be fanning out as far as three or four miles from the vessel's track or "cut." At the end of this run, the men could be ten or twelve-miles away – far enough to be in serious danger if there was an unobserved sudden change in the weather, meaning that experience was at a premium. If caught out on the ice, the men might have a punt (for crossing leads or hunting seals in the water) to use as a wind-shelter, or ice could be piled up in a wall, even a fire built (with shavings from a gaff for tinder and whitecoat fat for fuel); but with every care and the custom of always working in pairs and larger groups, men might be lost or frozen before they reached a rescuing vessel, especially if forced to remain on the ice overnight.

The sculps could not really be towed to the vessel for such a distance, and were piled in "pans" on the floes designated by the masterwatches and marked with 9-foot flags bearing the owner's insignia. When the vessel returned, the sculps were "whipped" or winched aboard – if bad weather or night had not set in and made the pans unfindable. Pans could disappear for several causes, including theft by another vessel. Sealing masters were highly competitive, and although there were certain customs, such as not cutting in front of the probable corridor of another ship as it headed into the ice to deposit its sealers, some captains were by no means adverse to following the wake of a successful rival, and a deserted pan of sculps was an easy temptation.

When the men reached the vessel they scrambled to get aboard, often having to jump for the sides of a moving 300-ton vessel carrying "side sticks," sturdy 5"x5" timbers chained along the sides to facilitate boarding. The desirable procedure was to get all the sealers aboard before the vessel "burned down" for the night. The sculps, once taken aboard, were left on deck to cool, but if it was necessary to send them below, they would be packed in ice to retard spoilage. When the vessel was "swatching," or hunting mature seals in the leads, only the punts (not enough to hold the entire crew) were overboard, with four-man crews selected by each foregunner.

Obviously there were hardships, and danger, and unappealing conditions, but for all that there was also the sense of excitement, of newness, of adventure. The men shared a feeling of cameraderie as members of a group dedicated to one purpose – "bloody decks and a bumper crop" as the sealers' toast had it. On the other hand, there was clearly also a competitive atmosphere, for it was the unflagging veteran who took the most seals or towed the most sculps (five or six made a 300-pound load) who had the greatest credit in this enclosed society. Ship's officers had to be strong men to command and to keep control, if necessary by the use of force. David Lindsay, surgeon on a Dundee sealer, the *Aurora*, in 1884, tells of a man who complained to the captain about the duff, breaking it in half and asking him to look at it. "But the Captain was a man of action, so he planted a blow between the man's eyes and asked him to look at that; the man dropped back dazed and the trouble came to an end at once."[20]

By all accounts such methods seldom were necessary, even though crews might be very mixed lots of Protestant and Catholic, Irish, English or Scots, religious and ethnic mixtures which on land might be explosive. It helped that liquor was very seldom allowed on board: such leisure as the men had was most likely to be spent in song or story

sessions, with a bit of gambling thrown in. Fights there were on occasion, but these took place more often before sailing as the men sorted
each other out or after unloading to pay off old scores. The common
feeling, aside from hunger and exhaustion, was enthusiasm, which a
shrewd master played upon. Sam Blandford, who when he found the
patch called his men together to address them, was in the habit of
pointing out how hard they had worked to get that far, how necessary
it was to continue to work hard for wives, children, owners, and the
ship herself. There was money to be made; "My God, men, it is like
picking up two-dollar bills off the ice; every half-hour you can get a
tow of seals worth eight dollars, every seal weighing fifty pounds. This
cannot be done in any other part of the world," or so Nicholas Smith,
who sailed with Blandford in the *Neptune* in 1888, recalled.[21] Blandford, of course, failed to mention the rather unequal distribution of the
rewards.

In fact, in the steam era, less of the eight dollars for a tow (four
sculps, presumably) went to the sealer's pocket than in the days of sail.
The record $303 was normally an unobtainable dream; Smith's return
from his voyage with Blandford was, at $77, unusually high. While
profits may have become greater and more regular with steam, wages
at first held their own and then declined from the standpoint of average
share per man. This was so not least because owners soon (though the
date is unclear) decreed that owing to the greater investment required
by a steamer and high outfitting costs their own share henceforth would
be two-thirds, not the one-half customary in sailing vessels. Since steamers
were more likely to find seals and clean off the main patch before
schooners could get to it (even though as of 1873 steamers had to give
sailing vessels a five-day head start),[22] sealers had to accept the new
distribution, knowing that while a fully loaded schooner paid more per
man, the key phrase was "fully loaded."

At first steam led to increased catches overall, though still never at
the level of the 1840s and of course varying widely from year to year.
Chesley Sanger has calculated five-year high-low ranges of averages
(not maximum-minimum figures) which show a clear decline to the
end of the century after the mid-seventies surge provided by the new
technology was passed:[23]

1870–4	$55–90
1875–9	20–45
1880–4	10–40
1885–9	8–35
1890–4	10–40
1885–1900	5–25

The hope was always there for a substantial return. In 1892, for example, sixteen vessels sailed, five making second trips: it was a good year for the sealers. An experienced hand aboard the ss *Diana* who made both trips earned $184.30, but a man aboard the ss *Eclipse*, a Dundee whaler-sealer which was bought for the 1892 season, earned a low $13.04. It is no wonder that given a choice sealers went with the lucky ship and the lucky captains, the real "jowlers" who were more likely to be "highliner" of the fleet, first to return "log-loaded" with all sails flying. There was no guarantee, though the chances were perhaps better with one of the famous northern families of captains, such as the seven Winsors, eight Barbours, nine Keans, or ten Bartletts.[24] In 1893, a bad year, no vessel made a second trip. The ss *Newfoundland*, first in, paid $68.00 per man to top the year, but the man on the *Diana* who made $184.30 in 1892 would have earned only $17.48 the next year.

The *Diana*'s voyage of 1892 demonstrates the profitability of a particularly good voyage: 40,904 seals in two trips, including so many in the second trip that 3,000 mature seals were brought home on the ss *Falcon*. The crew's share of $184.30 paid to each of the 224-man crew totalled $41,283.20, meaning, on the basis of a 1/3–2/3 division, a return to owners of $82,566, not counting any other charges to the sealers. Outfitting costs might run from $15,000 to 20,000, leaving at least $60,000 profit from this single voyage as return on investment. The cost of the *Diana* to Job's is unknown, but the comparable *Eclipse* (the *Diana* was 290 tons, the *Eclipse* 296) was valued at £12,347 in 1868, or about $60,000 (the Newfoundland dollar was worth 4/–); thus the entire cost of the vessel might be recovered in a single good year or two. On the other hand, the *Diana* brought in 7,263 seals in 1893, paying her owners $10,738.40, certainly less than outfitting costs.[25]

It is no surprise, therefore, that owners were careful to collect other charges from the sealers when they could. Berth money was apparently not assessed on sealers in steamers, perhaps because a much smaller percentage of the crew actually worked in punts and thus the distinction between gunner and batsman was less meaningful. But at some point (as with the allocation of shares, the precise date is not known), probably quite soon, owners found a substitute in "coaling charges," making the claim that while in the days of sail the men helped make and rig the sails and did much of the preparation for departure, now, with steamers, they arrived to find the vessel already fully loaded by dockhands at the owner's expense (no matter that the men would rather have done the work than pay; the owners could always use time as an argument, since the men normally arrived only shortly before sailing). At the end of the voyage, also, the vessel had to be cleaned for laying up or alternate employment.

By the turn of the century "coaling money" was a common $3 a man. Contemporaries called it "berth money," seeing little to distinguish it from the old system. Still another charge was for "crop"; owners argued that the men need not take it, though any man who refused might find himself without a berth. Then there was tare or "back money" – a few cents on each sculp charged to the men for meat adhering to the fat, whether there was any or not. All this could amount to a considerable sum which the owners were diligent in collecting. After all, 200 men paying $3 each would return an extra 1 per cent every year on the same $60,000 investment – a form of dividend which cost the owners nothing.

The sealers, on the other hand, saw their return steadily fall. They were aware of the vagaries of the hunt and the wild fluctuations of return to the men on individual ships in any given year, but they also knew that the basic source of their income, the price of young harp seal fat per hundredweight, had dwindled steadily in the mid-nineties depression years, reaching bottom in 1899 at $3.00, rising only slightly to $3.25 in 1900 and the following year. This figure – set by the St. John's sealing owners – was already very low, and a rumor that the price would be lowered again to a new record bottom of $2.40 in 1902 touched off the one significant twentieth-century labor action in the Newfoundland sealing industry.

The strike was important, but so too was the context. The issue was not merely a low price for 1900 and 1901 and a still lower one for 1902, but the long steady general slump in Newfoundland's economy which, along with slipping world demand for seal products of the sort exported by Newfoundland, had prompted the merchants to cast about for available remedies, including lower dockside prices for sculps. But the heart of the economy was still cod, and the cod fishery in the 1890s was in great trouble from which it had not escaped by the time of the sealing strike. The export price for salt fish fell substantially (32 per cent in the years 1880–1900), with production similarly falling some 20 per cent in a downward spiral. In the era 1884–1911, roughly 30 per cent of the 60,000-man labor force left fishing to go into other areas, a fall which includes out-migration (Newfoundland lost a steady 1,500 per annum, 1884–1935). The decline of the fishery, it should be stressed, led to emigration, not vice versa; although on the whole the population of Newfoundland rose about 2.5 per cent per annum during the nineteenth century (1805–1914), cod exports rose in the same years less than 1 per cent a year.[26]

The issue is complicated and largely falls beyond the scope of this study, but essentially the Newfoundland cod fishery was losing ground

1 St. John's from the Narrows (late nineteenth century)

2 St. John's from the air, looking toward the Narrows (early twentieth century)

3 Schooners off the south side of St. John's harbor (early twentieth century)

4 Typical Newfoundland schooners (late nineteenth century)

5 "Wooden walls" ready to depart for the ice, St. John's, c. 1900

6 Sealing works and schooners on the south side of St. John's harbor, c. 1900

7 A.B. Morine addressing the sealers, St. John's, 1902

8 Sealer being towed out from St. John's through the ice by her crew

9 Sealer's crew cutting her through the ice

10 Sealers over the side, breaking ice

11 "Wooden walls" racing out at the opening of the season

12 Capt. A. Jackman, "The Old Scorcher"

13 Capt. Samuel Blandford

14 Sealers readying gaffs

15 Sealer's punt for hunting adult seals among the floes

16 Scattered seals among the floes

17 Removing sculp from whitecoat

18 Towing in sculps

19 Panning sculps

20 Attaching whiplines to panned sculps

21 Preparing sculps for winching aboard

22 Loading fresh sculps

23 Sealskinners at work, St. John's (late nineteenth century)

24 Loading packed skins for export, St. John's

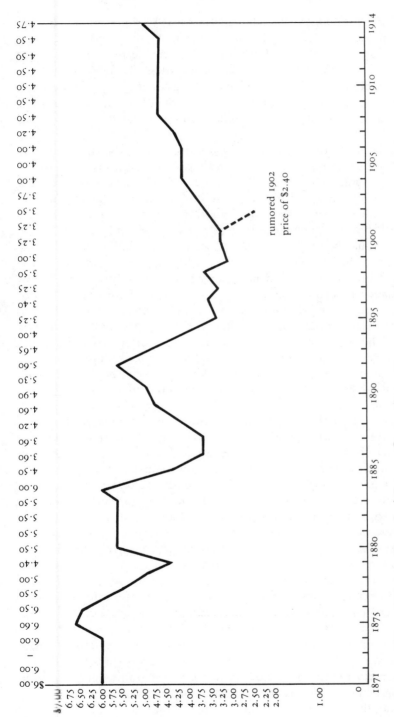

FIGURE 4 · Price of Young Harp Seal Fat per Cwt., 1871–1914

rumored 1902
price of $2.40

Source: Chafe's Sealing Book (1924).

to a better product from Iceland and Norway (seriously undercutting, in the process, the argument that Newfoundland's poverty was based on skimpy resources, for Iceland and Norway were if anything less naturally endowed, although Iceland had a longer fishing season). Unified marketing procedures in both competing countries permitted greater quality control than was practised in Newfoundland, where the credit system tended to discourage modernization of catching and curing methods.

But a decline of the cod fishery not only raised the cost of public relief, it also meant construction and repair of fewer vessels, thus affecting the entire remaining shipping industry. Fewer vessels meant less sealing, aside from the big steamers, making those steamers and the berths they offered all the more important – just at a time when their number, too, was falling due to declining demand, and experience was showing that between 3,000 and 4,000 men produced the optimum catch per man. (In 1902 the total according to Chafe was 3,978 men sailing from St. John's and northern ports.) In the meantime, Newfoundland plunged into an enormously expensive railway project, designed in part to offer alternative means of development in such areas as mining and forest products (though the big pulp mills came after 1905). The costs were huge, the immediate expectations wildly overoptimistic, and the result a steady slide towards bankruptcy under the twin burden of relief and railways. The merchants borrowed in the hope of better days, and, simply put, a serious bank crash in 1894 was the result.

Depressed times are not normally conducive to labor aggressiveness. There were, too, many other factors which made the strike unlikely – or even impossible, one might have said on the eve of the event.[27] Some are general, such as the religious and ethnic diversity of the population and its physical dispersal in 1,200 settlements along 6,000 miles of coastline, remote from all but telegraph and maritime communication. These communities were homogeneous in ethnic and religious content but were not self-reliant institutionally. There was no municipal government, for example; even school boards in a colony without state-run educational institutions were confessional and chosen by the churches. There were no district-level governmental operations; the only intermediary between outport and St. John's government was the Assembly district. Relief payments, medical treatment, communication development, sealing berths – all depended in a major way upon the relationship of the community with its Assembly representative. If that link broke, there was no other means of political expression. Everywhere party politics were slow to develop; only in St. John's and the Avalon Peninsula, where the population density was higher, was there anything like a meaningful political opposition.

St. John's, with a relatively stable population of 30,000, was another matter. Here the dominant mercantile elite was predominantly English and Protestant, but the lower class largely Irish and Catholic. It is no surprise that when in the course of the 1902 strike the owners unsuccessfully suggested arbitration the proposed committee would have included both Catholic and Anglican bishops for proper balance.[28] Such balance was perpetuated in the upper house, the Executive Council, to which appointments were made for life in such a way as to satisfy the denominational interests. But the real power was held by the Assembly, and that body – save for an occasional well-known sealing or Labrador captain who was given a quiet backbench position – was run by Protestant English professional men, lawyers, doctors, smaller manufacturers. In close alliance with the larger fishery-controlling merchants, whose viewpoint on fishery matters was seldom challenged, they insured that the highly centralized and very conservative system remained intact. These facts explain Newfoundland's long-standing opposition to federation with Canada: the oligarchy feared for its position and the Catholic element for its independence in a mainly Protestant (and English) Canada – French Catholics of Quebec had little attraction for the Irish of Newfoundland. For the outport fishermen, therefore, there was almost no chance for active political expression. Had there been it is arguable whether it would have been used, owing to the low level of literacy (30 per cent of the population over the age of ten were literate at the end of the century). David Alexander had singled out just this wide-spread illiteracy as the principal cause of Newfoundland's failure to keep up with such comparable areas as Iceland and Sweden.[29] Alexander realized that the causes of illiteracy are complex; his point was that it was not enough to cry "poor resources" as the city did or "exploitation" as the outports would have it. Illiteracy, economic dependence in the credit system, political dependence in the constitutional system (technically that of 1855) – these and the general economic situation all argued against militant labour action.

These conditions explain why the Newfoundlander is so often accused of displaying a sense of superiority, of acting as the hardy, independent survivor – an appearance which disguises his awareness of an inability to alter his situation. "It is a short intellectual journey from the view that the world is essentially debased to the view that nothing is worth changing," as Peter Neary has put it.[30] It may be, as Richard Gwyn, author of a perceptive biography of Joey Smallwood, Newfoundland's most famous politician, has written, "The struggle for survival exhausted ambition and creativity," but it would appear that there is more to it than that.[31] Smallwood himself was responsible for

Gwyn's remark, for he wrote much the same thing in an uncharacteristic moment of doubt.

It is a fact that for centuries we have lived by *killing* cod and other fish; by *killing* seals in the water and on the ice, and animals on land; by *cutting down* trees. Has all this developed in us a trait of destructiveness, or narcotized what ought naturally to be an instinct for creativeness? Is it not true that we have been intensely, bitterly individualistic, each of us preferring to paddle his own canoe and turn his back upon the other fellow? Have we not failed almost completely in the one virtue that the modern world has made an absolute essential, the ability and the desire to achieve a commonly desired end?[32]

It is not necessary to agree with Smallwood's analysis: after all, the Christian ethic has for centuries, and with only rare exceptions, dictated exploitation of resources for the glory of God and man, but the pessimism and the bitter distrust of what the outside world may inflict is, and was, real enough to be considered here.[33]

On the other hand, that very sense of separateness may have played a role in the strike. The outport men tended to stand together as outsiders – mainly from northern districts – in St. John's. They also had to aid their solidarity a history of intermingling, in the seal fishery itself, on the Labrador, in the off-season logging camps. This intermixture was far less common in the south, where the standard fishing method was more often the two-man dory on the banks paid by the day's catch, not the vessel's total haul – a system which pitted man against man.[34] All sealers, too, shared distrust of the owners, whom they considered responsible for the decline of prices, cod or seal, and the decreasing number of berths, however involuntary the decrease had been.

There was also some history of labor relations in Newfoundland, though it would hardly seem so from the record of labor action. As George Allen England wrote in 1924 (he went to the ice in the *Terra Nova*), "I have never known a country whose employers enjoyed such a sinecure as in Newfoundland. Labour, there, had hardly begun to dream that it has any rights."[35] The sealing incidents of the 1840s and 1860s were not very forceful precedents. There were unions in Newfoundland at the turn of the century, but they were skilled craft organizations such as the shipwrights (first to organize in 1851) and the sealskinners (1854), with the normal limitations of elite labor organizations. The skinners did strike Bowring's in 1900 to protest the introduction of skinning machines and disabled some machines at Job's the following year. This may have contributed indirectly to an atmosphere sanctioning labor activism but the skinners and the sealers were from different environments and social strata.[36]

More influential was a strike at the Bell Island (Conception Bay) iron mine against the Nova Scotia Steel and Coal Company – Dominion Coal Company operation.[37] From 11 June to 24 July 1900, the miners struck in an attempt to raise their pay by 40 per cent, an impressive figure until it is realized that their pay was 10 cents an hour. The situation became dangerous when the 1,600 miners attempted to stop the company's use of imported longshoremen, and later clerical staff, to load ore. A substantial share of the Newfoundland Constabulary was brought from nearby St. John's, and calm soon prevailed – but the top three officials of the "Wabana Workmen & Laborers Union" spent a week in a St. John's jail. The workers emerged with a compromise, not the 14 cents they asked, but 12.5 cents for skilled and 11 cents for unskilled laborers. The Bell Island strike ended without serious violence, but it left a bitter taste and a long-lived memory, proving at least that Newfoundland was not totally removed from the problems of labor-management relations common to the rest of North America. It remained true, however, that on the whole Newfoundland was behind in this development. Working-class political activity, let alone socialism, was minimal, but it served nevertheless as an adequate target for those seeking a scapegoat for the strike. As the *Trade Review* put it, experience in the mines (whether Bell Island or in Nova Scotia), or on the railroad, or in the United States, did its part, and "the home schoolmaster did the rest."[38]

The sealers' strike, like that at Bell Island, ended peacefully, but if anything its impact was greater. The events as they unfolded say much about political reality in Newfoundland, despite the fact that the evidence is limited to a few government documents and newspaper accounts.[39] How it began, alas, is a fog-shrouded question; it is not clear, for example, whether the movement had any planned preliminary organization. What is clear is that the authorities had absolutely no warning of its approach.

In 1902, the sealing vessels were to sail from St. John's, as the law specified, on Monday, 10 March. On the 7th, as was customary, the governor, Sir Cavendish Boyle, visited the ships in the morning and then entertained most of the captains to lunch, where in the genial camaraderie there was never a hint of trouble. But the next morning, at about 9 o'clock, the men of ss *Ranger* marched east along Water Street from Rendell's Wharf in a procession headed by a Union Jack or two. Along the way, men poured out of the various vessels so quickly that presumably there was some prearrangement. Leaders pushed up to the head, particularly one Simon Calloway.

Each steamer's crew did fall in line
While cheers out loudly rang,

Led in by one brave Calloway
The leader of the gang.
Free berths, it was their motto boys,
And no man would give in,
A fight for death or glory boys,
This victory to win.[40]

Reaching Government House, the crowd demanded to see Boyle, perhaps realizing that there was more likelihood of sympathy from him than from Premier Sir Robert Bond, who represented the Liberal Assembly majority. Boyle, only recently arrived from his previous post in British Guiana, was understandably alarmed and annoyed to be told that a crowd of some 3,000 sealers accompanied by perhaps 1,000 bystanders was outside demanding to be heard, when the previous day not a word of trouble had been whispered. When Boyle spoke with them, it was clear what they wanted. The 1901 price of $3.25 had been low enough, but the sealers would never accept $2.40 and in fact demanded $5.00 or they would not sail.

At Boyle's suggestion, sealers and owners commenced discussion of the issues. The owners were represented by Walter Baine Grieve, who controlled four of the fourteen vessels scheduled to sail that year from St. John's (Bowring's had as many, but Edgar Bowring, head of the firm, was in New York), supplemented from time to time by James Baird of the firm of that name, and William C. Job. While the sealers managed to choose a committee of half a dozen to look after their interests, it is indicative that they chose another to do their talking: A.B. Morine. Morine was a qualified lawyer and King's Counsel, with his own firm, and many times had been returned to the Newfoundland House of Assembly from Bonavista, the home of so many sealers.[41] He had already been Newfoundland colonial secretary, as well as holding several other posts, and had run, unsuccessfully, for the Canadian House of Commons from his native Nova Scotia (Newfoundland did not join Canada until 1949). Morine was a man with wide influence, probably the ablest speaker in the Assembly, in which he headed the Conservative opposition to Bond. As legal consul for R.G. Reid, concessionaire for the Newfoundland Railway, Morine was clearly part of the power structure. He was also impulsive, genial, impersonal, and widely distrusted for his extra-Newfoundland ambitions and implication in an apparent conflict of interest (as Newfoundland minister of finance he had negotiated Reid's railway contract while still Reid's solicitor).[42] Morine had, even worse, an attitude of condescension towards Newfoundlanders – "too green to burn," he is supposed to have said of them; the latter part of his career was spent in Toronto, not St. John's. The sealers had little choice, however, and Morine was colorful and able; his pince-nez and walrus mustache were well known to outport men in his district. Morine wisely refused any fee from the sealers, who appear to have offered him 10 cents a man (but there was no guarantee for the $300, and much credit to be earned by the refusal).[43]

The subsequent discussions, carried on over two days, were observed by the press and are interesting as a reflection of the economics of sealing. The owners maintained that coaling charges did not pay for vessel preparation expenses, especially important now that owners' risk was greater as the price for fat fell on the international market. The sum of $20,000 – argued Grieve, at any rate – was necessary to pay expenses, and the owners were lucky if their two-thirds share was adequate to "square the yards": of the 1901 crop of 400,000 skins, only 100,000 had been sold by the time of the strike, and the rest remained warehoused in St. John's, England, or the United States. Even with coaling money and the like, Job for his part claimed that his sealing ships over the last decade had cleared on average only 10 per cent profit.

Though negotiations were promising, neither side would give way, and when the dawn of Monday, the day for sailing, came, the men – who had slept aboard their vessels through the weekend – scampered ashore, along with most of the firemen. A few vessels limped out to mid-harbor, but the sealers had won the first round, to the owners' considerable surprise. Negotiations that day made some progress but reached no solution, and the sealers found what shelter they could in various halls and private dwellings, with perhaps 1,000 left in the open in March weather. The poorer districts of St. John's in particular seemed to offer sympathy, but there were familial and associational ties in operation as well.

Throughout the strike, the situation remained generally calm, though at first light on Tuesday a colorful scene took place when Capt. Arthur Jackman arrived at Bowring's Cove to inform a crowd of some 200 that his *Terra Nova* would sail in ten minutes. "There was not a sound or word of objection," reported a newspaper, and a lane was opened for Jackman who went aboard. He was, after all, the "Old Scorcher," and he had been taking seals since his first trip as master in 1871.[44] He was Bowring's "ship's husband" – marine superintendent – responsible for maintenance and preparation of all its vessels, but when not at work or at sea he was a fixture in the back bar of Strang's liquor store near Bowring's. A few officers and men in his wake were blocked in their approach, for the strikers held them in nothing like the same awe. But Jackman did have some two dozen St. John's men aboard, enough to raise steam to head south to obtain a crew from Cape Broyle and the south shore.

Sam Blandford, equally famous (well over half a million seals since his first command in 1874), was unwilling to sit quietly while Jackman put to sea and attempted about 8 A.M. to get his *Neptune* under way. When the strikers learned of his intention, a crowd estimated at 1,000 took a heavy hawser from the vessel and hauled it up Prescott Street. "The older the crab, the tougher his claws," runs a local saying,[45] and Blandford put on full steam: either the line snapped, or Blandford took an axe and personally sundered the hawser, but in any case, to the jeers and catcalls of the crowd the *Neptune* steamed away for Bonavista. Two dozen of the Constabulary had been present but had not interfered, thus avoiding the incident that all wished to prevent.

This struggle over the hawser, and a scuffle at Grieve's gate later that morning, where some part of a crowd was forced away by mounted constables, were as close to violence as the strike came, though there were many moments when an open clash was possible. One reason for the avoidance of conflict may have been the awkward dilemma in which the sealers were placed, far from home and badly in need of the work, well aware that delay could mean the loss of an entire season's catch. Perhaps the pressure of the more mature men quieted the younger hotheads, though a group which the press called "extremists" did its best to keep the strike going. The support of a few men influential in northern bays, such as Morine, helped, and it seems likely that a magistrate's order to close all the saloons (probably issued on Saturday) had the same effect.[46] Finally, there was steady progress toward a settlement, with concessions made out of fears felt by both sides.

In the end, the sealers won a price of $3.50 and no coaling charge, a most significant victory though not the $5.00 for which they had first asked. By late Tuesday afternoon, the men had all returned to work and the ships had sailed for the ice, so the season was actually delayed for less than two full days. But while the strike was over, its impact was not. Boyle in particular was worried. "It may not be always the case," he reported to the colonial secretary, Joseph Chamberlain, in London,

that so much forbearance, as on the present occasion, will be exercised: a very slight loss of temper on either side, and especially on that of the owners and of those working with them, would have led to consequences of a most serious nature. Since the forenoon of the 8th until the evening of the 11th the town was at the mercy of between three and four thousand men, who were practically homeless, and, in many instances, without food, and who showed their power by preventing anyone from going on board the vessels, from which they were themselves, to all intents and purposes, deserters.

The lesson, he concluded, was the need to increase local forces through the establishment of militia or volunteer corps, since a few hundred trained men would have prevented the recent "impasse ... as discreditable to the Government of the Colony as it was disturbing to the peace and well-being of the community."[47]

Increased garrison or new militia might be long in coming, and Boyle also wanted faster interim action. On 18 March he asked his Executive Council to appoint a "Commission of Enquiry" into the strike, its causes, its effect on capital-labor relations, conditions in the sealing industry, and the preservation of law and order in Newfoundland in the future. The next day, the council declined to act, on the grounds that the inquiry would "revive interest in a dispute which has, fortunately, been brought to an end, and the recurrence of which can be easily prevented by the ship owners another year."[48] To reopen the matter now might jeopardize the cod fishery – a decisive argument. Reluctantly, Boyle gave way, chiefly because he was unwilling to risk going against the opinion of his ministers that the inquiry might endanger the status quo, including that of the sealing industry; and the Water Street merchants who dominated the government in council and in the Assembly were not about to undertake any searching self-examination.

Nevertheless, the strike had thoroughly alarmed the owners who, like Grieve, had appealed in some panic for troops and who now realized that there was a point beyond which the sealers would not be pushed. Along with the Bell Island strike, the sealers' strike had one further impact in the subsequent development of Newfoundland's most important early twentieth-century labor movement, the Fishermen's Protective Union (FPU).[49] A year after the St. John's sealing strike, the FPU's founder, William Coaker, entered Newfoundland labor history by establishing the Telegraphers' Union. Coaker, son of a Twillingate carpenter and masterwatch, grew up in St. John's where, among his early experiences, he witnessed a strike of young fishhandlers at Job's. He became a clerk, then an outport branch manager of a St. John's firm, later buying out the branch – only to be wiped out in the 1894 crash. Background and personal experience bred in Coaker a deep distrust of Water Street. He became a farmer in isolated Herring Neck Island. Patronage, however, won him a telegraphist's and postmaster's job as a loyal Bond supporter and led him into the organizational activity which would make him one of the two important populist leaders the island has produced (the other is Joey Smallwood, who idolized him).

Coaker, a strange, magnetic, gifted, melancholy, austere man, had a greater vision in his mind than another craft union: an organization combining all those workers who formed the heart of Newfoundland labor, the fishermen, loggers, and sealers. The same men often did the same work, and shared the same "community of unrest."[50] In 1908 Coaker organized his first chapter, and the movement spread rapidly.

Within a year there were thirty chapters in Notre Dame and Bonavista bays and soon a newspaper, the *Fishermen's Advocate*; by 1922, twenty local councils spoke for 20,000 men. Coaker's energy and stamina were of critical importance to the movement; its greatest strength would only exist under his charismatic leadership. Coaker soon saw that cooperatives and lodges, while important, were not enough. Much of his procedure – flags, banners, parades, uniforms – was lifted from the Orange Order of which he was a member, which helps explain why the movement was more successful in the Protestant northeast than in the less homogeneous Avalon, though fishing methodology, religion, and distance to St. John's all had a role in defining the geographical boundaries of the FPU.[51] In 1913, the FPU elected nine delegates to the Assembly, thus becoming the largest opposition party. The precise link to the strike is difficult to pinpoint: R. Hattenhauer's report for the Royal Commission on Labour Legislation in Newfoundland goes as far as possible in noting that the FPU "may well have been created or strengthened by the sealers' success of 1902."[52]

Coaker was now in a position to press for reform legislation, particularly since Bond had retired, his faction in shambles, and the remaining Liberal rump, also occupying opposition benches, was quite willing to cooperate with the FPU. FPU representatives in their guernsey sweaters and oilskins (badges of working-class membership which they wore with pride) represented a definite political threat behind the "Moses of the North."[53] The result was the sealing bill of 1914, for the oligarchy realized that it could not afford total alienation of the fishermen.

There had been sealing legislation before, such as that of 1873 regulating sailing dates for the fishery, but that act was an attempt to strike a balance between sail owners and steam owners and not to benefit sealers. The 1879 act was the first to adopt a conservation measure, since it prohibited the killing of "cats," or immature whitecoats of under 28 pounds (loads with high percentages of such cats meant fewer seals in the future, and also meant that delayed starting dates were being ignored); similarly, the act of 1887 limited, and that of 1898 prohibited, making a second trip for mature seals during the moulting period. The 1898 act included the first safety legislation of sorts, since it stipulated that no vessel was to carry more than three men for every seven tons, and no more than 270 in any case.[54]

This was a minimal measure at best, and the same might be said for the 1899 legislation[55] which stated that "no sealer shall be liable ... for supplies to any greater amount than his share of the proceeds of the voyage," meaning that if the ship came home "clean" the sealer could not be pursued for crop money, nor could sealing wages be attached for any debt other than crop money. The owners might have been persuaded that such minimal concession was fair, but the more likely cause was cement for political fences in fishing districts.

More extensive legislation had to await the coming of Coaker, and even then, the 1914 act[56] which he proposed was resisted staunchly by owners and their associates in the Legislative Council who managed to cut much of the meat from his draft, approved already by the vessel masters. Nevertheless, for the first time there was some minimum standard for food, for example: "not less than one pound of soft bread shall be served to each member of the crew three times each week; beef, pork, potatoes and pudding shall be supplied for dinner three times each week; for breakfast stewed beans and fish brewse shall be supplied alternately; onions, potatoes and turnips shall be ingredients in the soup supplied on Saturdays," and "fresh beef shall be supplied to each member of the crew once each week" (or canned if fresh was unavailable).

Food was hardly the only complaint, but further redress had to await the twin disasters of 1914: the loss of the ss *Newfoundland*'s 73 men separated from the vessel by a storm, and the 173 men who went down at sea without a trace with the *Southern Cross*.[57] The former might have been avoided had the *Newfoundland* been required to carry a radio and operator, but that requirement was not included in the subsequent 1916 act.[58] Still, in that year for the first time sealers were not to be left on the ice after dark, or "when the state of the weather is such as endangers life and limb" (masters faced a $100 penalty in each case provided complaints were lodged within twenty-four hours). Jail terms awaited the captain if disablement to a sealer resulted. Vessels were now required to carry signal rockets, send out search parties for missing men whenever possible, and so on – and compensation would be paid the injured.

Other terms of this act limited maximum catch to 35,000 (a good harvest indeed) or whatever legal load lines (a new requirement) allowed. Equally important, as a belated conservation measure, seals were not to be taken by gun; since only mature seals were so taken, and many lost, this was a useful conservation practice. Masters, second hands, and masterwatches were all to be certified as competent by a Board of Examiners. But no word was included on the issues of minimum wages, overcrowding, sanitation, required radio communication, or doctors on board. As it was, only tragedy and time had forced some minimal advances.[59] The general rule had always been:

The hardy sons of Newfoundland
Wait not for sealing laws.
Mid ice and snow they daring go
To try and grease their paws.[60]

By the time the 1914 and 1916 acts were passed, the fishery had undergone change in another direction: the introduction of the all-steel vessel.[61] Steel ships were no novelty, but it was long after the initial unsuccessful experiment that they came to be used in the seal fishery – in fact, only after it was realized that the bow of steel vessels had to be sharply undercut. While the bluff wooden wall might butt the ice to crack it, the steel ship had to ride up and crack it with its own weight. Altogether eighteen steel sealers were added to the fleet before World War I, comprising at that point the premier ice-breaking fleet in the world.

Harvey's was the first to build a steel sealer, the *Adventure* of 826 tons, going to the ice first in 1906 (at about the same time sealing vessels stopped carrying square sails). The *Adventure* was an immediate success, returning with over 30,000 sculps in her first voyage. Harvey's rivals were quick to follow, for these new ships made obsolete the old wooden walls. By World War I, the numbers were roughly equal: nine steel to ten or twelve wooden sealers. Some of the new breed were massive by older standards: Job's *Nascopie* was 1,004 tons, Bowring's *Florizel* 1,980 tons, but the giant of them all was the same firm's *Stephano*, registered at 2,144 tons, which was introduced in 1912 and sent to the bottom by a German submarine off Nantucket in 1916. Most of the steel sealers were lost in the war (four were turned over to Tsarist Russia for icebreaking and never heard of again). By the end of the war, only three had survived in a condition to go sealing, and the few replacements were nothing like the same size, for the industry had well passed its zenith.

The steel vessels had added a further technological development to the industry and a brief surge in catches of seals before 1914. But steel did nothing to alter the distribution of profits or any other aspect of the trade. By the war, the pattern was so regularized that, given stable prices and no wartime interference, the trade might have continued indefinitely, or at least as long as there remained sufficient seals. But whether that requirement could have been met is a moot point. When the figures for "seals per man" or "total seals taken" are charted on an annual basis, the fluctuations are so great in the industry that little pattern emerges, and the catches remain fairly high. When the figures are rounded off to five-year averages, however, and contrasted with figures for tonnage and men employed, it is clear that there was an optimum point for the seal fishery so long as the hunt was not crowded with the last available ship and the last available man.

The harvest had declined steadily since the 1840s, and the take of seals per man declined also, and quite substantially, when steamer crew sizes went very much over 4,000 total. So long as owners were satisfied to take 200,000 to 250,000 seals each year, roughly the average after the late 1800s, the herd seemed to survive. The inverse relationship of "total seals" and "men" curves is clear, but beyond that it is useful to restate the conclusions of J.S. Colman, who gave the figures careful study in 1937: "it is not easy to see any evidence during this period [1895–1904] which shows that the seals were being overfished, except perhaps the fact that the number caught per man failed to rise when the number of men decreased so markedly during [World War 1],"[62] which failure might well be explained by other factors. The postwar situation was to be different again, when seal oil competed not only with petroleum but with vegetable oils as well. Not until after World War ■ would sealing once more occupy public attention in a major way, but never again would it play its former role in Newfoundland's economy.

There can be little question of the importance of sealing to the history of Newfoundland. It is evident in the settlement of the northeast, the financial power of Water Street, even the rise of a colony-wide labor movement – however tenuous the explicit connections. The effect on the seals is hard to calculate, though compared to presealing days the herd had been hard hit: catches had gone down steadily save in times of technological innovation. But the impact upon Newfoundland's economy, politics, and philosophy of life is still evident even to the most casual student of her history. The same claim cannot be made for the relationship of Alaska to the northern fur seal, yet for the Aleuts of the Pribilof Islands the seals were everything – and it is to that story that we now turn.

PART THREE

The "Sea Bears" of the Pribilofs

Now this is the law of the Muscovite,
That he proves with shot and steel.
When ye come by his isles in the Smoky Sea
Ye must not take the seal.
                    Rudyard Kipling, "The Rhyme
                              of the Three Sealers"

CHAPTER FOUR

The Wealth
of Alaska, 1867–1892

Of all the isles throughout the sea
The very best for you and me
Is old St. Paul of fur seal fame
Although with little other game ...

On summer mornings thick and wet
Is wen we drive the holostaik
We club him first then with a knife
Take off his skin just lightening like.

Verse probably by Joseph Murray, James
Judge Papers, Oregon Historical Society

Every biological order has its peculiarities, and certainly this is true of the Pinnipedia – the possessors of flippers. But of all the Otariidae, or eared variety, *Callorhinus ursinus*, the northern fur seal, stands out in two respects. First, its migration pattern, from the Bering Sea to the central California coast, is matched in distance only by the harp seals of Newfoundland and exceeded among mammals only by the great whales. Second, along with elephant seals it is possessed of the greatest sexual dimorphism, or difference in size between adult males and females, of any species of mammal on earth.[1]

Amazingly, a vast herd of northern fur seals, still numbering some two million, breeds only on a few small rocks, the Pribilof Islands which provide some 80 per cent of the total, and Russia's Commander Islands and Robben Island off Sakhalin (Japanese Kaihyoto, Russian Tyulenoy, it has changed hands more than once). Least important are very small herds on a few of the northern Kuriles and, since 1968, San Miguel Island off California. But the seals spend only the summer breeding and pupping season on the islands, remaining at sea the rest of the year. They come ashore rarely if at all during that time, and then only in case of illness or exceptionally rough weather.

The annual cycle is complex, beginning anew early each summer on the eight miles of rocky Pribilof beaches which are the species' principal home. This is one of the few great populations of seals which does not breed on ice. First to arrive from late April to mid-June are the older "harem bulls," substantial beasts indeed, measuring 7–8 feet in length and weighing 400 to 600 pounds; after two months without feeding they will be much reduced before leaving the islands. The bigger they are, the longer they are able to fast. Northern fur seal males seldom breed before the age of ten and then only for a maximum of perhaps four or five years, though individual seals may live into their twenties (tagged females of twenty-five have been recovered, and an age of thirty is not impossible). These beachmasters quickly take up a territorial position, which they defend with considerable determination against man or beast. It is not safe to enter a rookery in breeding season, but bulls normally will not pursue intruders beyond the edge of their own territory and much of their angry display is bluff – though not to other bulls. The northern fur seal is fully capable of driving off an interloping Steller sea lion three times its size.[2]

In June the older pregnant females arrive, much smaller and lighter – 75 to 100 pounds on average, so that the bulls, some four and a half times larger, stand out very clearly (by comparison, the bull sperm whale, largest on the whole dimorphic scale, is only two and a half times the size of the sperm cow). The females are soon corralled in harems of from one to a hundred by the breeding bulls, with the larger males dominating the scene: the larger the animal, the greater the fat store to provide energy for aggressive dominance and reproductive activity. A majority of females of ages eight to thirteen arrive pregnant, give birth in a few days, and then come into heat to become pregnant again, although implantation is delayed until the winter (the gestation period is thus some eight to nine months).[3] Since females are nonterritorial, bulls have considerable difficulty in containing them and do not try after impregnation: any disturbance, such as human intrusion or a bad storm, may irrevocably disrupt and reorganize the harem pattern of a particular season, though reproduction and pup rearing continue apace. Indeed, some scientists argue that the entire concept of "harem" is misleading, particularly when the female population reaches a peak in the rookery, and the herd is in fact continuous.

The immatures arrive in July and August, sometimes even later, staying at the rookery only two or three weeks. They occupy marginal areas on the landward side of the rookery in groups called pods, maintaining access to the sea by only a few paths. Their presence on the land side makes them easier to separate from the herd, for, as we shall see, only subadult males of three or four years have been killed for some time, while the immature females have been left to contribute to the breeding stock. Although to the untrained eye there is little difference between immature seals of either sex, the experienced Aleuts are wrong only some 5 per cent of the time.[4] The immature seals, like the mature females, come and go from the rookery; the mature females will leave several times to feed at sea during the nursing season of three to four months. At the end of the summer, when the pups also go to sea (the ability to swim is inherited, though the pups appear to take to the water only reluctantly at first), they will weigh some forty pounds.

Mortality among pups can be high. The higher the rookery density (and the larger the rookery total population), the more local food resources are strained, and the less likely are the pups to be well nourished.[5] For the survivors, the average size will then decline. The principal killer of seal pups is an endemic hookworm, the larvae of which live on at the rookery even in winter.[6] The hookworm eats through the seal's intestinal wall and causes fatal hemorrhage. Trampling and smothering by restless adults can also kill pups, and until the hookworm was discovered in the mid-twentieth century, this movement, or some mysterious undetected disease, was thought to be the cause of death. Little can be done to counter the hookworm, since it is impracticable to treat the wild seals in such numbers and ecologically dangerous to tamper chemically with the rookeries (assuming a successful agent is available). In the worst of conditions, 100,000 dead pups have been left at the rookeries: 50 per cent of pups, it has been estimated, die before the first year in any case, and 70 per cent or more before the age of harvest, most of which occurs at sea due to natural conditions (storms, failure to find enough food) or to enemies, natural and unnatural.[7]

One in every 100,000 or so is, incidentally, born an albino; its chances of survival are lower than for ordinary seals, since its eyes are weak without protective pigment and its coat is a conspicuous beacon for every sort of enemy at sea, such as the killer whale. Enough examples have been observed to provide the source for Kipling's "White Seal" in *The Jungle Book*, though in other details Kipling's monogamous world-wandering wonder seal rather exceeds the bounds of natural reality.[8]

At the height of the breeding season – only a few weeks in late July and early August – the subadult males to be killed, each weighing 60 to 80 pounds, are separated out and processed. Soon the remainder head back to sea, more or less in the reverse order of their arrival, though the pups mingle with the last to leave. By late November, the rookeries are empty save for a last straggler or two as the seal population fans out across the Northern Pacific.

The precise behavior of fur seals at sea is really little known, for all the years of utilization by man and the shorter span of his scientific study. The main Pribilof population, it is clear, proceeds down the west coast of North America, with the females and immatures ranging considerably further than the males. While the former reach the Channel Islands off Santa Barbara and on occasion even further, adult males seldom go beyond the Gulf of Alaska. Whether range differences result from the dimorphism or whether the dimorphism is the result of distinct sex-determined peculiarities of feeding habits is just one of the mysteries of the fur seal's existence. Into the spring, migrating seals will range from Sitka to Monterey, averaging ten to ninety miles offshore, rarely nearer, and mostly concentrated in the thirty to seventy-mile range, especially on various "grounds" such as the Vancouver Grounds off the Columbia River and Vancouver Island and the Farallon Grounds off San Francisco, where food is particularly plentiful owing to upwellings from submarine trenches, sea mounts, or the rising slope of the continental shelf. The fur seal can approach quite high speeds in the water – bursts of 15 knots – and can dive to over thirty fathoms (180 feet) but much prefers the food to come to it, hence its frequenting of the several grounds.[9]

At sea, despite occasional reports of seal "herds," seals normally travel individually or in small groups, sometimes, as with otters, called "rafts." Many single seals and various-sized groups may be sighted during optimum weather conditions. From six to twenty may be found gathered where food is very concentrated. Seals of this species seem to prefer water temperatures in a fairly narrow range of 47°–54°, though perhaps this preference has nothing to do with temperature at all, but is due to the existence in water of that temperature of some choice items of diet.[10]

Fur seals are, on the whole, opportunistic feeders, and they browse among fifty varieties of fish and nine of squid, depending upon local availability across their range. Off the British Columbia coast in spring, for example, perhaps 65 per cent of what they eat is commercially useful (and 5 per cent the particularly desirable salmon: less than fishermen are wont to believe). Off the Japan coast, on the other hand, where Pribilof (6 per cent) and Commander (30 per cent) and Robben (63 per cent) seals mix in considerable numbers, studies show that over 50 per cent of seal consumption in winter and spring has been of relatively useless lantern-fish and another 20 per cent of squid, most of types not exploited commercially.

Fur seals seem to prefer smaller species, anchovy, capelin, herring, and always squid, feeding mainly at night, when such creatures rise nearer the surface to be met by the hungry seals. Fur seals have unusually large eyes – part of the attractiveness of the whole order to mankind – to assist in feeding in dark waters. Since the fur seal feeds mainly at night, the corollary is that it will sleep on the surface during the day – to its considerable cost. But does the seal eat what it finds, or seek out what is most desirable? Again, an unanswered question.

These migratory patterns have occurred regularly for thousands of years, and small depredations have been made for nearly as long by man, whether in the Pacific northwest or along the Japan coast and in the Sea of Japan. Excavations of Ozette village, Cape Alva, Washington, show 2,000 years of seal consumption until this particular village was abandoned about 1900 – just when the population of mature male seals taken by village hunters declined sharply, as testified by the midden bones (size shows sex, and growth rings in teeth testify to age). The decline was perhaps a reason for abandoning the village, although economic conditions and government policy were at least as important. Indians of the northwest continued to hunt seal as much for meat as for skins into the mid-twentieth century, but Karl Kenyon, a well-known seal biologist who went sealing with the Tlingits in the 1950s, knew as well as his hosts that he was witnessing the last of a long era.[11]

But it was for furs that the Russians and their successors harvested Pribilof seals. Although like other seals the fur seal does have a layer of subcutaneous fat, it is also gifted with a luxurious pelt of densely packed hairs – some 300,000 per square inch, roughly half the density of the sea otter's pelt. While not as fine, the seal's fur is yet fully adequate to keep it warm and waterproof in the cold currents which it prefers. The real danger to fur seals is heat, manageable in a rookery warmed by the sun, since the seal's unusually large flippers operate as heat releasing planes, but deadly in seals herded on land. With very little effort they can be brought to the 107° which will induce fatal heatstroke.

Russian exploitation began as soon as otters ran low and the several rival Russian companies in operation learned of the Pribilofs.[12] From 1786 to 1806, beginning with a take of some 40,000 the first year, the exploitation was totally uncontrolled, and even the new twenty-year monopoly granted to the Russian-American Company in 1799 did nothing immediately to alter the situation. By about 1803 the company found that it had vast sealskin reserves in its warehouses, simply rotting away; one considerably later report noted that in that year some 700,000 of a hoard of 800,000 had to be discarded as useless. These skins were dried for leather and not salted to preserve the fur until the 1850s, when attempts were made, with some eventual success, to sell skins on the London market.[13]

The personal investigations of Russia's grand chamberlain, Nikolai Rezancv, in 1805–6, revealed the costly waste, and he used his powers to halt all sealing temporarily. It resumed in 1808. From that year until the charter renewal loomed the impression remains of generally unrestricted killing. In 1821, however, orders were issued to stop the kill every five years to allow for recovery of the herd, to kill no more than 50,000 each active season, and to take no bulls or pups unless for food or oil.

After the introduction of the first *zapuski* (sparing, or moratorium) in 1822, the Russians continued with mixed success to regulate the harvest until the sale of Alaska in 1867, having learned some lessons from the otter story of the previous century. No seals were killed on St. Paul in 1822–4 and 1835–7, or on St. George in 1826–7. Nevertheless, the quota of 50,000 so optimistically set in the regulations of the 1820s could not be met. By the mid-1830s the take was down to some 7,000 a year, falling ever lower in the early 1850s. A key regulation of 1847–8 prohibiting the killing of the all-important females – the "sacred cows" – allowed some recovery of the herd since it meant that only immature bachelors would be taken. In the last decade before the sale of Russian America, the annual take was between 30,000 and 40,000. The company had by a narrow margin found its way by trial and error to what later would be called a "MSY" – a maximum sustainable yield, which was not the same thing as the maximum possible seal population. By 1867 and Alaska's sale, the seal population had probably returned to nearly the pre-exploitation level of roughly three million (only four-fifths of which, it should be remembered, breed on the Pribilofs), allowing the population to expand beyond the MSY figure.

Again the question arises of how many seals had been killed before 1867, and again estimates vary. Victor Scheffer, than whom there are few more knowledgeable about northern fur seals, estimated that more than 3,000,000 were taken in the period 1786–1828, an average of 71,000 each year. Victor Tikhmenev, however, in his study of the Russian-American Company, recorded 1,232,274 – an often-cited number – for the period 1797–1821, an average of under 50,000 a year.[14] Using for want of a better solution a compromise of 60,000 computed from the discovery of the Pribilofs to the War of 1812 (25 years times 60,000 = 1,500,000), and adding in the wasted surplus of 1,000,000 (probably justified by discards and losses at sea), the total is 2,500,000. To this should be added another 500,000 taken before the Pribilof discovery, primarily from the Commander islands, and yet another 830,000 tabulated by Tikhmenev for the years between 1821 and 1862 – which we round off to one million to account for the missing years 1812–21 and 1862–7, reaching a grand total of roughly four million seals taken in Russian America for commercial purposes until the time of its sale in 1867. Much of this calculation is only an estimate, but at ± 500,000, it cannot be far off; it is not possible to match the precision of James R. Gibson: "From 1786 through 1832 there were 2,178,562 fur seals killed (185 daily or one every eight minutes),"[15] but his total for those years is reasonably close.

Meanwhile, another quite different resource was being exploited: the Aleuts This maritime people, numbering 15,000 (± 2,000) at the time of Russian arrival, strung out along an island chain extending 1,200 miles ("the longest longitudinal [east-west] distance occupied by a single language and racial group in the world") in small settlements and villages, had a rich culture but few defenses against Russian firepower and the diseases of civilization. Murder, robbery, forced labor and resettlement combined with epidemics to reduce this population to a mere residue The Fox Island (eastern dialect) Aleuts, once the largest group with some 10,000–12,000, was reduced to a total of 1,900 by 1790. It was mainly Fox Island Aleuts who were sent to the Pribilofs to work the fur seals, though the total number on the islands at any one time was never great (in 1816, 379; in 1834, 182).[16]

The establishment of Aleuts on the Pribilofs altered their lifestyle in important ways, making them permanent settlers instead of the mobile hunter-fisherfolk they had been throughout the Aleutians, and concentrating their skills on the processing of fur seals: there was far less fishing off these two islands. The population here grew distinct from that on the Aleutians, not least in the decline of their life expectancy. Their lives were now strictly regulated by the Russian-American Company, and although recognized as citizens or at least subjects of Russia – a privilege which was not extended to the mainland natives of Russian America – this distinction meant little in practice, for they were in many ways comparable to Tsarist Russia's serfs, a parallel which did not escape the company's directors.[17]

On the other hand, the company paid its Aleut workers from 20 to 75 kopeks a skin (100 kopeks to the ruble; the 1860s ruble was worth about 75 cents), a form of compensation not normally extended to serfs, at least on the owner's property. Although the money was generally spent on company goods at predictably inflated markups, still it was individual disposable income. The money was actually distributed by the chiefs (themselves paid at a higher rate) according to the level of proficiency in the worker's various assignments. This essentially merit-based distribution pattern was traditionally Aleut. Overall, as Dorothy Jones, a long-time student of the Aleut condition, has put it, "the Aleut's economic status under company administration was part serf, part proletariat and part traditional Aleut."[18] That unique system was to provide an essential background for American rule over the islands after 1867.

If the Russians had problems to overcome in the diminishing supply of both Aleuts and fur seals, an even greater difficulty was that of preserving Russian America at all.[19] The challenge was easily visible: the steady advance of American interests in the direction of Alaska. By 1812, the colony had signed a contract for food with John Jacob Astor's Columbia River enterprise; though the contract did not survive the War of 1812, the portents were ominous regarding the colony's ability to sustain itself. After 1815, Yankee merchants were back in such numbers as to bring forth a Russian ukase (decree) in 1821 excluding foreigners from Russian America. The resulting pressure upon the American government from such leading mercantile houses as Bryant and Sturgis was substantial, and in 1824 Secretary of State Adams in effect obtained abrogation of the ukase: already protection of American interests in the northwest was an important political issue, and in drafting the Monroe Doctrine, Adams clearly had an eye on Russia in the northwest as much as on the Spanish and Portuguese in the southern hemisphere.[20]

Until 1834, American trade in Russian waters was alarmingly uncontrolled. Payment was often in fur seal pelts until the decline of the herd brought a shift after 1829 to letters of credit upon the Russian-American Company itself. In 1834, Russia again attempted to close its colonies to American traders and once again was unsuccessful: for the second time, Russia was given a hint of potential conflict over the question. Fur traders were by now joined with other interests, for in the 1830s American whaling in Russian waters had increased substantially and in the 1850s it proved the mainstay for an otherwise declining whaling industry. Whaling focused New England's attention upon Russian America. Meanwhile the Baranov spirit of enterprise had gone, and the company had become "a stodgy, overstaffed, and inefficient concern,"[21] entering a quarter-century of decline and retreat. In 1839, the Hudson's Bay Company was given a lease for Alaska panhandle fur-hunting, giving British interests full control over the land fur trade. In 1841, Fort Ross was sold to Capt. John Sutter, as the company withdrew to its Kamchatka and Okhotsk interests. At times ice, mines, fish, timber, and even whaling were temporary substitutes for furs, and in 1852, in an optimistic search for profits from the California gold rush, a Russian commercial agency and vice-consulate was established in San Francisco – only to be closed down in 1861.

As Russia retreated, the United States advanced. Above all, after the 1840s, new and substantial west coast interests based upon San Francisco looked northwards; gold was potentially available in places other than California's Sierra Nevada. In 1851, the American-Russian Commercial Company (not to be confused with the Russian-American Company) was organized to carry basic supplies to Russia's colony (the Hudson's Bay Company, successor to Astor's contract, having failed to satisfy Russia). Within three years this new company, which included a number of leading San Francisco entrepreneurs, had contracted to carry coal and ice from Alaska and to be the chief supplier in the other direction.[22] Ice alone was a major item; by the coming of the Civil War, the company was importing 3,000 tons a year. Western and eastern sectional interests in the United States were now ready to join – a combination personified by the connection between William Mc-Kendree Gwin, formerly Andrew Jackson's personal secretary and senator from California in 1849, and William Henry Seward, who as secretary of state would negotiate the purchase of Alaska. By 1854, the first approaches had been made to Russia to buy the entire colony.

Although the Civil War did delay negotiations, future American expansion to Alaskan waters already appeared likely. Despite some reluctance to sell in Russia, authorities there realized that if the colony were not sold it might be seized. Although Alaska was in better financial condition in the 1860s than it had been in the 1840s, Fort Ross had been sold and feeding the colonists remained a problem, the more so as fur resources (and thus income) were dwindling. The desire for American friendship, strengthened by American commercial contacts during the Crimean War, was coupled with the real fear that, if gold should be discovered in Alaska (as was predicted and probable), there would be no holding Americans back: it would be 1849 in California all over again.

In the United States, too, there was opposition, some of it substantial, towards expenditure of $7,200,000 for "Walrussia," but it has been argued persuasively by Howard I. Kushner that there were real and powerful interests in the United States in support of the purchase, interests diverted not at all by the Civil War. Though superficially "a case of fallibility of international intelligence,"[23] in which each power supposed it was doing the other a favor – the United States by buying an expensive liability, and Russia by selling a property much valued by the purchaser – in reality the vision of Seward and Gwin was shared fairly widely, even if a few palms had to be greased in the final congressional vote.

Nor were the fur seals forgotten. Already at the end of the Civil War, Lewis Goldstone, an American merchant living in Victoria, was thinking of establishing a company based upon San Francisco which would replace the Hudson's Bay Company as lessee of Russian land fur enterprises.[24] He had the ear of a group of important San Francisco men (most of them unconnected with the American-Russian Commercial Company), headed by Gen. John F. Miller, collector of customs, Louis Sloss, a leading merchant, and Cornelius Cole, who had read law in Seward's office and was now senator from California. Even before the purchase of Alaska, this group had at considerable expense sent two schooners to survey Alaskan prospects and had opened negotiations not only for the land fur concession but for all the fur op-

erations of the Russian-American Company, including the Pribilof and
Commander islands. The purchase of the entire colony by the United
States was no reason to abandon the idea; on the contrary, a rush now
commenced to be the first new "owners" – that is, American citizens
– to reach the Pribilofs and their seals.

There were two ways of going at this. One was to negotiate with
the outgoing Russian-American Company to take over their commercial property, for purchase of the colony did not include such capital
goods. The other was simply to set sail at once for the seals, without
a care for any Russian property rights. Both methods were tried with
varying degrees of success, producing a year of vast confusion regarding
use and ownership of the seals.[25]

The principal agent in the first, more diplomatic approach was Hay-
ward M. Hutchinson, a Baltimore businessman who had done well
selling cooking equipment to the Union Army, but now was seeking
new opportunities. His friendship with Gen. Lovell H. Rousseau, named
by Secretary of State Seward as American commissioner to receive the
transfer of Alaska at Sitka, brought Hutchinson to San Francisco in the
same vessel and no doubt directed his thoughts northward. In San
Francisco, Hutchinson reached an arrangement for financial backing
with Louis Sloss and Lewis Gerstle. Sloss, born in Bavaria in 1823,
came west with the gold rush and settled in Sacramento, where he
conducted a successful grocery business with his lifelong partner Gerstle
(an association cemented by the fact that the men married two sisters).
Moving to San Francisco in 1861 or 1862, the partners established a
profitable business in wool, hides, shipping, and other interests; though
Sloss was usually the negotiator for the firm, one could always speak
for both.

Sloss had already participated in the ice trade with Goldstone (later
a determined opponent of Sloss's Alaska interests). It is not surprising
that Hutchinson contacted him. When Hutchinson reached Sitka in late
1867, it was with considerable backing. In December, after Alaska had
been formally handed over, Hutchinson negotiated the purchase of most
of the Russian-American Company's property for $350,000. Less than
a year later, the Alaska Commercial Company was to assume the same
assets at a book value of $1,729,000. Prince Dimitrii Petrovich Mak-
sutov, the last Russian governor, probably had orders to sell and get
out as quickly as possible (since he had visited San Francisco in 1861
and 1864 in connection with the ice trade, he probably knew Sloss as
well).

Hutchinson's purchase did not include the fur seals or the Pribilof
Islands as such but only company facilities on St. Paul and St. George
– and presumably the "good will" of the Aleut inhabitants (as will be
seen, the term is a misnomer in this connection). Maksutov also had

80,000 pelts in storage for sale. First to buy was an Oregon fur trader, Leopold Boscowitz. Boscowitz, probably a British subject (though he had lived in San Francisco), with his brothers conducted a fur business second in the northwest only to that of the Hudson's Bay Company. He now bought the best 16,000 pelts at 40 cents each, selling them at Victoria for $2–3 each.

Boscowitz was not the only potential buyer. At Sitka he encountered Capt. William Kohl, a Victoria shipbuilder of Pennsylvania Dutch origin, California '49er, and shipowner on the San Francisco–Sacramento run before moving to Victoria in the 1860s (and presumably known to Sloss and Gerstle from his earlier days). Kohl had taken his coastal steamer *Fideliter*[26] to San Francisco in 1867 ready to receive the horde of immigrants who naturally would wish to rush to Alaska. None came; a disappointed Kohl steamed for Sitka to seek other opportunities. His shipping interest would merge nicely with Boscowitz's furs.

Still another player was Capt. Gustav Niebaum, a tall, blond-bearded Finn in his thirties who had captained Russian-American Company steamers for several years.[27] Hearing of the sale of the colony, he rushed to buy an elderly brig, the *Constantine*, from Maksutov, with every intention of bringing off the skins he knew to be on the Pribilofs. Making no secret of his intentions, he soon reached an agreement with Kohl and Hutchinson, though precisely when and how is not clear. His experience of the islands made him invaluable, and in return the others offered capital and contacts.

In November 1867, the *Constantine* was in Sitka; in mid-December, she sailed for the Pribilofs. Niebaum was allowed to land since he knew the chief Aleuts: "I was the first American who ever landed on St. Paul's Island – I claim the honor of doing that."[28] He retrieved 30,000 skins (the difference between Maksutov's 80,000 and Boscowitz's 16,000 is more than double that, an unexplained discrepancy which might have resulted from spoiled discards), left a man in charge of native hunters to carry on sealing operations the following summer, and departed to deliver his skins to fur dealer August Wasserman in San Francisco in March for forwarding to Europe.

By this time a formal merger of the several interests had taken place. In January 1868 the independent operations of Boscowitz–Kohl and Hutchinson joined together in establishing the Hutchinson, Kohl Company. Its six directors represented the various contributors: Hutchinson held the Alaska properties; Boscowitz had experience in the northwest fur business, and Kohl in shipping; Sloss and Gerstle represented San Francisco financial backing; and Wasserman possessed links with the world fur market. Niebaum, upon his later arrival, was also given a partnership – he above all knew the Pribilofs – bringing the total to seven. The arrangement was complex, however, since Sloss and Gerstle at the same time became partners in Wasserman and Company.²⁹ In any case, the holdings were entirely reorganized in October of the same year into the Alaska Commercial Company (ACC), now with five directors (then called trustees) representing a considerably larger number of shareholders as will be made clear: Sloss (president), Wasserman (vice-president), Gerstle, Kohl, and Boscowitz. The ACC at that time agreed to buy out Hutchinson, Kohl for the $1,729,000 over a thirty-month period (the agreement was later renegotiated).³⁰

The reorganization is explained by the need to take into account other competitors who were not prepared to permit Hutchinson, Kohl to control the main wealth of Alaska without a struggle. Niebaum found this to his cost, when, after returning immediately to Sitka in the *Constantine* in late March he hurried on to the Pribilofs in the *Fideliter* (more capable of mastering spring ice conditions than his own brig). To his surprised displeasure, he found that in his absence others had staked out claims on the seal beaches, though it was still too early for the seals to arrive. As Niebaum remembered the event many years later:

When we came up to the seal islands we found a little schooner had come ashore from Honolulu – had managed to get through the ice and run ashore there and they had succeeded in bulldosing [*sic*] my partner, a German, and got possession of the houses and one thing and another and had pulled up all of our permanent stakes and they were going to put him off the island and finally to save something he made a proposal to them and they promised to give him one quarter of the catch they would get there that year. When we came back about two weeks after that we had to lay out in the ice and we could not get in but when we did we of course smashed the whole agreement.³¹

The schooner was actually tender to the whaling bark *Peru* (190 tons), just arrived from Honolulu, and now standing off and on. Her arrival was no accident. The *Peru* was owned by Williams, Haven and Company of New London, an important firm in whaling and sealing.³² The head of this firm was Thomas W. Williams, who with his partner Henry P. Haven had opened up Pacific whaling for New London, even establishing an agency in Hawaii in the care of his son, Charles Augustus Williams (later himself senior partner and mayor of New London). When news of the impending purchase of Alaska began to circulate Williams senior and another partner, Capt. Ebenezer "Rattler" Morgan, long a successful whaling master (among other accomplishments, he had commanded the first steam whaler), recognized the great potential of the fur seals. Morgan and yet another partner, Richard H. Chappell, raced by steamer to Panama and by packet to Hawaii. Quickly and quietly outfitting the *Peru*, then in harbor there, Morgan rushed on to the Pribilofs, arriving before Niebaum's own early voyage from San

Francisco and Sitka (Chappell meanwhile headed for San Francisco to look after that end of Williams, Haven's operations). Morgan's nickname of "Rattler" was deserved; he was a millionaire when he died.[33]

Morgan was not the man to be dismissed as casually as Niebaum remembered. After the two parties had come to blows more than once over control of the critical beaches, they managed to reach a compromise division of the spoils, a compromise all the more necessary since with the seals there arrived still more human competitors. These were now warded off by Niebaum and Morgan in concert, though in some cases newcomers were absorbed into their own gangs. Most simply went off to St. George, such as the 100-ton schooner *Caldera*, owned by John Parrott and Company of San Francisco.[34] Parrott, a Virginian, had been U.S. consul at Mazatlan and during the gold rush founded the first important local bank in San Francisco, after which he expanded his interests in other directions. Parrott was, like the principals of the American-Russian Commercial Company, to be an ardent opponent of Hutchinson, Kohl – until he too was brought into the Alaska Commercial Company.

Through the summer season of 1868 the seals were slaughtered in numbers far greater than the Russians had taken for decades, to a total of some 250,000 on St. Paul alone. Despite a later claim that only bachelors were taken the slaughter was indiscriminate, males, females, and pups all killed as they were encountered. The scene was as lawless as any gold rush in its early days. The various parties had taken control of the Russian barracks, along with the prettiest Aleut girls, and were living in "debauchery," according to William Dall, a coastal survey employee who arrived on St. George in September 1868 in a schooner belonging to the Pioneer American Fur Company, another would-be San Francisco sealing concern. Pioneer was a bit late; Hutchinson, Kohl men had seized three Pioneer employees and "tied them neck and heels and left them all night in one of the salt houses" to reflect on the dangers of interloping.[35]

Obviously there was as yet no official American presence on the islands. A force of 150 troops had arrived at Sitka in October 1867 under a General Jeff Davis (no relation to the Confederate leader), but none were sent on to the Pribilofs; given their lawless behavior at Sitka, their participation would not have improved matters at St. Paul or St. George Only after the height of the killing season had passed did a United States Revenue Marine officer, Capt. J.W. White, arrive in the screw steamer *Lincoln* (which he had brought round the Horn in 1865). White also remembered events many years later, in particular his fear that if all the seals were killed – a likely prospect the way things were going – the natives would probably starve in the coming winter. "So to bring the natives to their senses and what I considered common justice, I took possession and stopped the killing of the seals on both islands, seized and destroyed all the whiskey, made everyone settle up with the natives, in provisions, flour, etc., then called the natives together and in the presence of the White people told them the result and why I had done this, then ordered them to arrest any white man found killing seals, by my orders and I would sustain them."[36]

The intervention was perhaps critical in saving the herd, since White, with the assistance of stop-gap congressional legislation, permitted no killing at all in 1869 except by natives for their own food. White spent the next three seasons on St. Paul, recommending to Washington in the meantime that the Pribilofs be made a native reserve which, properly administered, could pay not only for itself but also for the whole of Alaska for many years to come. Meanwhile the financial organization of the sealing had also passed to a larger stage.

Niebaum and Morgan both sailed for San Francisco after the season of 1868 was over. Perhaps they reached more than an ad hoc agreement on St. Paul; in any case, an offer was made to Morgan and Chappel, on behalf of Williams, Haven, to amalgamate their Pacific sealing interests with Hutchinson, Kohl. A similar offer was made to John Parrott and Company. Both accepted, and the Alaska Commercial Company was born. One reason for haste was the undoubted hostility of J.M. Oppenheim, the major fur-processing firm of London, which had a contract to process skins for the Russian-American Company, and which claimed the entire harvest of 1867 under that contract. A young agent, Emil Teichmann, had been sent from Oppenheim's New York office to Alaska in 1868 to guard Oppenheim's interests.[37]

Even if they knew nothing of Teichmann's mission (Teichmann reached Sitka, and they were probably aware of his intent), the seven original partners of the ACC, together with Morgan and Chappell, realized that the expanded amalgamation would facilitate the next logical step: to persuade authorities in Washington to yield monopoly control to their new concern, at once enriching themselves and preventing the otherwise certain total extermination of the Pribilof fur seals. To this end, a several-pronged attack was mounted.

One ploy was to bring in yet another partner, this time as president of the ACC, in the person of the collector of customs for the San Francisco district, Gen. John F. Miller, already encountered in the ice business.[38] Miller was a Civil War veteran (noted especially for heroism in the defense of Nashville) and a good friend of President Andrew Johnson and Vice-President Schuyler Colfax. In addition, Miller played a vital role in favoring the company's interests through his customs connections: customs regulations were the only American laws immediately affecting Alaska until Congress passed special legislation, and when Congress acted with the Alaska Customs Act of 1868 to prohibit the sale of arms and alcohol, it was the Treasury's customs officials who would do the policing. Miller now joined the company, and his former subordinates could hope likewise to share in the largesse, particularly if it came to the attention of the ACC that they had given the company preferential treatment. The ACC's board was also reorganized, with Miller as president (at $8,000, subsequently $10,000, per annum); Gerstle as vice-president, replacing Sloss who now looked after the firm's interests in Washington, D.C.; H.P. Haven and R.H. Chapell for Williams, Haven; and John Parrott. Other stockholders included Williams, Haven, Morgan, Tiburcio Parrott (son of John), Jacob Greenbaum or Greenebaum (Gerstle's brother-in-law), and S. and D. Willets, Quaker owners of a whaling and chandlery firm of New York which had long been associated with Williams, Haven.[39]

The 20,000 shares were divided after some shuffling, 56 per cent to Hutchinson, Kohl owners, 29 per cent to Williams, Haven, and 15 per cent to Parrott, each group dividing its share block to its own satisfaction. It is important to note that while a considerable share of the company's profits went to San Francisco owners, nearly a third went east to New England. Enemies of the company charged that it was predominantly Jewish, but in fact only five of the eighteen listed owners in 1872 were Jews. (The charge perhaps stemmed from a mistaken belief that because both Kohl and Niebaum had German-sounding names they were members of the same German-Jewish society as Sloss, Gerstle, Greenbaum and Wassermann.) This point is ably defended by Frank H. Sloss, a descendant and historian of the company, but Louis Sloss and Gerstle were the commanding figures in the company's management, and by the end of the century only Niebaum in addition to the Gerstle, Sloss, and Greenewald (Simon Greenewald was an early associate of Hutchinson, Kohl) families held any shares at all.[40]

Reorganization, even the inclusion of Miller, did not guarantee the company's future, and as soon as the ACC was established Hutchinson and Sloss travelled to Washington, D.C., to exert their influence, supplemented with cash gifts, to obtain monopoly control of the seal herd. Times were right in the "gilded age" of unfettered capitalism. In this case, widespread corruption was joined with righteous advice from all sides that some action should be taken to prevent the herd's extinction, though not all agreed that the ACC should be its guardian. But Hutchinson Sloss and Miller had the contacts (Sloss, for example, was close to William M. Stewart, Nevada's senator);[41] their company had obvious experience in the Pribilofs, a hold on the Russian-American Company's property, and claim to a five-year "contract" with the Aleut workers on the island.[42] The result, in mid-July 1870, was legislation authorizing the Treasury Department to grant an exclusive lease. More than a dozen bids were now tendered. A twenty-year lease was granted to the ACC, by no means the high bidder.[43] Clearly bids mattered less than influence; also, the company had offered to match the fees suggested by the highest bidder. The secretary of the treasury argued that he had exercised the discretion permitted by Congress: the act had not required that the top bidder be chosen.

The lease, dated 3 August 1870 and signed by Miller on behalf of the company, stipulated that no more than 100,000 seals would be taken each year (75,000 from St. Paul, 25,000 from St. George); seals would be killed only in the months of June-July, and September-October (in August the immature male pelts are worth less since the animals are moulting), none at sea, and none with guns – and only males over one year of age.[44] In return for this right, the ACC agreed to pay $55,000 for its lease each year, and $2.625 for each pelt shipped (compared to the 1869 selling price of $4.00 gold in San Francisco and $6.50–7.50 in London),[45] and 55 cents for each gallon of seal oil. The company also agreed to supply the Aleuts with 25,000 dried salmon (which soon became salted salmon, then corned beef in cans), sixty cords of firewood, and sufficient salt and barrels for the curing of meat. The Aleuts were also to have a school at company expense, but the company was to see to it that they would not be supplied with alcohol. Not specified in the lease, but soon company policy in any case, was the payment to Aleut workers of 40 cents for each skin for their labor (the money was paid into a common "seal fund" for later distribution, much as the Russians had done). The company also came to provide for widows and orphans, recognized the authority of the local chiefs, and gave free housing and medical care.

Though it was not the 99-year concession for which the ACC directors had hoped, a twenty years' lease on sealing was certainly a valuable right, and the company got to work at once. The take for 1870 was only 23,733 (the lease had been granted rather late in the season), but the average annual kill over the life of the lease was 93,090 seals taken for skins and another 9,727 for food, putting the total near the 100,000 limit (see table 2). Killing and counting were all done under the supervision of agents of the Fur Seal Service, a branch of the Treasury Department given responsibility for the herd under the 1868 act.[46]

The relationship between company and government was generally harmonious on the islands. The first treasury agent, H.H. McIntyre, was a firm believer in the ability of the herd to sustain an annual kill of 100,000 (the Russians in their last years had considered 75,000 the most which could be taken safely each year). McIntyre was told in his first year to permit the taking of seals only as absolutely necessary for food and clothing – a limit he interpreted most generously to the total of 85,901 skins, dividing the pelts equally between Williams, Haven, and Hutchinson, Kohl (White's recollection that no skins were taken

TABLE 2
Harvest and Dividends, Alaska Commercial Company, 1870–1892

Year	Skins Taken	Rental and Skin Tax	Dividend Per $100 Share (20,000 issued)	Total Dividend Paid
1870	23,773	$ 101,080	–	–
1871	102,960	322,863	($8.00 per share assessment)	
1872	108,819	307,181	$ 10.00	$ 200,000
1873	109,117	356,610	10.00	200,000
1874	110,585	317,495	17.00	340,000
1875	106,460	317,584	37.50	750,000
1876	94,657	219,156	17.50	350,000
1877	84,310	253,256	10.00	200,000
1878	109,323	317,448	45.00	900,000
1879	110,511	317,400	85.00	1,700,000
1880	105,718	317,595	100.00	2,000,000
1881	105,063	316,886	90.00	1,800,000
1882	99,812	317,295	47.50	950,000
1883	79,509	251,875	42.50	850,000
1884	105,434	317,400	45.00	900,000
1885	105,024	317,490	35.00	700,000
1886	104,521	317,453	65.00	1,300,000
1887	105,760	317,500	55.00	1,100,000
1888	103,304	317,500	50.00	1,000,000
1889	102,617	317,500	75.00	1,500,000
1890	28,859	–	–	–
	2,006,136	$6,010,566	$ 837.000	$16,740,000
		(lease cancelled)		
1890			115.00 paid	2,300,000
1891			70.00 from	1,400,000
1892			50.00 surplus	910,000
(no dividends 1893–1908)			$1,072.00	$22,450,000

$5,975,253 rental and skin tax
3,625,000 import duties on skins finished in London
  550,000 fees on skins taken in Commander Islands

$10,150,253 total paid U.S.A.
 (7,500,000 purchase price of Alaska)

Sources: Fur seals taken: Francis Riley, *Fur Seal Industry of the Pribilof Islands, 1786–1965,* 4; payment to U.S. Government: George Rogers, "An Economic Analysis of the Pribilof Islands, 1870–1946," 26; dividends: Gerstle Mack, *Lewis and Hannah Gerstle,* 41–2.

is rather at fault in this regard). It is no surprise to find that by his second killing season on St. Paul, McIntyre had left government employ and was now the local superintendent for the ACC, a position which he continued to hold until almost the expiration of the lease.[47]

McIntyre's several successors, for the most part, similarly supported company operations. They were, after all, dependent upon the company's people (normally about a dozen on St. Paul, with another four on St. George), for companionship and upon the company's transport for contact with civilization in Unalaska, Sitka, or San Francisco. The Pribilof agency was not the remote exile it might appear: it was quite well paid (the senior agent received $10 a day, his assistants $6–8) and wielded considerable power. It was a patronage post, and the agents over the years were a very mixed lot. All had in common a total lack of experience in sealing and an equally unfortunate dearth of instructions from their seniors. Dorothy Jones, in her several studies of the islands in the last half of the nineteenth century, blames the government, not the company, for the subservient position of the Aleuts: "In effect, while the company treated the Aleuts like other American workers, government agents treated them like Russian serfs in forcing them to labor without pay. It was the agents of a democratic government, not of a private profit-making company, that imposed this violation." Jones had overlooked the fact that company supervisors and government agents each needed the other to perpetuate the system, and without orders to the contrary, it was only natural for both to continue to operate in a way that the older hands on the islands advised was customary.⁴⁸

Little changed, for example, in the method of taking the seals. At dawn (or what passed for summer dawn in such high latitudes) of the chosen day, Aleuts ran along the beach between the seals and the sea, awakening and turning the confused animals away from the water. Bachelor males, grouped inland, were herded slowly to the killing ground. "Pods" were treated carefully to avoid overheating and allowed frequent pauses in the half-mile drive; they were not killed in the rookery itself to avoid disturbing the entire herd. Native foremen culled out and drove back those of the wrong age or with less valuable scarred skins: the company paid tax by the skin and therefore wanted the best 100,000 skins available.⁴⁹

The killing took place at 6 or 7 A.M. Each man was equipped with an oak or hickory club, 5–6 feet long. At the foreman's order, the seals were clubbed in the head, crushing the thin skull bones. They were then stabbed in the heart and allowed to bleed. Though other methods have been tried more than once, the use of clubs was – and is – regarded as the fastest and most humane method of killing, just as with harp seals. As David Starr Jordan, the well-known biologist, wrote in 1912: "The processes of driving and killing the seals are simple and humane, comparing, both in their nature and effect upon the animals, to the ordinary processes by which the domestic animals supplying food and clothing to man are handled."⁵⁰

The carcasses tend to heat rapidly after death and quickly become "stagy," that is, the hair comes out in patches; it is necessary to deal with the skins at once. The carcass therefore was laid out with flippers cut loose for the skinner to detach the pelt – a skilled operation requiring frequent whetting of skinning knives and taking about four minutes for each seal. Clubbing and skinning were done by the most expert first-class workers; "sticking" in the heart and "flipping" by the less skilled second-class or beginners; the classes were paid at different levels in the eventual distribution of the "seal fund."

Although occasional attempts were made to find a use for the carcasses, in general they were simply left where they fell, to rot – another reason for not killing the seals on the beaches – and a different killing ground was used each year on a three-year rotation. One observer explained the effect of 75,000 such bodies burning up through the snow at the Lukannon grounds: "the odor along by the end of May was terrific punishment to my olafactories, and continued so for several weeks until my sense of smell became blunted and callous to such stench by long familiarity."[51]

After a careful count, the skins were carried to the salt houses, placed in layers with the flesh side up, salted, and left to cure for a week, at which point they were uncovered, resalted (this time with the flesh down) and left another week or more. When properly cured, they were tied flesh side together in bundles of two (each bundle weighing 12–15 pounds) and stored. At the end of the season, they were taken by small boat to a waiting ship and transported to San Francisco, where they were packed in barrels for the overland journey to New York and onward to auction in London. Although after the 1820s some skins were processed for caps in the United States – whether of Alaskan or South American furs, purchased in London, is unclear[52] – the market for furs to be worked in the United States remained small, and by far the majority of skins were shipped to London until World War I. Well before that time a substantial demand had developed in America for fashionable sealskin wraps, but the dying and finishing was almost all done abroad. Oil was never an important factor: an oil works operated on St. Paul in the 1870s, but the absence of labor to operate it at the peak season when the blubber was available, coupled with customs duty on oil, transportation costs to San Francisco, and a low U.S. market price meant that the plant was insufficiently profitable to maintain.[53]

Sealskins, however, were profitable indeed. In its first year of operation, 1872, the ACC paid $10 per share in dividends, and soon went on to greater things (see table 2, averaging over an eighteen-year period (1872–1889) a dividend of $46.50 per face value $100 share. Explanation of this incredible rate of return is not hard to find. The price for Pribilof skins at the London auctions of the C.M. Lampson Company averaged $14.67 over the entire twenty-year lease, and brought $17.04 each in the last three years. The 1,840,364 skins sold by the company (the difference between sale figures and the harvest of table 2 is due to discards, etc.) thus produced approximately $27,800,000, from which had to be deducted fees (lease and royalty) to the United States government of just over $6,000,000 (21.6 per cent), wages to the Aleuts (under $1,000,000 or 3.3 per cent), and a small overhead composed of salaries of company people on the islands and in San Francisco, the firm's impressive headquarters at 310 Sansom Street, and – the largest figure – transport of the skins, all of which totalled under $1,000,000 over the twenty years. In other words, roughly 28 per cent of gross return went for expenses, and the entire remainder was profit. A low-investment, low-overhead operation produced net profits of $18,102,140, or just under an average $1,000,000 a year. The risk was small, the labor force immobile, the market assured, and monopoly enforced at law: even in the age of the "robber barons" this was success to end all successes.[54]

And there was more. In 1871, Hutchinson, Kohl, Maksutov and Company was created to obtain a Russian lease for the Commanders and Robben Island (Maksutov was the obligatory Russian partner required by Tsarist law). The lease, also for twenty years, required the company (after 1874, renamed Hutchinson, Kohl, Philippeus Company), to pay an annual fee of 5,000 rubles (about $4,000), a tax of two rubles a skin ($1.50), and wages of a half-ruble a skin. Since this company immediately made the ACC its sole agent for management and processing, the latter now had a monopoly on all northern fur seals save for the few which bred on the Kuriles (in 1875 Russia traded the Kuriles to Japan in exchange for the southern half of Sakhalin). The first few years of this contract were disappointing, since these islands had been badly oversealed between 1867 and 1871 – perhaps by disappointed rivals of the ACC – but from 1871 to 1891 they produced overall 769,863 skins – 38,500 a year. Though Russian skins commanded a lower price than the more luxuriant Pribilof skins (Russian skins were sold in a spring London auction, Pribilof skins in October), this subsidiary contract produced additional useful income, as did some 4,000 otter skins taken by the company, worth $60–100 each, together with sundry fox skins, land furs, etc., upon which to calculate the company's total profit.[55]

There was of course some payment to the United States. The tax of $2.625 per skin produced over the lifetime of the ACC lease about $6 million, not quite the equivalent of the purchase price of Alaska – though if to this figure is added the income produced from excise taxes upon finished skins imported from London (the U.S. government would have received this money in any case, whatever the original source of the pelts), more than the purchase price was earned back. Still, $6 million was roughly six times the cost of all the services and programs paid for by the United States, according to Henry Wood Elliot, a man of no small importance in the fur seal story as will be seen.[56]

But what of the Aleuts? As George Rogers has written, "This human resource was managed very much as another renewable resource, the objective being to maintain a suitable and efficient workforce to meet the needs of the harvest." The Aleuts were paid a basic piece-rate of 40 cents a skin with additional minute sums for bundling (1 cent a bundle) and other tasks, in addition to the lease-mandated supplies and seals killed for food. Although the total sum of wages paid of under $1,000,000 for twenty years was exceptionally small even for the era as a percentage of return, nevertheless the Aleuts compared to the average American worker did not come off badly, even though goods purchased with their pay had to be bought at the company store which was permitted a maximum 25 per cent markup over San Francisco prices. Aleut money not immediately spent after division of the "seal fund" was set aside in company-held savings, on which the ACC paid 4 per cent per annum: $40,000 stood in that account at the end of the lease. In addition, as promised, the company provided free housing and medical care.[57]

The company's policies, at least in the instructions issued to its agents, showed the best intentions. President Miller wrote in 1872:

It is important that the utmost care be taken to see that the natives are kindly and liberally treated; that friendly relations between them and all our employees constantly exist, and that no injustice, even in the smallest degree, be done them; that the free schools are maintained; that no interference with their local government or religion be practiced, and that they are constantly treated as people having the same rights, privileges and immunities as all citizens of the United States enjoy.

All efforts to elevate them in the scale of civilization should be encouraged.[58]

The picture is easily conjured up of eager Aleut response:

Through the years, the Company sold them increasingly American types of store goods. The men loved bowler and derby hats, beautiful ties and waistcoats. The women loved to study fashion pictures in Harper's Monthly at the stores and to order San Francisco gowns and hats and cloaks or materials to sew their own stylish garments.[59]

All true, but also all misleading if left at that. In fact, the Aleuts were a captive population, and some of the apparently altruistic concessions of the company were designed to keep them that way. Health services, for example, meaning particularly a doctor for each island, were an inexpensive investment to maintain a dwindling labor force, already too small for peak season needs. Between 1867 and 1890 the population of Aleuts on the Pribilofs fell from 422 to 303, a decline of 28 per cent, resulting from the excess of deaths (of which the rate was high) over births. The company blamed insanitary conditions and worked to build sanitary housing. But that housing was probably a negative health factor: the turf-roofed half-underground *barabaras* which the Aleuts had previously used were the result of long adaptation to the climate and resources of the islands. When the company, with government approval, destroyed these homes, it moved Aleuts into wooden houses which could no longer be heated by seal blubber lamps but required instead coal or driftwood, often in short supply. Nor was free housing much of a concession: had the company allowed the Aleuts to build or buy their own houses, a vested right or interest would have been created which, at a later date, might have proved to be very troublesome.[60]

Regulations of all types insured that there was little escape. The government controlled and often prohibited movement from the islands and allowed no visitors not equipped with advance government approval. Refusal to work for the company (sealing) or the government (construction, unloading of supplies) was punished by the agent. No freedom of trade was permitted; no vessels were allowed to touch at the islands. No free use of funds was allowed either; the Aleuts were paid not in cash but in credit, and the agent limited what each could spend. The government, like the company, had pledged noninterference in traditional institutions but soon found the Aleut chiefs tiresome and began to appoint and remove them at pleasure, insuring more amenable leadership. The company punished no one, but the government did whatever was necessary, particularly using reduction in pay grades for sealers as a penalty for idleness, absence from school, abuse (or use) of alcohol, even simple disrespect.[61] The relationship is reminiscent of that between the Spanish Inquisition – which burned no heretics – and the civil government, which always did what was asked.

Samples from the logs and other documents collected by Dorothy Jones demonstrate a wide range of sanctions, from leg irons or threats of exile to interference in the marriage arrangements of private individuals, or the preannounced search of the Aleuts' (actually, the company's) houses. Always the most common crime was the brewing of qvass, an alcoholic beverage made from anything which would ferment, and official logs are full of references to denial of molasses, sugar, raisins, preserves, even sugar cookies to the Aleuts in penalty for consuming one of the many formulations of the stuff. Bay rum and cologne were equally consumed until the company stopped their import. "Raw quas," wrote one agent, "takes rank as the most villainous compound that ever traversed the human gullet, making the drinker not only drunk, but sick also, and unfits him for work, even after the stupor has passed off."[62] This agent cut off the sugar supply, at which point the sealers in the midst of the June slaughter threw down their clubs with an ultimatum: "No sugar, no seals will be." The agent refused to make any concession and ordered another gang of men up from Unalaska, an action that broke the back of the strike.

Mr. Gavitt [St. George agent] ... explained to me that he found the story of quass under the floor [of Aleut homes] was not true, by going to the houses and tearing up the carpet and opening the trapdoor where it was supposed to be and he found none secreted there.

There had been one occasion when three of them went aboard a vessel and got some [alcohol] from a cook ... We had them up to the police court and tried and fined them $10 apiece.

Kerric Tarakanoff on the street drunk. Called him to government house and put him in irons.

Alex Galaktionoff ... goes to St. George to find a wife and with the distinct understanding that he is to find one before returning. If he gets married he is to return next spring – if not he is to seal over there.

Mr. Redpath [company manager] reported that Peter Krukoff was saucy to him. Peter was ordered to do some work and report to me.

Complaints made about Widow Popoff being saucy to a passerby. I sent word to keep her mouth closed or I would let her live in some other part of town more out of the way.

Shortly after I got there ... I called up the priest and the men and told them the state of affairs [alleged immorality] was very bad and they must correct it or that the government would send them to the Aleutian Islands to live on codfish; that this government could not afford to have such a black spot in existence.[63]

In short, the Aleuts in comparison to the inhabitants of the continental United States were not badly off economically during the period of the ACC lease, but were very badly off indeed from the standpoint of the theoretical rights and privileges of a democratic society. But from the outset it was agreed that the Aleuts were inhabitants of a preserve under government control, and this put them in limbo: neither "citizens" with the right to vote (in this Miller was misinformed), nor Indians and thus "wards" of the government. Though occasional petitions against these conditions emerged from the islands, in practice there was no real means of appeal. The only resistance possible was refusal to work or to attend school, and both had their penalties. The school represented a special problem; after initial enthusiasm, the Aleuts soon came to see it for what it was, at least in part: a determined attack on the Aleut and Russian languages and traditions, which were prohibited in the school.[64]

Though the Aleuts had little choice but to remain silent on the whole, the ACC had its critics from the very beginning, not least of which were the disappointed bidders for the lease. There was also still J.A. Oppenheim and Company. When Hutchinson, Kohl won the first struggle for the skins, Oppenheim's San Francisco agent, one Agapius Honcharenko (a Ukrainian and a Russian Orthodox priest) with the help of anti-ACC elements organized a critical weekly newspaper, the *Alaska Herald*, which for some years mixed fact with fiction in its incessant attack upon the ACC for profiteering, smuggling of seals over the legal limit, and so on. Parrott and Company, along with the Pioneer American Fur Company, petitioned Congress against Hutchinson, Kohl, and then in May of 1869 formed, with several other firms, the Alaska Traders Protective Association to mobilize the attack – though it was undermined by the subsequent union of Parrott with the ACC. The goal was to force cancellation of the lease and opening of the seal market to any and all bidders. Interestingly, a petition from the American-Russian Commercial Company to Congress was prescient enough in 1869 to warn that any monopoly company "would have a kind of feudal sway over the natives." Several California newspapers joined the hue and cry, but, given the quick financial success of the ACC, it was all to little avail.[65]

Still another major attack was mounted in the mid-1870s by a coalition of opponents formed into an Anti-Monopoly Association. Arousing national attention with its charges, the association managed to provoke a congressional investigation by the House Committee on Ways and Means, only to have the committee conclude that the charges were unsubstantiated. In the process, Robert Desty, chief writer for the *Alaska Herald*, felt obliged to retract all his criticism (perhaps because he was a French citizen and was promised – or threatened with – a speedy return to the land of his birth). Desty maintained that he had been duped by Hocharenko, who had dictated what he claimed were translations from the original Russian of various Aleut petitions and complaints. The *Herald* now closed, its life and death clouded in obscurity.[66]

The ACC lived in relative peace for another decade, until the 1885 appointment of Alfred P. Swineford as governor of Alaska. A political appointee from Michigan, Swineford launched an impassioned assault upon the company in his annual report for 1887: "Conceived ... in corruption, born in iniquity, and nurtured and grown strong and insolent on ill-gotten gains wrung from a hapless and helpless people, this giant monopoly, which lies like a blighting curse upon the progress

and welfare of this great Territory, should be shorn of its corruptly
secured, much-abused franchise with no more delay than may be absolutely necessary." Swineford modified his charges after his first visit
to the Pribilofs in 1888, but he did not reverse his opinion that the
lease should be cancelled. For the ACC, Swineford, who does not appear
to have possessed significant political influence, was simply another
storm to be weathered.[67]

Ironically, while many of the charges levelled against the ACC were
implausible – smuggling of extra skins, for example, would be hard to
conceal given the considerable number of people who counted them in
and out each time they were handled from slaughter to processing
(particularly at the annual London sales, the only major market) – yet
the very real basis for severe criticism was overlooked almost entirely
until after the lease had been cancelled. The issue was simple to raise,
nearly impossible to answer: how large was the seal herd? Though
many estimates were made, most remained no more than that until the
coming of Henry Wood Elliott to the Pribilofs in 1872.

Elliott, born in 1846, visited Alaska in the 1860s with Union Telegraph and U.S. Geodetic Survey expeditions, but he now arrived at
the Pribilofs as a special Treasury agent sent to study the seals. His
qualifications included a mysterious position in the Smithsonian Institution (he had certainly been a clerk and artist there, but he also claimed
to have been private secretary to Joseph Henry, the Smithsonian's first
director, although no such position is mentioned in the Smithsonian
archives). But from his first glimpse of the seals, he became committed,
and from 1872 until his death in 1930 remained a vocal, forceful,
emotional, and influential spokesman for the seals (but never for the
Aleuts, though he married one). Unfortunately, his effectiveness was
forever handicapped by his arrogant assumption of infallibility in the
one aspect that mattered above all – the size of the herd.[68]

In his census of 1872, Elliott had no trouble in assessing the herd,
though most men would have found the task as bewilderingly impossible as counting an active hive of bees from simple observation. Elliott
assumed that each seal required "2 feet square" (meaning two square
feet, not 2' × 2'), a very small space indeed, on rookeries which he
calculated to include a total of 6,300,000 square feet. This gave him a
calculation of 3,193,670 seals, to which he added 1,500,000 nonbreed-
ing seals, for a total population of at least 4,700,000. Elliott's pride, and
conceit became locked to this "imperative and instinctive natural law
of distribution."[69] As he wrote in a lengthy study of Alaska published
in 1887 (and of which the bulk actually concerns the Pribilofs):

There is no more difficulty in surveying these seal-margins during this week
or ten days in July than there is in drawing sights along and around the curbs

of a stone fence surrounding a field. The breeding seals remain perfectly quiet under your eyes all over the rookery and almost within your touch ... There is not the least difficulty in making such surveys, and in making them correctly.[70]

How a scientist could make such a claim defies explanation. Far closer to reality is the description of G.C.L. Bertram, written in 1950:

So close-packed are the few great rookeries, so great are the size differences between new-born pups and aged harem bulls, so continuous the confusion and movement of cows and young males to and from the sea, indeed so fantastic the crowding and biologically ordained commotion, that no one can do better than say that probably there are between two and four million animals in the herd.[71]

Elliott had indeed added in his 1887 claims that his calculations were "not systematic ... and my estimate for them is purely a matter of my individual judgment," though he meant this qualification mainly for his addition of 1,500,000 *holluschickie*, stragglers or bachelors. Twenty years later he was inclined to forget that admission: his figures, he wrote in the *North American Review* in 1907, "have never been successfully denied as to sense or accuracy by any man, and they never will be."[72]

"Never" is a word to be used with circumspection. Alas for Elliott's reputation, his estimate was greatly inflated. Seals need considerably more space, for the group is nothing so compact as Elliott assumed: six square feet per animal would be a closer approximation. Moreover, the rookeries, as studied by Karl Kenyon in 1948 (and in other studies), were more on the order of 3,500,000 square feet. Given his double error, Elliott's guess was about 300 per cent too high. Unlike Columbus, who first set sail for the Indies on the basis of a similar miscalculation, Elliott never saw cause to doubt his own figures through his periodic visits (1872–6), nor did the agents on the islands: the fourth census of 1886 reported a seal population of 4,768,430 with the astonishing precision which was now commonplace, though no actual, thorough physical enumeration had occurred.[73]

But in the summer of 1890, after a hiatus of fourteen years, Elliott returned to the Pribilofs, only to discover that the total number of seals on the rookery beaches were at best some 20 per cent of his earlier census. Outraged, he accused the company of destroying the breeding capacity of the males by driving them to the slaughter grounds – his assumption being that even those allowed to go free were so adversely affected that their capacity to breed was forever destroyed. This hasty and biologically absurd conclusion brought bewildered astonishment in Washington. In the opinion of his superiors, Elliott clearly had been smitten by some form of insanity – whether temporary or permanent made no difference – as evident in his numbers as in his causal reasoning. His report was quickly pigeonholed and withheld from publication, and he himself was sacked – but not silenced – in April 1891. Elliott had a fine capacity for making enemies in any case, and his marriage in 1872 to Alexandra Melovidov, an attractive and intelligent "Creole" (the term used for Aleut-Russians; her father was a Russian official), did him no good among racially biased administrators who frowned on any fraternizing with the Aleuts, let alone marriage.[74] Elliott, for his part, came fairly soon to admit that pelagic sealing – taking seals at sea – probably had also played a considerable role in the reduction of the herd, as indeed it had, as will be shown in the next chapter.

The most amazing part of the story, however, is the failure of anyone connected with the islands to report in any meaningful way the decline of the herd in the last decade of the lease to the ACC. For all his idio-syncratic methods and conclusions, Elliott did prove vital in awakening the Treasury Department to the danger of the collapse of the herd from over-exploitation, even without pelagic sealing. The last point must be stressed, for pelagic sealing thoroughly complicated the problem of measuring the herd's size and, therefore, the maximum sustainable yield (a term not then in use). One conclusion is unchallengeable: the com-bination of land sealing at the ACC rate and uncontrolled pelagic sealing doomed the herd unless something was done, and quickly.

The signs of the herd's decay were there for those who wanted to see; serious blame attaches to both company and government for failure to pay attention. In the last eight years of the lease, according to Gerstle Mack (who from the family association implied by his first name had access to private correspondence of some of the directors), the ACC was already culling seals from beaches not normally used for this purpose because of their distance from the killing grounds. Similarly, to the normal take of three- and four-year-old bachelors, the company was adding less valuable two-year-olds, even yearlings, to reach its quota. In 1889, half those killed were two-year-olds and 25 per cent yearlings – a development that was legal but ominous.[75] Leon Sloss, son of the director, who acted for the ACC on the island, testified in 1902 that the size of average pelts was declining steadily. A simple but indicative measure was the number which fit into a cask (always the same size) for shipment: 47.5 in 1885, 50.5 in 1886, 55.5 in 1888, and 60 in the last year of the lease.[76] And yet, astoundingly, agents on the islands continued to report that the herd was increasing. Not until the summer of 1889, the last under the ACC's lease, did agent Charles J. Goff send in a report so alarming that it inspired the Treasury to send Elliott as special investigator, with the results just described.

The ACC's lease was not renewed, and there was certainly just cause. Yet just cause had nothing to do with nonrenewal: Elliott's new and shocking census was not made until July of 1890. The previous March, when the lease expired, the Treasury Department negotiated a different contract with another concern – the North American Commercial Company, a new group not to be confused with others with similar names (one of the complexities of Alaskan history) discussed in this chapter. The ACC thus passes out of our story, though the company continued for some years to operate its other Alaskan interests.

The fur seals had created some rich men, and made others already well off richer yet. Though some, like Kohl and Niebaum, preferred a quiet if luxurious ranch life, others continued to be prominent in Washington, where Hutchinson died in 1883, still the company's representative, and where Miller similarly expired three years later during his term as senator from California, or in San Francisco, where Sloss and Gerstle turned their attention to such enterprises as the San Francisco-San Joaquin Railway, the North American Navigation Company (whose steamers ran to Panama in competition with those of Southern Pacific), the Union Trust, and the California Hawaii Sugar Company. Both served as treasurers of the infant University of California (Berkeley). Williams and Haven and their New London Associates will be encountered again in connection with the elephant seal industry. Commercially, the ACC had been an outrageous success: it remained to be seen whether the NACC would have such great good fortune.

CHAPTER FIVE

Pacific Rivals,
1890–1911

*And oh, the marvel of it! the marvel of it! That
tiny men should live and breathe and work and
drive so frail a contrivance of wood and cloth
through so tremendous an elemental strife!*
            Jack London, *The Sea Wolf*

Given the ACC's miraculous production of massive profit from minimal
investment, its position should have been unassailable as it set about
applying for renewal of its lease in 1890 – unassailable, that is, so long
as the problem of seal numbers did not arise. But 1890 was not 1870,
and despite the wealth and influence of Gerstle and Sloss, there were
other men of wealth and influence equally interested in the furs of Alaska
and even better placed to obtain preferential treatment in 1890. The
North American Commercial Company did, it is true, tender a higher
bid that year than the ACC, but a higher bid only partly explains the
sudden abandonment of the ACC's twenty years of proven experience.
The country as a whole was fast developing, and such prizes could no
longer be the province of a few San Francisco merchants; the game was
much more likely to be played out at a national level.[1]

Two of the half-dozen dominant figures in the NACC were of that
caliber. Darius Ogden Mills, NACC vice-president, was exceedingly
wealthy by virtue of various California interests and was father-in-law
of Whitelaw Reid, a powerful leader of the Republican party (he would
run for national office in 1892 as vice-presidential candidate to President

Harrison). Reid owned and edited the influential New York *Tribune*, and was a close associate of James G. Blaine, sometime Speaker of the House (1869–75), secretary of state (1881, 1889–92), and an unsuccessful candidate for the presidency in 1884.

The second important stockholder was Stephen B. Elkins, Blaine's financial adviser and campaign manager; his future included appointment as Harrison's secretary of war (1891–90) and subsequent election as senator for West Virginia (1895–1911). Blaine, Elkins, and Harrison, and their families, were business associates and social friends, and the result was a chain of connections which led from the Pribilofs to the State Department and the White House through Reid, Mills, and the man who actually ran the NACC, Isaac Liebes. Liebes, like Sloss and Gerstle with whom he was sometimes linked in various enterprises (but not the ACC), was also a member of the German Jewish merchantocracy of San Francisco and in particular the fur business.

Precisely how Liebes came to be associated with Mills and Elkins in the NACC – incorporated in December 1889 with a capital of $2,000,000 in time to bid on the lease – is unknown, but most probably through Mills's own San Francisco connections, including Rothschild banking representatives.[2] The prospect of replacing the ACC must have seemed both intriguing and possible, with the help of some highly placed friends. After all, not only was the ACC's lease expiring, but after 1888 Harrison was in the White House. Just how much any of this mattered in early 1890 when the NACC's higher bid was accepted is not provable. As Charles Campbell has explained in his study of the Bering Sea negotiations of 1890–2, however, it was most probably a factor of considerable significance in the development of American foreign policy regarding that area.[3]

In any case, the lease was granted to the new company on 12 March 1890 (the ACC's lease expired on 30 April) for a similar twenty-year period, and was signed by William Windom, secretary of the treasury, and Liebes on behalf of the NACC.[4] Basically, the terms were similar to those given to the ACC, except that the annual rental was raised to $60,000, the tax on each skin to $9.625, and the quota made a variable figure set at the discretion of the government (but not to be over 60,000 for the 1890 season). A number of concessions which the ACC had made of its own volition were now required specifically of the lessee, including twenty tons of coal each year, the maintenance of educational and health facilities, a church, "the necessaries of life for the widows and orphans and aged and infirm inhabitants," along with "a sufficient number of comfortable dwellings" for the Aleuts, to whom would be paid "a fair and just compensation," to be set by the secretary of the treasury. Overall, the company agreed "to contribute, as far as in its power, all reasonable efforts to secure the comfort, health, education, and promote the morals and civilization of said native inhabitants." The NACC purchased the ACC's entire Pribilof plant for $67,264.82 and moved in, ready for the 1890 season.

Ostensibly, the relationship of government, company, Aleuts, and seals had changed little, except that the government's fees had increased. In actuality, a rather important alteration had occurred in the role of the treasury agent on St. Paul (in 1903, the Pribilofs were transferred to the new Department of Commerce, and, in 1909, to the latter's Bureau of Fisheries). Since the agent now was responsible for the timing and quota of the harvest, he had to concern himself actively with the problem of numbers as never before. This responsibility would have benefited from some biological training, but unfortunately agency jobs were "too good to be kept from the faithful,"[5] as one assistant agent, dismissed so that another man could be put in his place, remarked, and these posts remained political plums. Moreover, the wage structure would now be set by the agent acting for the Treasury and not by the company, so that he had now to concern himself, presumably, much more with comparative compensation rates, food prices, and the like, and not simply insure that the government received its rightful share of profits as landlord.

As a result, the situation of the Aleuts became the responsibility of the federal government far more than in the days of the ACC. Unfortunately, the agents were caught between the needs of the Aleuts (to whom they might or might not be personally sympathetic and over whom they still presided as prosecutor, judge, and jury), and the obvious decline of the seal herd which was nevertheless expected to provide for the basic needs of the Aleuts through the payment of wages for sealing work performed. The basic wage was increased to 50 cents a pelt in 1894 and 75 cents in 1906 (the latter primarily to offset increased prices at the NACC store, the profits from which, the agent in 1905 calculated, were enough to match the company's entire Aleut wage bill for that year).[6] Over the life of the lease, 1890–1909, the average income of St. Paul Aleuts (213 in 1890, 193 in 1909, to which should be added St. George's 90 and 87, respectively, to arrive at the islands' total population),[7] including sealing wages, supplemental pay for government work (normally 10 cents an hour, or 15 cents for unloading cargo), money from the sale of fox skins, goods donated by the company under its lease requirements, and a government subsidy to alleviate poverty, totalled some $25,000 a year – about $500 a year for each of the fifty adult workers on the island.

Roughly comparable with continental United States levels at the start of the lease, this income represented considerably less than the average by 1909, though since Aleuts worked only about two months of the year in the formal sense such comparisons are a bit misleading. The real problem was that costs on the island were very high, and the per capita income, given the higher average size of Aleut families, below general United States levels. Seldom, in fact, did average Aleut income rise above subsistence levels, and the level of personal consumption of purchasable goods was visibly low. Agents might worry about extravagances ("Nor do I want *fine*, thin, toothpick shoes, nor dancing slippers for either men or women under any circumstances"),[8] but in fact there was seldom surplus money for a second pair of shoes.

By 1892, it was clear that wages simply were not adequate to sustain the population, and the Treasury began paying a supplement which over the life of the lease averaged more than $15,000 per annum and totalled $310,863 for both islands. It was given out in food and clothing, not cash. Because the harvest in the NACC era only once exceeded 30,000, and generally averaged little more than half that, the Aleut's sealing wages for both islands were about $9,000 a year. Relief, in other words, came to be the principal source of income. On St. Paul, the average total annual income over the lease for the island's entire population was $7,583 for sealing, $694 for miscellaneous labor and fox skin sales, $7,075 of donor goods from the NACC, and $9,226 from government appropriations. It is no surprise, therefore, that income from whatever source came to be regarded as charity – a conception all the easier to adopt since wages were not paid in cash but only credited to the individual's account at the store, where a very limited range of items was obtainable at relatively high prices.[9]

Walter Lembkey, a St. Paul agent who protested against the altered pattern, was clear on the overall effect:

Heretofore the native could expend his earnings as he pleased. After the [relief] appropriation, however, the earnings were sequestered by the agents, and the natives had no voice whatever in the expenditure of the money for which they toiled. Each native was allotted articles of necessity to a certain amount each week payable from his wages and after the latter were expended the appropriation was drawn upon at that same rate until another sealing season intervened … This plan of compensation … is highly objectionable when considered from a sociological standpoint, its weakness being that it reduces all to a common level. It prevents the progress that accrues from the cultivation of superior skill or greater self-denial and makes a virtual almshouse of the Pribilof reservation by dealing with the inhabitants as indigents.[10]

By 1909 the Aleuts had become wards of the government, and the islands a "reservation," not by any legal definition – no Act of Congress had made them so – but by administrative fiat through the vast powers of the chief agent on the islands and his deputies.

In the actual seal harvest, on the other hand, changes were few.[11]

25 Killing fur seals, Pribilof Islands (late nineteenth century)

26 Skinning seals, St. Paul's, c. 1890

27 Fur seal carcasses, St. Paul's, c. 1939

28 Gatling gun used for defense of the Pribilofs, c. 1900; agent's wife demonstrates correct position

29 Sealing fleet, Victoria harbor, c. 1905

30 Victoria sealers under sail

31 The *Casco*, a famous Pacific sealer

32 Capt. Victor Jacobsen, Victoria

33 Pelagic hunters, *Thomas F. Bayard*, 1908

34 Indian sealing crew, Victoria schooner

35 Indian sealers with their gear, Victoria

36 Spearing seals

37 Aleut *umiak* and crew

38 Shooting elephant seal, Macquarie Island

39 Stripping blubber from elephant seal, Macquarie Island

40 Rafting empty casks for oil ashore, Macquarie Island

41 Rolling casks up the beach, Macquarie Island

42 Leopard seal

43 Shooting walrus

44 Chopping out walrus ivory

45 "Headhunters" with their catch

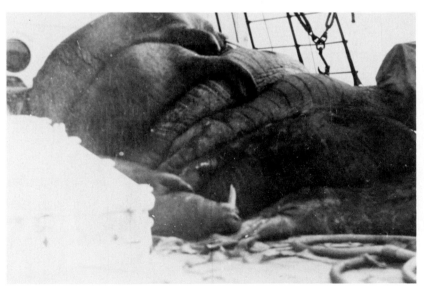

46 Dead walrus stacked aboard a Coast Guard vessel
(to be used as food for Eskimo villagers)

47 Sea lions killed on the Oregon/Washington coast

48 The *Sea Lion Hunter*, with male organs of seals drying in the rigging

49 The *Sea Lion Hunter*, a typical California "pick-up" vessel, moored at San Pedro, 1927

TABLE 3
Pribilof Land Sealing, 1890–1910

1890	28,859	1902	22,386
1891	14,406	1903	19,292
1892	7,509	1904	13,128
1893	7,390		
1894	15,033	1905	14,368
		1906	14,476
1895	14,846	1907	14,964
1896	30,654	1908	14,996
1897	19,200	1909	14,368
1898	18,047		
1899	16,812	1910	13,586
		Total	359,462
1900	22,470		
1901	22,672	Average	17,117

Source: Francis Riley, *Fur Seal Industry of the Pribilof Islands, 1786–1965.*

Modernized transport (a marine railway, winches, and so forth) and a new method of skinning were the principal improvements. The date of the latter innovation is unclear, but in essence it proved easier to make a cut around the head and flippers down to the stomach, pin the skull to the ground with a sharp iron bar, and simply rip the skin off entirely with clamps. The skin had still to be separated from the fat and processed for shipping. Biologists – then more likely to be called naturalists – attempted to add one stage at mid-season by which females could be so marked as to be undesirable to pelagic sealers. Several experiences in branding with hot irons or shearing a strip of hair on female pups proved expensive, impracticable, or both. To the frustration of those in charge, the seal population continued to erode. That being the case, the government permitted the NACC to take only small annual harvests. The company objected to the point of legal action, with the result that in 1893 the annual $60,000 rental fee was henceforth reduced on a percentage basis when the crop did not approach the 60,000 annual expectation implied by the quota set for the first year of the lease.

The average number of skins which the company was permitted over the life of the lease (see table 3) was approximately 17,000 per annum – a far cry from the 100,000 taken by the ACC. The Aleut workers, paid by the skin, were the first to suffer. The United States government, however, was also paid by the skin, though its $9.625 was rather more than the 50–75 cents paid to the Aleuts. Its seal income fell so quickly that when Aleut relief and Bering Sea patrol operations against pelagic sealing were added to other administrative costs the Pribilofs were no longer a profit-making matter. The net deficit to taxpayers did not tend to encourage material progress on the islands.

The company was naturally displeased to be denied larger quotas, but its profits remained high – though not at ACC levels – due to the fact that demand for sealskin in the last decades of the nineteenth century surpassed all supply, and of the various sources of skins Pribilof pelts were the best. Sale prices for skins taken by the NACC averaged, over the life of the lease, $27.39 each.[12] Even after preparation and transportation charges and deduction of the $9.625 tax, a high profit margin remained; had the price remained at the $14.00 of earlier years, the NACC would probably not have continued in operation with the small quotas allowed.

The dramatic increase in demand is attributable to one cause: fashion. Before the middle of the nineteenth century, Pribilof skins were likely to wind up as leather or felt. Between 1850 and 1870, however, a technological revolution of sorts occurred which permitted faster and better dehairing of guard hairs and dying, producing overall a better product. At that point – the early 1870s – seal fur became fashionable, which it really had never been. Furs had long been known and appreciated as an item of decoration associated with rank (in medieval Europe, use of furs was strictly regulated by law), but now sealskin was all the rage, the most preferred material for outer garments such as capes and coats for those who could afford them.[13]

The first modern "fur coat," made entirely of fur and not simply trimmed with it, may well have been made of sealskin. Such coats were popular equally among men and women. Those who could not afford the real item (selling in the 1880s for 30–40 guineas or $150–200), made do with second-class "musk sealskin" (7 guineas or $35) or "seal plush" (3 guineas or $15) which was not fur at all. In the nineties, fur fashions became even more elaborate, with sealskin trimmed with sable, ermine, even – incongruously – lace. In 1900, high fashion could demand a full-length seal coat, with collar, cuffs, and lapel edging of sable, and ermine lining – adding in the next decade sealskin muffs and hat to match. By this point, however, sealskin was most expensive and difficult to obtain – just in time, synthetic dyes of a new high quality capable of transforming rabbit magically into seal were becoming available – though seal furs continued to be popular as late as World War I.

Consumption figures tell the story. The fashion world of Europe and America made use of 10,000 skins a year in 1860, 20,000 in 1865, but 150,000, in the 1870s and 200,000 (if they could be obtained) in the 1880s. In the NACC era there simply were not enough skins to meet the demand, which explains two important developments: the continued high return to the NACC despite low quotas and the growth of an extensive pelagic sealing industry, despite the fact that pelagically taken skins were always worth less – averaging for the same era (1890–1909) just over $14 – for reasons which will be explained.[14]

Though figures differ on the number of seals taken overall, particularly those taken pelagically, the extent of the NACC's profit is clear. For the period of the lease, the company took in $9,700,000 in harvest income (including $235,000 from fox skins); its costs came to $4,200,000, including taxes ($3,400,000) and transportation and supplies ($800,000). After a further deduction for Aleut wages ($700,000), office expenses in San Francisco and Washington, depreciation, and the like ($400,000), the net profit was $4,390,917 – or 45.2 per cent of income – and this on a quite small investment base in vessels and property on the islands and in San Francisco. George Rogers has estimated that the U.S. government subsidized the company's operation to the extent of $2,400,000, through Aleut relief, Bering Sea patrol, and the cost of the agencies over and above return to the government from skin taxes and lease payments.[15] If that sum had also been paid by the company, it would have brought the profit margin down to 20.3 per cent, perhaps a more justifiable income than the 45 per cent it did receive. It is no wonder that the NACC continued its Pribilof sealing operations as long as possible despite the limited quotas.

It is also no wonder that others were eager to share in the spoils. Though the NACC held the Pribilof lease, the seals roamed the seas and were available for the taking. Indians of the British Columbia coast had done just that for generations, working their canoes off Vancouver and the Queen Charlotte Islands. Not until 1866, however, according to the distinguished northwest historian Judge F.W. Howay, did a larger vessel owned by whites take seals at sea. This was the trading schooner *Ino* of Victoria, 20–30 tons, owned and captained by Hugh MacKay. Indian hunters had suggested to him that he could take seals offshore by carrying out Indian canoes and their hunters and launching them when seals were present and the weather cooperative.[16]

MacKay found the *Ino* too small to carry more than two or three canoes (traditional Indian canoes required three men, but it was found a better use of resources to carry smaller canoes for only two men). William Spring, MacKay's partner in the trading business along the coast of Vancouver Island (a business which included Indian-prepared dried sealskins), purchased or chartered the *Surprise* for the 1868 season, adding the *Alert* the following year. Spring enlarged his fleet over the years and by the early 1880s owned and operated seven schooners obtained from various sources; the *Favorite*, for example, added in 1873, had been built in 1868 as a codfisher for the Okhotsk Sea based on Hawaii, and then later saw service as an Acapulco packet. MacKay also did well, and had two schooners of his own, *Onward* and *Alfred Adams*, by his death in 1881; his interests were taken over by William Spring and Company.[17]

Spring and MacKay were not without competitors. In 1870 the sloop

Thornton was taken sealing by her owner/master, J.D. Warren. Warren was destined to be a power in the business, in part because his partner was Joseph Boscowitz, whose brother Leopold has already been met as a founder of the ACC. Boscowitz operated stores trading for furs on Vancouver Island and elsewhere (some independently) and thus at least by 1870 was dealing in pelagic sealskins, though dried, not salted. Since dried skins could not meet the competition of salted skins from the Pribilofs and could only be used for leather, these traders turned to pelagic sealing when they saw it might pay. Boscowitz, who owned shares in several schooners, was also one of the first to send out Indians with their canoes on his vessels, although initially they hunted only in nearby waters for a week or two.

By the 1880s, the Warren-Boscowitz interests were converting schooners to steam, but Spring did not find this necessary. According to the testimony of his son Charles, the need never justified the expense, since in these waters dead calm conditions were rare. Though several master/owners also operated schooners of their own, making the Victoria fleet roughly a dozen in the early eighties, the two companies of Spring and Warren dominated the business, still using native hunters with spears. In 1884 a major change occurred when the first vessels were equipped for sealing by white hunters – beginning with the *Mary Ellen* under the command of Capt. Dan McLean, probably the first Canadian master to venture into the Bering Sea (the American *City of San Diego* was first in 1883).[18]

Although profitable, pelagic sealing was slow to develop, for several reasons. First, the really major demand did not build until the 1880s; second, the methodology had to evolve over time to combine Indian techniques with larger "mother" vessels to carry the canoes, and the time and place of seal migration routes had to be learned. Third, the west coast of Vancouver Island and the Queen Charlotte Islands, bases for many Indian crews and adjacent to sealing grounds, were themselves slow to develop – indeed, remaining largely wilderness country even today. Victoria, which was to become the principal northern sealing port, was surveyed only in 1837 and founded in 1843 as a Hudson's Bay Company trading post. Vancouver and Seattle, both of which were to surpass Victoria in size and industrial output (but never in sealing), were not founded until much later – Seattle in 1851 and Vancouver in the eighties. Finally, capital was required, and it was in limited supply; as it was, much early investment in sealing seems to have come from San Francisco, of which Victoria in many ways was a commercial outport.[19] The effects of the trade were varied, however, and even the Indians of Neah Bay (near Washington's Cape Flattery) became schooner owner/operators on a limited scale. Sealing's impact upon these peoples remains to be studied, but James Gilchrist Swan, local customs agent in 1873, hinted at the result in a diary entry even at that early date. Noting that the Makah Indians no longer bothered to hunt the plentiful whales in the vicinity, he concluded that devotion to sealing was the cause: "I think the business as now conducted is a positive detriment to these Indians. They neglect all other avocations during the sealing season from January to June, and the money they receive for the skins they secure is either gambled away or is spent for flour, bread, sugar, &c, is distributed in potlaches to their friends."[20]

Indian or white, in time the fleet expanded to match the demand for skins, and when it did so, the seals were doomed by their habit of feeding by night and sleeping by day. Fur seals, unlike hair seals, tend to break clear of the water when swimming, a peculiarity which makes them far easier to spot; but the easiest way to take them was to find them sleeping, often in peculiar attitudes familiar to sealers. A "dog-sleeper," for example, was easiest to catch, since with her nose just awash she seemed less easy to arouse. A "jughandle" slept on her side with one flipper over her head; a "breaster" crossed her flippers, while a "finner" with one flipper extended was often only half asleep. "Travellers" moved at a good speed, but "moochers" did not – but always seals were best taken asleep, and from upwind. To land-bound hunters this point may seem surprising, but sealers could drift down upon their prey from upwind, and seals were commonly more alert to danger through hearing – a canoe downwind would have to paddle up to the seal – than through sight or scent.[21]

Most seals taken below the Gulf of Alaska were females, since males did not migrate so far; most were also pregnant, unless under bearing age. Occasional surprises were in order. Frederick Schwatka, shooting a seal from his vessel, saw the seal disappear, only to be seen floating nearby in the next morning's calm. A boat was lowered; Schwatka rolled up his sleeve, put his arm round the seal to lift it aboard – only to find that he was embracing a suddenly very active unwounded seal which had merely been sleeping.[22]

Indian hunters, who operated with one paddler (often the hunter's wife), used spears with detachable harpoon heads which could be thrown 15–30 feet. The goal was to hit the water a yard or so away from the seal; sliding along, or just below, the surface the spear was thus more likely to hit the seal than if aimed directly at the animal. Once the seal was pierced, the task was to get to it quickly before it sank as, unfortunately, most did; though estimates vary from five or six to ten seals lost for every one retrieved, no hunter with experience in pelagic sealing ever contested the point that more were lost than recovered, whether shot or speared. Skins thus taken were always marred by a hole or rent, nor could they always be skinned immediately or preserved with the same care taken on the Pribilofs – all of which explains the considerably lower prices for pelagic pelts.[23]

White hunters used a special sort of sealing boat, double-ended, some 18–19 feet long and 4 feet 6 inches wide; they were good boats for open water, for they rode seas easily and did less splashing than other small craft – an important point when going up to a sleeping seal. If built in the northwest they were clinker-built, the planks overlapping; these sailed and rowed a bit better but turned a little harder than the carvel-built (smoothsided) models preferred in the South Atlantic. The boats normally carried three men: a hunter who stood in the bow to shoot, a puller or oarsman who sat on a middle thwart facing aft, and the boatsteerer aft who faced forward like the hunter, often standing to push his oars and place his boat in proper position for the hunter to make his kill. The boat would be equipped with some rations and water and both short and long gaffs (the latter 17 feet, three-pronged) for retrieving sinking seals.[24]

White hunters, who were in charge of their boats, used either long-barreled rifles – in the early days, Kentucky rifles were favored for their accuracy, though the powder smoke might obscure the target when speed in retrieval was essential – or, somewhat later, double-barreled 12-gauge shotguns. Hunters were paid by the pelt, so they were eager to work. Max Lohbrunner, a famous Victoria sealing master, in later years recalled some of these men: "Many of the best all-round shots I have come in contact with were hunters on some of the Victoria schooners. Snap shooting with a rifle from a fast moving boat in a heavy sea requires some skill. I have seen hunters who could hit a jumping salmon or shoot a bird in the air with a rifle."[25]

Shotgun shells were loaded aboard, and this was probably the most dangerous aspect of the voyage, at least to Carl Rydell, who was a navigator on the vessels of the H. Liebes Company (brother to Isaac Liebes of the NACC): "The cabin where shells are loaded is about twelve feet long by ten wide, and it contains a stove. As many as six hunters with a few helpers will be in the cabin loading shells. Twenty-five pound kegs of powder, some of them open, stand about the floor; the stove is going for melting lead to cast bullets; cigarettes and pipes are being smoked; and everyone is in a hurry, as there may be but a few hours for sleep at best. Under such conditions accidents will happen occasionally."[26]

Although different masters favored different combinations, the standard procedure was to recruit five to ten white officers and men at the ship's home port, then proceed to a coastal Indian village, load as many canoes and crews on board as could be carried (normally eight to twelve; at times competition was fierce for experienced Indian hunters), and proceed to the sealing grounds, where both white-manned boats and Indian canoes would be set out. The seals presented no danger, but the weather did. Both the Kuriles and the Aleutians could be dangerous as breeding grounds for storms when cold Siberian air masses encountered the moisture-laden, warmer atmosphere of the Japan current. Low pressure systems pass quickly in succession over the entire area, and sudden storms and fogs are facts of life. Gales could overturn schooners, and rip tides and treacherous rocks send them quickly to the bottom. Sealing here was no sinecure, and a good many men were lost, as the Russians had lost Aleut hunters before them, though sealers rescued others in trouble whenever possible. Max Lohbrunner:

Sailing from Victoria in the first part of January, we sealed mostly from 50 to 300 miles off the Oregon and California coast as far as Mexico ... If it was fit to lower a boat, even though the weather was pretty heavy at times, we were out anywhere from twelve to twenty hours a day, quite often longer than that, pulling and sailing as much as 90 to 100 miles some days looking for seal, with the schooner following the boats if the weather was clear and if it became foggy the schooner hove to but the boats still kept on hunting. The schooner after heaving to, would fire a small cannon called a bomb gun, every hour, also a hand-operated fog horn was used to assist the boats, to find the schooner on their return. We carried a five-gallon cask of water, a large box of hardtack, three loaves of bread, canned salmon and corned beef, in case we got lost and that happened quite often in foggy weather. When lost, we would try to find the schooner for the first few days, then if the weather was getting too tough and the schooner could not be found, we would start for the nearest land. As a rule we would not see land for months at a time but always had an idea in what direction the land bore by compass, and that might be anywhere up to a thousand miles or more. If the weather kept fine and you were lucky after some days to reach the shore and could make a landing, the next thing to do was to look for grub. As a rule we always saved a couple of seal carcasses if we were lost. Some of our boats have been out a couple of weeks before being found. The charts we had in those days were not like they are to-day and there was very little traffic on the Coast in some places. Many small sealing boats, especially in Northern waters and off the Japanese Coast, never made the land and foundered in heavy weather making a run for it.[27]

Usually in March the schooners headed north again, up the California and Oregon coast, calling somewhere on Vancouver Island to replenish stores and fresh water and offload skins for shipment to Victoria. Then the vessel was on her way, sealing along the Queen Charlotte Islands and the Alaska coast, calling at one of the Aleutians for water (usually in June), making for the Japan coast and Kurile Islands, then back to the Bering Sea where she would stay until the latter part of October or November, just before ice set in and the weather worsened. She called again at one of the Aleutians for fresh water and more ballast, "to drive her homeward bound to Victoria," as Lohbrunner put it, "to get a little beer before the other schooners would get in ahead of us and drink it all up." A month in Victoria to spend pay and display Japanese silks and gewgaws, and it was time again to head south.

Such lengthy voyages were demanding, even more so when catches were small and wages low. These were tough men, and battles among the crews were not unknown. Some masters preferred Indian crews as more quiescent, but among Indians too mutiny was not unknown.[28] If the attractions of civilization were near, it was hard to rein in either whites or Indians. Monterey, California, was a particularly favored spot. Capt. Matt Mathieson, master of the *Umbrina* in 1907, found a canoe missing while hunting fifteen miles off Monterey in a thick fog. When he called at the harbor the next day, he found his Indians in the local jail. They had found the town by the sound of the whistling buoy outside the roadstead and with some advance pay in their pockets had soon discovered many new friends. Mathieson found himself the purchaser of a new drugstore window and sundry receipts for judicial fines to a total of nearly $500 before he had his two men back aboard.[29] For some years, so many sealers frequented Monterey that a regular trade developed in the illegal sale of "gin" (alcohol, sugar, and water) to the Indians. In 1906, the *Ainoko*'s Indians gave a war dance in the local opera house, attended with interest by the crew of the *Penelope*, who collected, and then decamped with, a public subscription in honor of the *Ainoko*'s dancers.[30]

Monterey, Sitka, Hakodate, the Bonin Islands (home of skilled otter hunters and boathandlers), Victoria – the sealers knew many corners of the North Pacific. Losing an eye to a harpoon's point did not mar Victor Jacobsen's recollection of the exuberance of youth: "We had good times, too, lots of fun and excitement at the northern ports when we put in, dancing and music and feasting. I've seen some beautiful women among those Russian half-breeds, and they were the finest dancers in the world. All our schooners carried a musician or two. Rum and whisky were plentiful at the settlements. We were all young and devil-may-care. Great days!"[31] Russian ports, however, were visited only involuntarily, when a boat was seized by a patrol vessel. Similar problems could arise in Japanese waters, especially when the Japanese organized their own pelagic sealing industry.

Above all, however, the wise mariner stayed clear of the Kuriles. A cold arctic current sets from the northeast along this coast, inside the boundaries of a stronger, warmer northerly Japan current. The result is a terrifying conglomerate of tidal rips and fogs. Heavy gales are frequent; only the presence of deep water inshore gives scope for some maneuverability in case of trouble – warning of which may come from kelp beds (always near shore) or sea lions, whose persistent barking is a sure sign of rocks and thus gladdened no sealer's heart in a fog.[32] Nevertheless, the presence of otters and seals brought from long distances hunters, encouraged by the fact that throughout the nineteenth century the islands really were governed by no one (whatever their legal status might be), neither sealers, nor Russian (to 1875) or Japanese officials, nor local Ainu residents.

Just when Victoria fur traders were expanding their operations to include hunting at sea, the last remaining otter hunters from California – by mid-century a mere handful of unique characters, akin to the mountain men who opened up the fur trade of the Rocky Mountains (some of whom had become otter hunters) – often turned their attention to the Kuriles. According to Capt. H.J. Snow, that rare mixture of experienced otter hunter and sealer who was also an articulate writer, a Captain Kimberley of Santa Barbara in the small schooner *Cygnet* was responsible for rediscovering the otters of the Kuriles in 1872 (they had been hunted by Russians in the previous century). Kimberley, according to Snow, was searching for fur seals and iced-in whalers and had moved to the Japan side to take advantage of the customary earlier break up of ice there.[33]

Kimberley did reasonably well hunting for otters, then bringing $80–90 a skin; crewmen he left in the Hokkaido port of Hakadote, however, spread the word and a minor otter rush followed. Snow himself took the lead, buying the schooner *Swallow* in Tokyo for $2,000 and fitting out for ottering though he had absolutely no experience in their pursuit. Knowledge did not come easily; he was first dismasted in a storm and then wrecked in the Kuriles. But luck, fortitude, and the backing of a western firm operating in Japan kept him hunting otters and seals from Japanese ports until 1890.

Otter hunting required persistence, skill, and teamwork. The otters had to be chased down by three boats working in tandem, with four to six men in each (one schooner normally carried only three otter boats). The otter, once sighted, was kept constantly surrounded by the boats, sometimes over a long haul of three or four hours; smooth water was essential, otherwise the otter, whose nose was likely to be the only thing rising above the surface, could not be followed. Even sunny weather was a handicap due to glare. Though otter hunters could and did take seals, their boats were different, being worked by slender sweeps, not oars, and normally carrying no sail. Few sealers took otters, however, for normally they lacked the necessary experience. Hunters considered sealing the harder work, though masters who had to navigate in and about the Kuriles probably would not have agreed.

Snow estimated in 1897 that some fifty vessels from America and

Japan had hunted otters in these islands. Five were seized by the Russians, thirteen lost with all hands, and another seventeen wrecked. When fourteen quit after lack of success, only a handful were left, several of which were Japanese owned and manned. Snow was inclined to blame the Japanese for the rapid exhaustion of the otters and seals here, remarking that they were incapable of exercising control yet refused to lease the harvest to a company which might have done so.

The presence of western sealers inspired a Japanese sealing industry which, while little faster to develop than that of Victoria, came to play a major role in the issue of pelagic sealing, and not only in Japanese waters.[34] When Kurile otter/seal hunters moved on to the tempting quarry of the Commander and Pribilof islands, the Japanese soon followed, though American masters were the first to risk direct raids upon the beaches. The intrepid master of the *Cygnet* once more was in the forefront, raiding the Pribilofs in 1874; the *Ocean Spray* was next in 1876. Raiders landed to take seals, but that was not necessary; often it was enough to send a boat or two close inshore to fire off guns and thus drive the frightened seals into the sea.[35]

The ACC, first to be affected, was not prepared to accept such interference, whether on the Pribilofs or on its Commander Islands leasehold, without a struggle. It equipped its rookery guards with guns – on St. Paul even two Gatling guns and three light cannon.[36] Some seals were taken by the raiders, but some raiders lost their lives. Three of Snow's men never returned from a raid upon Copper Island in the Commanders, and Snow himself had a narrow escape with five bullet wounds (and thirty-four holes in his boat).[37] It was just this sort of story which intrigued Rudyard Kipling – who heard the tales at the Yokohama Club while passing through Japan in 1892 – and gave him inspiration for his dramatic account written soon after, "The Rhyme of the Three Sealers."[38] The raids continued into NACC days; five Japanese lost their lives on St. Paul in 1906.[39] Seal losses in such raids were small, however; it was sealing at sea which was increasingly costly to the herd after the first schooners began to hunt in the Bering Sea in 1883 and 1884.

Given the easy availability of seals, and the rapidly rising demand for skins, it is no surprise that the sealing fleet grew dramatically in the mid-1880s.[40] The precise number will never be known, since vessels came and went without record, particularly the small, locally owned schooners of British Columbia, California, and Japan. Others simply vanished, changed profession (going to cod or salmon fishing), or altered name and designation because of a sale or legal problems with the Bering Sea Patrol. At a minimum, however, 300 individual vessels, almost all sloops and schooners (only one square-rigged vessel is known to have gone sealing), some equipped with auxiliary engines, went out

TABLE –
Pelagic Sealing Vessels, 1868–1911

1868	2	1891	88
1869	2	1892	124
1870	–	1893	107
		1894	113
1871	4	1895	116
1872	4		
1873	5	1896	91
1874	9	1897	59
1875	8	1898	40
		1899	26
1876	6	1900	36
1877	3		
1878	5	1901	39
1879	9	1902	36
1880	4	1903	23
		1904	22
1881	19	1905	18
1882	21		
1883	18	1906	16
1884	20	1907	13
1885	31	1908	8
		1909	4
1886	38	1910	4
1887	48		
1888	50	1911	4
1889	59		
1890	68		

Sources: See below, p. 290n40.

for at least one sealing season between 1868 and 1911. Fore and aft rig was necessary since only it could be managed by one or two men – very likely the captain and a mate or the cook – while all the boats and canoes were at sea (the rig was also more maneuverable in island waters).

Before 1880, not more than a dozen sealing vessels operated in any one year, but then the number of vessels for which there is definite record rose rapidly. The 18 to 20 of the early 1880s became 50 in 1888, nearly 100 in 1891, and – the top year – 124 in 1892, after which the fleet fell off slightly, to 91 in 1896 and then, the following year, began to decline rapidly due to regulations against American pelagic sealing (these numbers are likely to be somewhat lower than those of vessels actually involved, but records are skimpy). By 1911, only four schooners were still in operation, though many were laid up in various ports awaiting either freedom to seal pelagically or monetary compensation for the inability to do so.

Most vessels actually made few trips for seals. Of the 260 whose detailed career is at least partly known, 65 went only once, while 165 went five times or less. Fifty vessels, however, made ten trips or more, and a selected fifteen schooners went fifteen times or more, among which were some famous names in the sealing fraternity, such as the *Anna Beck* (sixteen voyages), the *Dora Sieward, Triumph,* and *Libbie* (seventeen each), the *Mary Ellen* and *Teresa* (eighteen), *Umbrina* (nineteen), *Penelope* (twenty), *City of San Diego* (twenty-two), and the apparent record-holder, Spring's *Favorite,* with twenty-nine voyages, spanning nearly the entire era of pelagic sealing. Since aside from Snow's informal calculations on the Kuriles at least fifty ships were lost, foundered, or wrecked, the *Favorite*'s lifespan surely defied prediction.

Of the 270 of which ownership by nation is known, 116 were American, 94 Canadian, and 11 changed national colors at least once (the remainder are unknown, save for four listed as German, Japanese, or Mexican). The place of construction is unknown for too many (86), but of those for which records exist about 65 were built in Washington or Oregon and another 65 in California (51 of which hailed from San Francisco Bay). A handful – fewer than a dozen – were built in Japan, Shanghai, or Alaska, but 32 came from British Columbia, mostly Victoria. Though most vessels wound up sailing from Victoria, the figures demonstrate that the shipbuilding industry of the western American states benefited most from the industry: only a third of the sealers built on the west coast came from Canada.

Nearly fifty vessels, however, were built on the east coast of the Americas, including twenty-five which came from Nova Scotia, Grand Banks schooners for which more lucrative careers were now foreseen. These ships came in a steady stream around the Horn (in some cases, the Cape), generally in the 60–100 ton range which was found to work best for sealing. But tonnages for all the fleet varied widely, from the tiny *Alexandra* (8 tons) to the *Maggie Mac* (196 tons; a three-masted schooner). Eighty-eight fall in the 60–100 ton range, but only twenty-eight are over 100 tons (fifty-seven are unknown).

Even the best-built "Bluenose" Nova Scotia schooner was not inevitably bound for luck in the Bering Sea once safely around the Horn. The *Fortuna,* a 97-ton Nova Scotian built in 1888, arrived from Halifax in 1895, took 200 skins, and then disappeared the same season with all twenty-three men aboard. The *Maggie Mac* from Jordan River, N.S., left only sparse wreckage, but not a trace of her similar-sized crew, on Vancouver Island in 1892. The *Carmolite,* sent out late in 1890, battled the Horn for fifty days, had one successful season in 1891, then was seized and sold by the Russians: the *William McGowan* arrived in 1892 from Cape Breton only to be similarly seized and sold at Vladivostok with no catch at all.

Some were famous. Jack London sailed on the three-masted *Sophie Sutherland.* Robert Louis Stevenson chartered the *Casco,* built as a luxurious private yacht in 1878, for a South Seas voyage in 1888 at the princely sum of $500 a month before she became a sealer; her long career ended in 1919 when she was abandoned by gold prospectors on the Bering Sea's King Island. The *Halcyon*, alias *Vera*, was a notorious Far Eastern opium smuggler. But among sealers, it was achievement that mattered, such as being first to the Bering Sea or winning through in a bold raid on the "Smoky Isles" – feats which became part of the story of the *City of San Diego* (built in San Francisco in 1871, she was 67 feet long by 20 wide, and just under 50 tons). It was not origin or even size, however the later might limit manpower, which made for a successful vessel but her sailing qualities and the experience and wisdom of her master. Mathieson's *Umbrina*, a consistently profitable schooner which rounded the Horn in 1891 (built in Shelburne, N.S., in 1888; 85 x 22.3 x 9.7', at 98 tons), "would lay hove to for hours with the wind four or five points on the bow, under a double-reefed foresail and a reefed trysail, taking anything the Pacific had to offer with dry decks."[41]

What worked in one context was not necessarily successful in another. The *Thomas F. Bayard*, an unusual vessel built in Brooklyn in 1881 as a pilot boat, combined oak, teak, and copper fastening with a heavily sparred schooner rig. Capt. Hans Blakstad took her sailing in the 1907–11 seasons, with Mathieson as his mate; Mathieson reported her more likely to fall off when hove to, bringing the wind abeam, and, with it, the sea aboard to wash off even heavily lashed canoes. But however appropriate the vessel, luck was very useful, as in Victor Jacobsen's account of the ordeal of the schooner *George W. Prescott*, which managed to survive a 360° flip-flop under water but snapped off her masts, losing two men in the process, and sank shortly after the rest of the crew were taken off by a mail steamer.[42]

Owners and men would not have run such risks were there not a decent living to be made. A good catch for a year's work for a pelagic sealing schooner was 2,000 pelts; 1,000 was adequate, and often an unlucky vessel took less. The *Umbrina* brought back 25,950 seals in her nineteen seasons (until cut in half by a steamer off the Oregon coast in 1911), an average of 1,366: her high trip was 2,801 in 1894, her worst was 648 in 1907. In the 1880s, though London prices might average $14, the Victoria dockside price was more likely to be $6–8.[43] An average catch of 1,400 skins would, at $7, return roughly $10,000 gross, from which wages and outfitting costs had to be deducted. As it happened, the *Umbrina* was valued at $10,000, a bit higher than the average, in the 1890s, of about $8,700.[44]

Some idea of economics is given from Victor Jacobsen's experience. Jacobsen bought a derelict steamer in 1902 for $250, and rebuilt her into the three-masted schooner *Eve Marie*. Total reconstruction from a rebuilt stern to new masts and spars cost $9,400; outfitting from scratch, including provisions and ammunition, another $5,300. Total expenses were about $15,000. Jacobsen claimed that the *Eve Marie*'s cargo of 837 seals in 1905 made $30,000 – at first sight an error in calculation, but by 1905 skins were indeed bringing $35. Even at $14 (Victoria price, not London sale price), which is closer to the average between the low rates of the 1880s and the fierce demand of the last years of pelagic sealing, his cargo would still have brought nearly $12,000.[45] Sealing could indeed be lucrative.

The hunters, of course, had to be paid. Jacobsen in the early nineties paid his Indian hunters $2–3 a skin (depending on where they were taken).[46] Later they were paid at the end of the voyage on a sliding scale: $2 each for the first fifty seals, $3 for 50–100, and $4 each if 100 or more were shot (and retrieved). Some famous hunters were paid even more. Rowers and boatsteerers were paid 50–75 cents per seal, and an additional wage of perhaps $15 per month; the hunters, who worked strictly on piecework, could be expected to keep the other boatmen up to scratch. Overall, $5 a skin is a fair estimate of labor costs. The net return, radically simplified, might be as follows:

1,000 skins at $14		$14,000
less outfitting costs	$5,000	
and labor costs	5,000	10,000
Profit to master/owners		$4,000

Put another way, the return might be roughly the same 40 per cent on a vessel worth $10,000 as the North American Commercial Company enjoyed in its lease, though for rather less risk.

Many vessels did not enjoy return on this scale. Jacobsen and other masters testified, however, that $2,000–3,000 a year was not an unusual income for themselves, though finding seals was tougher in the later years; Rogers estimates that the Victoria Sealing Company – a combine formed of the various owners when times turned hard – averaged just under 9 per cent profit over the years 1901–3 . But some might make that 40 per cent, and the prospect was available to all for the dreaming.

The masters to whom such a gamble appealed were a tough and polyglot group, from all parts of the earth.[47] Spring, for example, was born in Russia, where his father, a Scots engineer, married a Russian; H.F. Sieward was German; Owen Thomas, as might be expected, hailed from Wales; others were mariners from Sweden, Norway, Denmark, or, like Victor Jacobsen, Finland. A few learned their sealing in the harp seal hunt: Frederick Cole was born in Newfoundland, W.H. Whiteley came from a well-known Labrador family. But the biggest single group came from the Maritimes – Prince Edward Island (James D. Warren; Laughlin McLean), Cape Breton (the famous Alex McLean – generally regarded as the model for Jack London's *Sea Wolf* – and his brother Daniel, and above all from mainland Nova Scotia, birthplace of at least a dozen well-known captains, such as William Munsie, William O'Leary, Robert E. McKiel, John G. Cox, Luke McGrath, William D. Byers, and G.W.S. Balcom, each of whose careers spanned the bulk of the pelagic sealing years.

It was uncommon for these men to stay long with one ship, since they soon moved on after a wreck, an offer of higher wages, or investment in their own vessel. Alex McLean arrived in the Pacific, aged twenty-nine, in 1880 as second officer on a clipper, shifted to steam, and in 1883 turned to sealing. He and his brother Daniel brought the *City of San Diego* north from San Francisco to be the first into the Bering Sea. Alex then commanded the *Favorite* in partnership with William Spring until the latter's death, and then took the *Mary Ellen* for two seasons, turning next to the *J. Hamilton Lewis* (alias *Grace*), in time to be seized off Copper Island by the Russians and languish for four months in a Russian jail. His later vessels included the *Rose Sparks*, the steam bark *Alexander* (at 294 tons, too expensive to stay long in this trade), and the *Bonanza*, in which he is reported to have brought back the largest catch landed at San Francisco – 2,189 skins.

His older brother Daniel, after prospecting along the Alaska coast, worked the *City of San Diego* and then commanded the *Mary Ellen*, bringing in the highest single-season cargo on record for a pelagic sealer: 4,268 skins in 1886, a cargo worth at the prevailing price of $6.50 approximately $14,228. Daniel then returned to Halifax to bring out the *Triumph* and sailed again in the *Mary Ellen* and later the *Edward E. Webster*, making several "visits" to Japan and Copper Island in the 1890s.

Such flexibility was typical, as was the familial relationship among masters. George Heater, to give one more example, commanded at various times the *Vera*, *Ainoko* (a famous vessel; Heater commanded her, 1892–1900), *Markland, Penelope, Rosie Olson*, and *Jessie*; his brother William was master of the *Penelope, Beatrice, Arietis, Libbie* and *Dora Siewerd*. But Victoria sealing was never a closed system built over generations as in Newfoundland; there was always room for a determined newcomer with luck, and many of the famous men of the Victoria fleet were loners – Robert McKiel, Laughlin McLean, Luke McGrath, Hans Blakstad, Max Lohbrunner, and Victor Jacobsen being the most famous cases.

Since only the capital necessary to buy and outfit a schooner was required to enter the trade, no one firm or small group dominated – a situation recalling the outport days of Newfoundland sealing – nor did North Pacific sealing ever have its "wooden walls," which would not have worked here. But some firms do stand out. Spring and Company, operated after William Spring's death by his son George, owned at least seven schooners at various times and chartered others (it also operated several trading schooners). William Munsie was an enterprising Nova Scotia captain who took up sealing in 1884 and a year later brought round the Horn the *Pathfinder* (alias *Pioneer*) which he bought in Halifax – probably the first such import. Profits enabled him to add whole or partial ownership of another half-dozen vessels, and William Munsie and Company became an established Victoria concern, with a share in the Shawnigan Lake Lumber Company.

James D. Warren, another early entrepreneur, was the first to send out steam sealers, and at one point his concern ran eight schooners or steamers (one of which, a trading vessel, shows his commercial associations in her name, the *Barbara Boscowitz*).[48] He was hard hit by the seizures which touched off the Bering Sea controversy between Canada and the United States. In its prime, however, Warren's fleet had a high reputation: "the proudest thing that ever put to sea from Victoria," said Victor Jacobsen, who knew what he was talking about, "fine big vessels, brand new, snowy white canvas – oh, a pretty sight!"[49]

Not all leading owners were masters or even sealers themselves.[50] E.B. Marvin, a Victoria shipchandler, worked in partnership with John Graham Cox, a well-known master, to acquire a fleet of eight or nine vessels. Herman Liebes owned at least two. Edward G. Prior, owner of an ironmongery (originally a Yorkshireman, Prior was at various times inspector of mines, sheriff of Nanaimo, mayor of Victoria, MP for Victoria, and lieutenant-governor of the province, 1919–20; his house was, of course, "The Priory") invested in schooners. One firm, that of Robert P. Rithet, accumulated a considerable number of vessels near the end of the trade. Rithet first made his mark in San Francisco, then controlled a major import, shipping agent, and insurance firm in Victoria which owned, among other things, the Rithet wharves – the most prominent in the town – and an impressive office building/warehouse which still stands on Wharf Street as the home of "Tourism British Columbia" (fully restored, it is worth a visit). Rithet's investments were many, his position in the community unassailable (in addition to being colonel of the militia, he was Victoria's mayor in 1885, sometime member of the colony's legislature, president of Victoria's Jockey Club, and not least member of San Francisco's elite Pacific Union Club);[51] he stands as typical of the merchant aristocracy of Victoria which at the end of the century came to occupy a predominant position in the elite of the city, previously confined to Hudson's Bay Company agents and associated clerics, landowners, naval officers, and colonial administrators.

While these firms and individual owners could and did compete for cargoes, they were not above combining in adversity. In response to

United States pressure to cease pelagic sealing (discussed below), public meetings of protest were held in Victoria, beginning in 1889. Out of these grew a combine known at first as the British Columbia Sealing and Fishing Organization, formally established in 1891, with leading owners such as Rithet, Boscowitz, Prior, and William Munsie (a captain who himself owned several vessels including the *City of San Diego*) as its officers. Initially this group did its best to counter American seizures, not least by hiring a steamer, the *Coquitlam* (each member was assessed $50 a vessel owned to cover the lease) to collect pelts from the vessels at sea. In time the group, better known as the Victoria Sealing Company, included the great majority of sealing ships. It found that such an association could serve other purposes, such as setting hunter's wages or reducing the charges made by Indian chiefs for contracting their men to the steamers (bonuses, "potlach" ceremonies, and the like).[52]

Towards the end of the sealing industry, the company was principally a lobby to persuade the Canadian government to greater efforts in its search for just compensation. As such its shares became increasingly concentrated in the hands of a few of the larger owners, who assumed mortgages of the lesser holders. Rithet, for example, who owned at least eight schooners of his own (*Favorite, Marie, Libby, H. Sieward, Carrie C.W., Dolphin, Ocean Rover,* and *Teresa*), held mortgages or indentures on many others. Not all masters, particularly those who chartered their own course like Victor Jacobsen, were happy with such arrangements or the company's policies: Jacobsen claimed to have lost $30,000 by cooperating with the company.[53]

Until a thorough study is made of the Victoria Sealing Company much of the detailed financing of the fleet will remain obscure. What is clear is that sealing made its mark upon Victoria. In part that contribution was through shipbuilding, but as has been seen the lion's share of the construction of sealing vessels was taken by ports further to the south. But outfitting, including the hiring of white crews, provisioning, and transshipment, was a Victoria speciality in the 1880s and 1890s, bridging the important hiatus between the decline of the Hudson's Bay Company enterprises, land furring, and supplying the Russian colonies, and the era of the Klondike gold rush and the final ascendancy of rival Vancouver. Hotels, saloons, ship chandleries, all catered to the sealing fleet. In the late nineties, when American vessels could not openly sail as sealers and took off-season refuge in Victoria, the hundreds of vessels moored in rows were an impressive sight, as old photographs testify. Victoria was then the world's premier sealing port, St. John's wooden walls (which only sailed for a few weeks in the spring) notwithstanding – if one neglects to compare total harvests of harp seals with northern fur seal skins.

And yet the total investment was never great, nor was the permanent, as opposed to seasonal, labor force, ever substantial: far more men had regular employment in the sealskin industry in London than ever did in Victoria. It must be remembered that a majority of the vessels were not Canadian-owned, and thus much of the profit was siphoned away.[54] For San Francisco, on the other hand, the sealing fleet was less important, though it did give some work to the shipbuilding industry (seventeen sealers alone were built in Matthew Turner's facilities at Benicia, modern Oakland). But given San Francisco's far-flung and expanding interests, it cannot be claimed that pelagic sealing ever made more than a marginal difference, except to those who owned the vessels involved. Except where an occasional prominent name such as Liebes, Boscowitz, or Rithet emerges (all closely linked to the fur business and Victoria), it is impossible to trace the profits of pelagic sealing. These profits were never of the magnitude of those of the NACC, let alone the ACC; we are dealing here with small business, the hardest form to identify and trace. The principal result for San Francisco interests was of a different sort: the determined hostility of the NACC to the whole concept of pelagic sealing, which was visibly robbing the leaseholder of current and future profits.

The question again arises, how many seals were taken pelagically? The answer is anything but clear. Table 5 shows part of the problem, the differing estimates of seals taken and sold. The three columns represent a close approximation to official United States figures (from Riley's 1967 compilation, based on congressional hearings), a 1976 calculation by Rogers of skins specifically sold on the London market, and catch figures from British Columbia archives dating from 1889, which are very precise for all vessels returning to Victoria year by year. If the first were more than the second, and both more than the third, there would be no problem: not all skins arrived at London auctions, and not all skins auctioned were from the Pribilof herd, still fewer from British Columbia. The extent of the problem can be gauged by the British Columbia figure for 1893, giving a total of 142,712 (70,592 brought to British Columbia, 21,793 to Hakodate, and 50,279 to San Francisco), to compare with the 30,812 of which American officials were aware and the 121,618 sold in London.

For the period to 1880, the differences are not critical. Riley's figures for 1868–80 produce an average of 6,700, which seems adequate for the number of vessels operating in all North Pacific waters in those years, including the high years of 1871 and 1879. Similarly, an average of 13,500–15,000 produced by both Ridley and Rogers for 1880–5 seems comparable. In the succeeding decade, however, the figures show wide disparity, with averages of 42,000 and 64,000 respectively. After 1897 the calculations close again, as American sealers were forced to operate only in Japanese or Canadian ports; both calculations average

25–30,000. The totals contain a difference of 300,000 over forty-four years: 981,744 vs. 1,310,728, with averages ranging between 22,312 and 32,768.

In one sense, such calculations are an exercise in futility, except for the hunters and owners who brought in the pelts. In counting seals, at least in relation to the state of the herd, the vital issue was how many were lost for each shot and retrieved. The testimony here shows only agreement that the loss rate was high. Those who knew best were the boatsteerers and pullers; hunters were likely to claim a high return, proving their skill, and masters normally remained in the mother vessel. Thomas Brown, boatpuller: "I don't think we got more than one seal out of six that we killed." Louis Cutter, boatpuller: "[white hunters] … do not save more than 2 or 3 out of every hundred shot." Arthur Griffen, boatpuller and steerer: "A good hunter will often lose one-third of the seals he kills. A poor hunter will lose two-thirds of those he shoots. At an average, hunters will lose two seals out of three of those they shoot."[55]

The pelagic hunt returned approximately 1,300,000 skins for sale over the forty years, 1870–1910, or 25,000 a year (but since the bulk was taken between 1885 and 1905, the average is misleading). A fair estimate should at least double this number for seals shot and lost; probably a better guess would be 1:2 – two lost for each one retrieved. Using this ratio, the average kill each year would have been 75,000, and the total closer to 4,000,000. But even 50,000 annually from the mid-1880s onward put an unbearable strain upon the reproductive capacity of the herd in those years. Therein lies the explanation for the herd's decline and the American response to pelagic sealing.

Precisely how the Bering Sea controversy began is quite clear; its causes, in the detailed sense, remain clouded in obscurity.[56] In the 1886 season the Revenue Cutter *Corwin* was issued orders – on whose ultimate instigation and authority is uncertain – to seize all vessels found sealing in Bering Sea waters. The schooners *Thornton, Onward*, and *Caroline*, all Canadian, were gathered in and their masters and mates jailed and fined – $500 and $300 respectively – for killing fur-bearing mammals without license. The intent was to stop pelagic sealing and the legal justification the claim that the Bering Sea was American territorial water (since the schooners had all been taken many miles offshore, the normal three-mile limit did not apply).

The claim was absurd from the viewpoint of international law, especially since the United States in general contested such claims elsewhere in the world and in particular had opposed similar Russian claims in the Bering Sea before 1867. Nor was this a claim which either Canada or Great Britain – as sovereign power in respect to Canada's external relations, Britain was responsible for negotiations with Washington –

TABLE 5
Pelagic Sealing, 1868–1911

	Riley (1967)	Rogers (1976)	British Columbia Archives
1868	4,367	–	
1869	4,430	–	
1870	8,686	–	
1871	16,911	–	
1872	5,336	1,728	
1873	5,229	40	
1874	5,825	5,071	
1875	5,033	2,224	
1876	5,515	3,104	
1877	5,210	722	
1878	5,540	2,698	
1879	8,557	14,609	
1880	8,418	13,501	
1881	10,382	15,887	
1882	15,581	22,886	
1883	16,587	8,704	
1884	16,971	19,357	
1885	23,040	10,148	
1886	28,494	49,079	
1887	30,628	39,419	
1888	36,389	30,285	
1889	29,858	39,884	35,310
1890	40,814	25,746	43,695

was prepared to admit at any time. Indeed, only the display of wisdom at critical points and the general desire of Britain to remain on good terms with the United States in the last tension-filled years before World War I kept the Bering Sea crisis from reaching truly dangerous proportions.

The negotiations which followed were complex and fall more properly under the heading of general diplomatic history than of the history of sealing. They were, moreover, linked with other issues of Anglo-Canadian-American relations, such as the Alaska Boundary and North Atlantic Fisheries disputes which cannot be considered here but without which the diplomatic story is incomplete. Since these questions have been fully discussed by others, only the briefest summary of fur seal diplomacy is needed here.

Despite immediate protests, no change was made in American policy, and in 1887 six Canadian and six American schooners were seized. Others were inspected and released if they had no seals on board or could prove from their logs that their skins had been taken outside the

TABLE 5
Continued

	Riley (1967)	Rogers (1976)	British Columbia Archives
1891	59,568	48,093	50,445
1892	46,642	61,858	48,793
1893	30,812	121,618	70,592*
1894	61,838	119,980	95,044
1895	56,291	104,724	73,614
1896	43,917	73,665	55,678
1897	24,332	43,082	30,410
1898	28,552	47,033	40,889
1899	34,168	44,769	35,346
1900	35,191	40,480	35,523
1901	24,050	32,189	24,443
1902	22,812	20,582	16,725
1903	27,000	22,513	14,701
1904	29,006	39,744	14,646
1905	25,320	22,205	14,177
1906	21,236	21,251	10,385
1907	16,036	14,356	5,397
1908	18,151	23,107	4,954
1909	14,139	16,460	3,742
1910	12,111	795	4,452
1911	12,671	139	2,673
Totals	981,744	1,310,708	**
Averages	22,312	32,768	17,119

Sources: Francis Riley, Fur Seal Industry of the Pribilof Islands, 1786–1965; George Rogers, "An Economic Analysis of the Pribilof Islands, 1870–1946" (pelagic skins sold at London); Seal Catch Statistics, 1889–1911, Provincial Archives of British Columbia, I/BS/SEi.
*1893 estimate (the only year thus given) gives 70,592 landed in Canada, plus 21,793 in Hakodate and 50,279 in San Francisco for a (slightly inaccurate) total of 142,712.
**No total is given since BC figures have been found only from 1889 onward.

Bering Sea. While the more cautious masters were reluctant to risk their vessels others followed the seals anyway, trusting to luck and weather to permit them to catch their seals and escape in good time, and knowing in any case that there were seldom more than two or three revenue schooners in these waters at any one time. Nor was a seizure easy to accomplish without putting a prize crew aboard and thus depleting the cutter's own strength. Schooners which were ordered to Unalaska or Sitka had a habit of disappearing, only to live again with a new name and papers. Jacobsen, seized in his schooner *Minnie* in 1889, simply set sail for Victoria despite the protests of the prize crew put aboard by the cutter *Rush*. But those vessels safely seized were likely then to rot in harbor or more commonly be put up for auction. It was not unknown for a captain to reappear as master of his former vessel, having in effect paid a substantial fine in this way to the United States government for the privilege. To prevent the escape of others, yet save the prize crews, cutters developed the tactic of seizing all weapons aboard, including spears, to make hunting impossible; masters naturally took to secreting weapons aboard, and Indians had a knack of manufacturing impromptu spears.

President Grover Cleveland permitted the arrests, and the cat-and-mouse game, to continue in 1887, but his secretary of state, Thomas F. Bayard, who believed the seizures to be illegal, quietly ordered them stopped in 1888 while he negotiated with Britain and Russia for a closed Bering Sea season. The succeeding Harrison administration took a different, harder line, and the seizures were resumed in 1889, just as serious negotiations initiated by Bayard began. Talks in Washington in early 1890 quickly broke down due to the impatience of the new secretary of state, Blaine, and the refusal of the Canadian minister of marine and fisheries, Charles H. Tupper (not to be confused with his father, Sir Charles Tupper, Canadian high commissioner in London at the time), to admit the need for a closed season at all. To Tupper, the whole affair, as he wrote six years later, was an attempt "to drive our flag off the ocean in the Northern Pacific," when in actuality the industry was "a proper and natural opportunity for the exhibition and encouragement of British pluck and daring."[57]

London, aware of the depth of feeling, had in this matter to avoid further irritation to a Canadian public opinion already inclined to perceive British policy as altogether too self-serving where American relations were concerned. Should care not be exercised, further support might be aroused for total Canadian independence, even union with the United States. On the other hand, the prime minister, Lord Salisbury, simply did not believe Canada's argument that pelagic sealing did not ruin the fur seal herd. Temporizing, Sir Julian Pauncefote, British minister in Washington, in April 1890 submitted to Blaine a proposal for a joint commission to study and report on proposed regulations, while in the meantime a partial closed season should operate on both land and sea. Blaine rejected the proposal, above all the idea that land sealing should be halted. The month was May 1890; in March, Blaine's friends in the NACC had been given the lease precisely for such land sealing.

Blaine's rejection of the proposal, which had originated with Salisbury, brought the risk of conflict only too near, though this was not necessarily perceived by all participants at the time. Not only did Salisbury protest against American policy with considerable vehemence, but also he ordered – in secret – that four British warships be held ready, two in Yokohama and two in Esquimalt (British Columbia's naval base, part of the Victoria harbor complex), at the same time giving Pauncefote discretion to use these warships as a threat if necessary.[58] Blaine, hearing of British preparations, wisely ordered the cutters to seize no Canadian vessels, and conflict was avoided for 1890.

Embarrassingly, just at this point in November 1890, Elliott's report blaming land sealing for the diminution of the herd was submitted. Elliott recommended that killing be prohibited on land for seven years (the chief agent on the Pribilofs, Charles J. Goff, had already made a similar recommendation and had ordered the NACC to take no more than 21,000) and totally at sea. Elliott and Windom, secretary of the treasury, discussed this report with Blaine in November and again in the following January (1891); it was all most awkward, for if Elliott was right, Blaine had no reason to resist the proposed halt to land sealing, and his friends were likely to be the poorer. Elliott's report, as we have seen, was buried, though justice was not entirely upon his side, considering the legitimate issues raised regarding his computation methods over the years.

Blaine was nevertheless forced to reach agreement with Britain, for he found navigation impossible between the rocks of confrontation with British naval strength and the reefs of charges at home of favoritism, jobbery, and – if Elliott's report ever came to light – callous disregard of the herd's survival. After tortuous negotiation and considerable policy-making confusion – and the intervention of President Harrison – Blaine proposed a *modus vivendi* which was not far from Britain's own proposal. NACC lobbying, however, had been fierce upon the Treasury (now headed by Charles Foster), as well as upon Blaine, who was in close touch with Mills (now senator), Elkins, and probably President Harrison himself.[59]

The result, in April 1891, was agent Goff's removal and replacement by an inexperienced man and a (secret) permit to the NACC to kill 60,000 seals in 1891, apparently in the certain expectation that Canada would reject the *modus vivendi* proposal. When Canada did the unexpected and accepted, a distressed Washington administration had to compromise. Meanwhile Elliott, who had learned of the secret permit, made angry charges in the press, to which he released sections of his suppressed report – and was sacked from his Treasury post as a result.

But these embarrassing disclosures forced administrative surrender, and the agreement with Britain was proclaimed in May 1892. Neither power would permit its subjects to take seals pelagically, and the United States pledged to limit land sealing to 7,500 the following year (several thousand more actually were taken for Aleut food supplies), during which time a joint commission would investigate means to protect the seals and the issue would be presented to the Paris Tribunal for arbitration. Elliott's disclosures and the need to avoid confrontation had done the trick, but there were many disgruntled parties: the NACC, which was barred from sealing in the desired numbers; the pelagic sealers, similarly affected; and authorities in Canada generally, who had not been consulted with regularity by Salisbury and who had not even been promised the requested compensation for Canadian pelagic sealers.

The major crisis was over, though the negotiations were not. It was necessary to renew the *modus vivendi* once more, in 1892, before the matter could actually be taken up in Paris (March–August 1892).[60] The tribunal had no difficulty agreeing that the United States had succeeded to the rights of Russia, but it denied the claim that the Bering Sea fell under the heading of "territorial waters." This being the case, the arbiters set about framing regulations to govern the seal fisheries outside territorial waters (a clause of the Arbitration Treaty had stipulated this procedure). The result was to prohibit sealing within sixty miles of the Pribilofs and all pelagic sealing in the North Pacific or Bering Sea from 1 May to 31 July; in addition, the weapons and vessels to be used were specified, and the issuing of licenses and keeping of records of the hunt regulated.

Unfortunately for clarity, the tribunal did not decide whether seals should be seen in the same light as whales, the property of the catcher, or, as the United States argued, rather like cows straying beyond the pasture in which they were born; that seals were not quite United States citizens, entitled to every protection on the high seas, was clear nevertheless. The American seizures of earlier years were held to be illegal (of Canadian vessels, at least), and the question of damages now remained to be settled. The claims presented by Britain totaled $542,169.26; the United States offered $425,000. The lesser sum was accepted, but Congress refused to pass the appropriation, on the grounds that American owners would then have to be compensated. Only after considerable delay, and the judgment of a judicial tribunal, did the United States eventually pay $473,151.26 in June 1898.[61]

Not surprisingly, the tribunal had failed to harmonize the peculiar habits of the fur seals with international law. As one result, Canadian authorities made no effort to force vessels to stop taking seals, partly because the judgment concerned only the eastern Bering Sea claimed by the United States, and not any Japanese or Russian waters. Not only had those powers not been party to the agreements, but also Japan had entered the sealing business in a substantial way once the United States and Britain ordered their own nationals to cease sealing. Finally, American authorities were plainly not adequately equipped to patrol the entire Bering Sea at once, even up to 1 July, after which all pelagic sealing was perfectly legal.

A considerable number of Victoria and San Francisco vessels dutifully trekked across to the far side of the Pacific in April 1894 for the early season, but others simply paid no attention to the regulations, which, in any case, left plenty of scope for a successful voyage, provided there were still seals to hunt. The tribunal's regulations were in effect for five years, during which time there were many small incidents involving sealers and the Revenue Marine. Some of the revenue cruisers had intimidating reputations, particularly the *Bear*, a Dundee "wooden wall" (Walter Grieve and Company), bought by the U.S. government as an arctic rescue ship and subsequently engaged in a career of forty years and more of Alaskan service. Still, there were no more of those controversial seizures and confiscations.[62]

Negotiations on the larger issue of a permanent solution to the problem of the herd's survival (meaning, the end of pelagic sealing), were slow to reach a fruitful conclusion. When the United States prohibited its citizens from sealing at sea, American vessels simply moved to Japan or Victoria for a new base of operations. Without full Canadian co-operation, pelagic sealing would continue. Canada's position, however, was consistent in the ensuing years: for such a concession, not only would Canadian sealers have to be compensated, but Canada as a whole should be given some equivalent concession in one or another of the outstanding Canadian-American issues.[63]

By the first decade of the new century, all officials involved agreed that pelagic sealing should be ended, but there was little agreement on the price to be paid. Eventually, Canada's leverage was undermined by separate settlement of the fisheries issue and the steady reduction of the Alaskan fur seal herd. The Canadian sealing fleet declined substantially, victim to the rapid growth of a Japanese sealing fleet which had no legal need to observe the same regulations which Canada had accepted in 1893.[64] By 1909, the choice was to accept a 1906 American offer of 20 per cent of the land-harvested Pribilof skins, or see the seal herd totally exterminated. Several suggestions had already been made by responsible parties in the United States towards just that end out of sheer frustration. Theodore Roosevelt, despite his reputation as a conservationist, remarked in the *Metropolitan Magazine* (1907): "In case we are obliged to abandon the hope of making arrangements with other governments to put an end to the hideous cruelty now incident of pelagic sealing, it will be a question for your serious consideration how far we shall continue to protect and maintain the seal herd on land with result of continuing such a practice, and whether it is not better to end the practice by exterminating the herd ourselves in the most humane way possible."[65]

Fortunately such measures were unnecessary. In January 1909, Secretary of State Elihu Root proposed to Britain, Russia, and Japan a conference to shape a multilateral settlement. Though the question of compensation was a complication, in February 1911 an agreement was reached at last. Canada was guaranteed 20 per cent of the seal harvest which in turn was guaranteed not to fall below a fixed annual minimum. In addition, the United States advanced a sum in lieu of future annual instalments out of which compensation could be paid to sealers (the advance was $200,000). In July 1911, with Canada now appeased, the four powers signed a treaty banning pelagic sealing in the northern Pacific for fifteen years, to become permanent should no opposition be registered.

Canada now would receive 15 per cent of the American Pribilofs' catch, 15 per cent of the Russian take on the Commanders, and 10 per cent of the Japanese harvest on Robben Island. Japan was both compensated (15 per cent each from the Pribilofs and Russia's Commanders) and compensator (10 per cent each of the Robben catch to Canada, Russia, and the United States). As an early and successful multipower essay in conservation, the 1911 Fur Seal Treaty was a historical landmark. It was also a landmark in Anglo-American-Canadian relations, part of a general settlement over the years 1905–11 of a host of complicated international issues, during the negotiation of which Canada emerged as a far more independent state than ever before in her relationship with the United States (though the Canadian government which signed the several agreements was thrown out of office on charges of a "sell-out," especially in the Boundary Waters Treaty of 1911).[66]

For Victoria's fleet, however, the future was all in the past. Less than half a dozen vessels brought back skins in the last three years, 1909–11. The Victoria Sealing Company now had forty-odd vessels on its hands, but few sailed. Yet still dividends were paid, even as the herd was disappearing and the fleet rotting at its moorings, since prices remained extremely high for the few pelts brought in. The five schooners sailing in 1910 accumulated a total catch of 4,652. At the going Victoria price of roughly $35, the pelts returned over $140,000. As always, the prospect of high profit kept sealers and vessel-owners in the business, though now the latter were mainly the wealthier men who could afford the overhead, and the gamble.[67]

But the heyday of profit was never to return, and after 1911 the fleet vanished. Some simply disintegrated on Sitka's aptly named "Rotten Row." Others became sodden hulks in "the Arm," a neglected corner of Victoria's inner harbor: "a dismal, stagnant sort of spot, dedicated to garbage, and dead cats, and snags, and sawmill refuse, and mouldery slipways, and sinister leaning boathouses that seemed waiting for murders to be committed in them."[68] A few schooners found useful employment, such as the *Casco*. Renovated and fitted with a gasoline engine in Seattle, she turned to halibut fishing, though the carved and decorated wood which had graced her interior cabin now adorned the home of her new owner.[69]

Similar use was made of the *Markland, Carlotta G. Cox, Borealis, Jessie,* and the famous *City of San Diego.* The *Aurora* used a new gas engine for a northern Vancouver Island mail route; the *Vera,* one of the best, was sold at auction for a mere $2,100 and went to Ottawa owners to serve as an Arctic exploring vessel. Only one survived long into the twentieth century, the *Thomas F. Bayard,* which was well built to begin with and entered the trade late, but when rescued was a forlorn hulk. She is now to be fully restored and enshrined in fitting splendor in the Vancouver Maritime Museum – a bit ironically, because Vancouver's role in the sealing industry was never great.

When the trade was all over, total claims for losses by the owners of the fleet came to the staggering sum of $9,800,000. The British government had in hand to pay these claims a total of only $200,000, received in October 1912. It proved more than enough; since there was in fact little legal basis against the government for any claims at all, by early 1916 only $60,000 had been paid out, giving – out of "fairness" – the value of the vessel and the equivalent of the last year's return to masters and hunters (70 per cent to able seamen, boatpullers and boatsteerers, and only 50 per cent to the Indians, on the ground that they were still permitted by treaty to seal by their ancient methods, meaning without schooners!). Thus had the claims been reduced in scale, cancelled, or forgotten.[70]

For the NACC, the final settlement should have led to joy unbounded, so long had their harvest been constrained from reaching the level they desired. Table 3 shows their total catch of 345,962, which, rounded to 350,000, we may now add to the total northern fur seal harvest:

Seals taken to 1867	4,000,000
Seals taken 1868–9	250,000
ACC lease, 1870–89	2,000,000
NACC lease, 1890–1910	350,000
Pelagic sealing, 1868–1911	1,300,000
Total	7,900,000[71]

For a total of seals killed, as opposed to "harvested," however, to this approximately 8,000,000 should be added the pelagic sealing estimated loss rate of 2:1, which would bring the total to well over 10,000,000. But the NACC was not in a position to celebrate their potentially greater contribution to these statistics, for, by an ironic twist, the company had lost its lease. There was no successor company; the United States government had now assumed sole responsibility for operation of the Pribilof fur seal industry.

Two reasons for the demise of the NACC stand out. One dates from

the origin of the company and the jobbery associated with its lease –
and to charges made early in the leasehold that the Liebes interests were
involved in pelagic sealing, as in fact they were. Pelagic sealing was
illegal for Americans in the nineties, and in any case the lessee was
prohibited by the contract from killing seals in any other than the
customary land harvest. The charges were at first denied, though a
better response would have been to separate the various family branches,
but the issue was still very much alive in the last years of the lease.

A more important reason was the need to stop sealing altogether if
the herd was to survive and recover. Elliott, in his suppressed report,
saw a "scant million" left, with only 100,000 young males in the entire
herd, and that was at the beginning of the lease.[72] Yet pelagic sealers
took a total of 142,000 seals (most of which were females) in 1892,
probably the high point of pelagic seal catches. All agreed that this
could not continue, and even while high-level diplomatic negotiations
were producing the ineffective Paris tribunal regulations, London and
Washington agreed to send separate missions to study the population
problem, thus introducing to the Pribilofs yet another important participant in the fur seal story, David Starr Jordan.

Jordan was an eminent biologist, but his fame as a university ad-
ministrator was even greater: he was president both of Indiana University (1885–91) and of Stanford University (1891–1913; chancellor,
1913–16). Known in scientific circles for his numerous authoritative
works on ichthyology, after his work on the Pribilofs he became as
tireless a defender of seals as Elliott, writing in all seventeen studies on
the subject, from official government reports to *Matka and Kotik: A
Tale of the Mist Islands*, a book ostensibly for children which focused
in a romantic way upon the plight of Pribilof seals (Matka, a female,
dies with a spear in her throat; Kotik, her pup, starves to death). Published in 1897, it was reprinted in 1910 and 1921.[73]

Jordan and Elliott had common objectives but shared little else. Elliott
distrusted Jordan as an interloper sent to discredit his own calculations;
Jordan, for his part, found Elliott's work "marred by unbalanced as-
sertions, many of which seem to show a lack of both honesty and
sanity," and protested to S.P. Langley, the secretary of the Smithsonian,
against Elliott's glib assumption of Smithsonian authority, calling himself "Prof. Elliott, of the Smithsonian," and giving its offices as his
return address. (Langley disclaimed any such link for the past twenty-
five years, and told Elliott to halt the practice.)[74]

But Jordan agreed that pelagic sealing had to end. "It is not a question
of who shall have the skins, but of protecting and saving the race of
animals, thus preserving to the United States and the world its industry." Herein, however, lay a fundamental disagreement. Elliott was
determined to stop *all* sealing: "You must curb the land butcher as well

as the sea butcher: they have the same instincts: they are impelled by the same motives: and have the same tender regard for the seals," he wrote to Jordan in 1897. Jordan refused to agree. "Whatever the instinctions and motives of the 'land and sea butchers' are, their interests are wholly different," he replied; "the interests of the Company and of the Government are identical."[75]

Once on the islands in 1896, Jordan soon discarded Elliott's computations and set about preparing a revised census – a task vastly more easy than in Elliott's day because there were now far fewer seals to count. Jordan based his figures upon a method devised by Treasury agent Joseph Murray which counted dominant bulls and average harem size, and then was extended to produce a total for the herd. Four informal censuses had been made in 1895; their results for the total herd ranged from 200,000 to 65,000. Jordan's estimate for 1897 was probably closer to the mark at 400,000, but this figure still provides a striking contrast to Elliott's original 4,500,000.[76]

Of the 400,000, some 10,000 dead pups were counted after 1 August and the opening of the legal season for pelagic sealing (though all sealing within ten miles of the islands was prohibited, it was hard to control, and females in any case went considerably further in their search for food); the pups were starving because their mothers were taken at sea. Each female taken pelagically, to put it another way, might cost three seals: the female, her pup ashore, and the unborn pup of the following year, for by this time she had been impregnated again (given natural mortality and missed pregnancies, however, the average figure would more probably be just over two). The herd totalled 402,000 (actually Jordan's count was 402,850), yet the land quota for that year was 19,200, to which pelagic sealers added at least 40,000.

Jordan's calculations did not bring him to predict immediate extinction of the herd, but obviously it was in dire trouble. The British investigation produced a similar result, except that its report continued to blame land sealing for the herd's decline. The United States, in response, prohibited sealing by its nationals without license on land or sea, and barred the importation of pelagically taken skins – i.e., all skins which could not be demonstrated to originate at the Pribilof land kill.

Though that regulation hurt Canadian sealers, since there was by then little market outside the United States and Canada for skins prepared in Europe, still the killing continued, under British and Japanese flags, and with all the more determination when the Paris award expired in 1898. But so too did the land harvest continue while international negotiations languished. To bring that killing to a halt, Elliott mounted yet another effort, this time in Congress, to force a moratorium upon the NACC. Only as the lease approached expiration did his efforts bring results, by which time he had recruited some influential supporters, such as Senator Joseph M. Dixon, chairman of the National Resources Committee, and William T. Hornaday, director of the New York Zoo. Since Hornaday was also a director of the Camp Fire Club of America (and chairman of its wildlife committee), he was in a position to mobilize considerable public opinion.[77]

Since the advisory boards of scientists created in 1909 for both the Fur Seal Service which administered the island and the Bureau of Fisheries also favored a moratorium, the only remaining resistance was that of the NACC and its friends. When the lease did expire, the growing voice of public opinion insured that it was not renewed. Congressional legislation now required that the government assume direct control of the entire process, which was done in April 1910. Nevertheless, no holiday on killing was ordered by George M. Bowers, commissioner of fisheries (and manager of Elkin's West Virginia campaigns);[78] in the same year, 12,920 seals were harvested (with seals for food, the total was 13,586) of a herd which now was estimated to total only 130,000.

Elliott was not finished. He now brought charges against the NACC of misconduct, and the result was a congressional investigation under Representative John H. Rothermel of Pennsylvania, chairman of the House Committee on Department of Commerce expenditures, which continued into 1914. Its findings agreed with some of Elliott's points, noting particularly that the lease should have been canceled in 1890 when Isaac and Herman Liebes outfitted the *James Hamilton Lewis* for pelagic sealing under "one Alexander McLean, known as a notorious British pirate." (Without the detailed accounts of that vessel, her precise ownership will never be known, though she certainly was owned by the Liebes interests.)[79] No formal charges were ever brought against the NACC or its owners, but meanwhile, in 1913, a five-year moratorium on the killing of seals (save for Aleut food needs) was ordered: Elliott had won at last, but by the narrowest of margins.

With the new *zapuski*, or moratorium, and the establishment of government control, the history of Pribilof sealing entered a new era which will be considered in the final chapter. The hunt for the northern fur seal without question had been of influence: in the development of interest – Russian, Canadian, American – in Alaska; in the growth of the maritime industries of the entire west coast through construction and outfitting of sealing schooners; and in the distribution of profits from the furs. In the larger world of international relations, the Bering Sea controversy further complicated the tangled relations of Canada, Great Britain, and the United States, proving, if new proof was needed, the importance of commercial interests in the formation of foreign policy.

From the conservationist standpoint, while the seal herd had gone a considerable way down the road to extinction, at least its administrators had emerged by World War I, under the prodding of men like Elliott and Jordan, with new determination to study and understand the biology of seals. The Murray-Jordan census method provided a much better technique for conserving the herd (Jordan was not infallible, however; he remained convinced that the Pribilof and Commander Island stocks did not mingle).

Finally, the literary impact should not be forgotten. Though Jordan's *Matka* may not warrant an exalted place in the annals of fiction, it was a popular book in its day, and Jordan's name was one to conjure with. " 'I think I have read something about them,' " reflects Maud Brewster, the heroine of Jack London's *The Sea Wolf*, " 'Dr. Jordan's book, I believe.' "[80] Kipling's "White Seal" of the *Jungle Books* and his "Rhyme of the Three Sealers" fall into a different literary category from *Matka*, as does London's successful novel. It is seldom remembered that, when Kipling and London are added to Cooper and Poe, more authors of lasting fame wrote of seals than of whales, though no sealing tale can match the place in literature held by *Moby Dick*.

For the independent sealers of North America, the North Pacific had certainly been the place to be in the 1880s and 1890s. As the herd fell off in the early twentieth century, sealers came again to think of other waters. As will be seen, Victoria had already sent some of her fleet back into the South Atlantic to hunt the fur seal of that hemisphere, so long forgotten. Meanwhile, since the early nineteenth century, a specialized trade had developed, following its own cycle of boom and decline, in the taking of elephant seals, not for fur or leather but for oil.

PART FOUR

The Wilder Shores

Because I see these mountains they are brought
   low,
Because I drink these waters they are bitter,
Because I tread these black rocks they are barren,
Because I have found these islands they are lost;
Upon seal and seabird dreaming their innocent
   world
My shadow has fallen.
                    Kathleen Raine, "Eileann Chanaidh"

CHAPTER SIX

Elephants
of the Oceans

On board of some ships they have plenty to eat
But it is here they put a stop on our ration
It is work for nothing and find your own grub
And starve your self to death on Desolation
Journal kept on *Ocean Rover* 1859, quoted in
Pamela A. Miller, *And the Whale is Ours*

No seal species is without its unique characteristics, but one, the elephant seal or "sea elephant," stands out at once if only because it is the biggest.[1] Indeed, it is the biggest living representative of the order of meat-eating mammals (Carnivora), let alone the earless seals (Phocidae) family to which it belongs or the eared seals (Otaridae). But there are two elephant seal species, which never mingle. *Mirounga leonina* is the southern version, found principally near to but usually north of the Antarctic convergence, with separate herds based upon the islands of South Georgia, Kerguelen, and Macquarie. *Mirounga angustirostris*, on the other hand, breeds along the coasts of Mexican and American California and ranges as far north as Alaskan waters. Each had its role in the sealing story.

The two species have slight but recognizable differences. The northern elephant seal has a longer head, but more noticeable is its longer "trunk" or proboscis – a feature which elephant seals share only with the much smaller hood or bladdernose seal, already encountered in the Newfoundland hunt. The trunk, well developed only in males, was thought to serve as a resonating chamber. Since the sound is now known to come from the larynx, the trunk's purpose, other than display, is obscure. Nevertheless, on its rookery beaches in mating season the bull elephant's roar can be heard a considerable distance.

It is not clear whether its elephant-like snout or its vast bulk gave the elephant seal its name. But there is no question about the size: at the height of the hunt for elephants in the nineteenth century, bulls were taken measuring 22–24 feet in overall length, with a circumference of 12–15 feet at the shoulders – a size which defies the normal image of the friendly performing "seal" at the local zoo, who is probably a sea lion from quite a different family. It is often asserted that northern bulls are the smaller of the two species. Charles Scammon, an experienced whaling master, recorded a bull taken off Santa Barbara in 1852 which was 18 feet in length[2] – a notable size then and an exceptional one today when 16 feet would be closer to average. The average length of mature bulls killed in South Georgia in the mid-twentieth century was 18 feet.[3] But comparisons are difficult, given the disparity in population sizes. The northern species currently numbers about 70,000, roughly a tenth of the 750,000 estimated for the southern variety.[4] In either species, however, a bull could often weigh a ton or more and it could take six men with poles to turn one over for skinning. A claim of four tons may be exaggerated, but three tons is certainly possible, and the biggest bulls tended to be selected first by the sealers. Females measure eight feet at maturity, and weigh some 1,300 pounds; though the sex/size dimorphism is great, it is not as great as that of the northern fur seal.[5]

Although *Mirounga* can move fast enough on land when necessary, hunching along like a caterpillar, it prefers not to have to do so. Whether north or south, the elephant chooses beaches, not rocky coasts. That the beach may be composed of rocks makes no difference: it is not sand this species seeks but a place to haul out where progress of an animal of such size does not require leaping obstacles or climbing cliffs. Like the fur seal of the Bering Sea, the elephant seal is seasonally migratory and solitary at sea. Elephants do not range as far, but some branded at South Georgia have appeared at Marion Island, a distance of 1,600 miles, and Graham land in Antarctica; some bred along the California coast occasionally reach the Gulf of Alaska.

The bulls, or "wigs" to the sealers, arrive first at the beach in August, the cows next a few weeks later, with the breeding season lasting until October (December to March in the northern hemisphere). Once landed, the herd displays much activity: dominant males appear to establish harems (actually, elephant seals rather rely on the cow's gregariousness to form themselves into groups, then strive to keep other males away), savage fights occur among rivals (grimly colorful but seldom fatal), and the females give birth and are again impregnated. The scene is most reminiscent of the Pribilofs with the difference that fur seal males are territorial while elephant seals compete for social rank.[6]

Bulls do not return to sea while the breeding season lasts, nor do the cows for the entire four weeks or so of lactation. Since the female must therefore be miserly in conservation of body fluid, the milk is low in moisture but amazingly high in fat content (over 50 per cent). The pup, 4–5 feet long and weighing 75 pounds at birth, will often triple its weight during a month of feeding. Only then, following its moult of a thick black natal coat, is it ready for sea.[7] Adults also moult each year beginning in May along the California coast, December in the southern latitudes; this is a characteristic by no means found in all seal species. Seals come ashore to moult, if possible finding mud to wallow in to ease the irritation, or at least constantly flippering sand onto themselves. Once moult is complete, the seals again return to the sea.

Pupping, breeding, roaring, fighting: while the seals are present, the scene is fascinatingly alive. Pup mortality is fairly substantial, however, owing to disease (generally a trivial number, but one which may increase with herd density), starvation (loss of the mother), or the crushing weight of a giant bull; pups can take a great deal of punishment, but some inevitably perish. Southern species have another peculiar danger: a period of sunshine after a snowstorm will melt the snow under a pup, who then sinks into a pit from which, once the surrounding snow freezes again, there is no escape save starvation. At Año Nuevo Island off California, pup mortality is about 14 per cent per annum but may increase to 70 per cent in a bad storm year; despite ice pits, the rate is reported to be considerably less at South Georgia.[8]

Pups that survive now take to the sea to feed. In the northern species, about which more is known than about the southern elephant seal, adult seasonal migration may be substantial, up to 600 miles, apparently in search of food. Northern elephants feed primarily on pelagic animals that live far offshore in deep water over the continental slope, but above all several varieties of squid. Of the approximately thirty prey species upon which they feed, the majority are squid, but this seal also feeds extensively on Pacific hake and sharks and rays. No other pinnipeds of the North Pacific eat squids and sharks to the same extent and in the same variety. On the other hand, larger sharks, above all the great white, and killer whales are the main enemies of elephant seals in these waters. In the southern hemisphere, where the elephant seals also feed on squid, the voracious leopard seal should probably be added to the list of enemies.[9]

Among the several separate herds, there are other differences, often the result of human interference. The age of sexual maturity, for example, depends upon population density, which in turn depends not only upon available space, but also upon human depredation. In South Georgia, where, as will be explained, there was a steady, controlled harvest throughout the first half of the twentieth century, females are sexually mature of the age of three, while males breed at four. At Macquarie, where there has not been sealing for some years, maturity is not reached until the fourth or fifth year in females and the fifth in males, though it is the same species as that on South Georgia. In elephanting, mature bulls, not bachelors, are taken first, so there is more opportunity for younger males to establish themselves sooner in the rookery; any significant depredation in the herd size, whatever the cause, by decreasing competition will commonly invoke younger female reproduction.[10]

From the standpoint of the seal hunters, what really distinguishes the elephant seal is the amount of oil which may be obtained from its blubber. For the most part, everything but the fat – three inches thick in cows, six to eight in bulls, depending upon condition – was wasted. Skin, meat, bones, all were simply discarded unless the hunter required fresh meat, in which case he would save the tongue ("more luscious than a neat's tongue," reported Edmund Fanning),[11] flippers, heart, even the trunk (boiled in salt water for hours, then served stuffed and roasted, and known to devotees as "snotters").[12] The oil itself, though whiter than the average whale oil, was all too often listed as "whale oil" in various summary returns. In fact, different grades of whale oil commanded different prices, and elephant seal oil was sometimes worth more than average-grade whale oil.[13] A big bull was worth taking: Scammon's 18-foot specimen produced 210 gallons of oil.

Better-grade oil, whale or elephant, burned with a clearer flame in the lamps of those who could afford it, and also had wide use in lubrication; the more industry developed, the greater the demand. Even after the mid-century initiation of the commercial petroleum industry in the United States, sperm whale and elephant oils were still wanted for these and other purposes (soap, paint, oils for tanning and finishing of leather). Provided costs, especially labor, could be kept low enough, elephanting, like whaling, could still be a successful business.

As with fur seals, it is impossible to say when man first recognized the value of elephants for oil. The first visitors to Falkland and other South Atlantic beaches certainly found and took elephants while searching for the more valuable fur seals. The first American vessel specially fitted out as an elephanter, according to Charles H. Stevenson, an authority on oil and skins, was the *Alliance* of New Bedford, which returned from Patagonia in 1804 with a full cargo of oil.[14] That coast was continually exploited for the next decade and a half, after which the focus shifted to more southerly islands.

Oil was always saleable, in addition to being useful to the mariners themselves, but it was not so valuable that the last elephant seal was hunted to extinction in the Falklands and along the mainland coast. Indeed, since the parent population for the Falklands herd is at South Georgia, until that island too was sealed the Falklands could always be recolonized. In the meantime, elephant oil was often a cargo for disappointed fur sealers or even successful ones wishing to fill the last corner of their holds before heading home.

South Georgia itself was certainly visited by sealers before 1800. When Fanning was there for the 1801–2 season, he hoped to meet the crew of another vessel, the *Regulator*, but found that her captain had disposed of her skins for cash to "an English elephant oil ship," a strong implication that elephanters as well as fur sealers were here before the eighteenth century was over.[15] When the South Shetlands were discovered in 1819, elephants were taken there too; Weddell estimated that 940 tons of oil were taken along with 320,000 skins in the years 1821 and 1822.[16] However, the same South Georgia parent herd was involved here too. Of all these dependencies of the Falklands, South Georgia was the most productive, providing a generation's worth of mixed cargoes. The Schooner *Pacific*, for example, took 256 skins and 1,800 gallons of oil from South Georgia in 1829, at least three decades after the first sealers arrived.[17] Elephant sealing became a regular trade which, though fur-sealers and elephanters were often the same men, developed its own methodology.

The lethargy of the beast on land and its unwillingness to move unless forced was the key to the business; there was no fear that the elephants would make for the water at the first sign of danger, as was very possible with watchful fur seals. The elephants could be herded away from the water and killed in convenient groupings, particularly where beachside cliffs blocked further escape. Bulls blinded in one eye would keep their harems intact, yet not react to men passing by (the one good eye was kept on the females).[18] Fanning explains how it was done at South Georgia:

In taking the younger, a club is commonly used, and for the old ones, a lance; yet in order to overcome the largest bulls, it is necessary to have a musket loaded with a brace of balls; with this, advancing in front of the animal, to within a few paces, they will rise on the fore legs or flippers, and at the same time open the mouth widely to send forth one of their loud roars; this is the moment to discharge the balls through the roof of the upper jaw into the brains, whereupon the creature falls forward, either killed, or so much stunned as to give the sealer sufficient time to complete its destruction with the lance. They are frequently discovered sleeping, in which case the muzzle of the piece is held close to the head, and discharged into the brain. The loudest noise will not awaken these animals when sleeping, as it is not unusual, though it may appear singular, for the hunter to go on and shoot one without awaking those alongside of it, and in this way proceed through the whole rookery, shooting and lancing as many as are wanted.[19]

Bull elephant seals were not entirely without defenses. Capt. Joseph Fuller, a leading New London elephanter at Kerguelen in mid-century, had a word of advice on this:

You have to be cautious in killing the elephants; don't get too near them for they will give a sudden spring towards you and if you have no way of retreating you are gone up the spout, for if they once get you, or any part of your anatomy, between their powerful jaws nothing will save you. At times when you are trying to lance the critters they will grasp the lance between their teeth, and if you keep ahold of it, like as if by magic, you will pick yourself up about twenty feet distant, and when you come to look at your lance, you will find the tip of it gone.

"Sometimes," he added, "it takes as high as ten bullets to stop them and then a good spearing to kill them."[20]

Once the animal was killed, the blubber, with skin attached, was cut into strips 5–10 inches wide, depending upon the thickness of the blubber and the size of the animal. The method of cutting might differ; Fanning notes that the fat was slit from head to tail, but photographs taken a half-century later at Macquarie show circular cuts around the body.[21] Once taken off, the blubber was washed clean of blood, then cut into smaller pieces two or three inches square at a mincing table, where as much as possible of the skin was detached, then tryed out in kettles. For fuel, sealers used scraps from the trying-out process, but anything flammable might be burned, including penguin carcasses, for penguins, which also have fat, burned well once the fire was hot enough. Trypots abandoned at South Georgia often had one flat side, indicating that they were used in pairs at the same fire; it was common to have one pot higher than the other, allowing the oil from the first to flow to the lower for a two-stage boiling process.[22] Unlike whalers, sealers preferred to try out their oil on land; otherwise all the blubber would have to be carried to the vessel, often through dangerous surf.

The oil was then placed in casks, though as Fanning explains this process was not simple:

... a new cask after being filled with the boiling oil, is then started [emptied out] and coopered, necessarily, from not being fully shrunk requiring to be filled again with the boiling oil, and even a third time, if it has not done shrinking after the second filling, which can readily be discovered; this course being

particularly attended to, it may finally be coopered and stowed away in the
ship's hold, to be filled up by the hose, and will remain tight for the voyage,
in all climes, nor require wetting for any length of time, or lose a gill of oil
by shrinkage.[23]

Whale oil was treated in the same way; the problem here was the new
casks. But the treatment of elephant seal oil, in addition to the double
boiling, had yet another complexity. Even after boiling, a good deal
of oil remained in the scrap pieces before they were consigned to the
fire as fuel. Elephant oil did not "run" as freely from its blubber as did
whale oil. Among specialized elephanters, therefore, it was common
to boil the oil for ten to fifteen minutes, then put the scrap pieces through
a special "scrap press."[24] Once through the trying-out process, the oil
was well worth the effort, for it was very rich: the rule of thumb was
"a cask of oil for a cask of blubber."

Though no doubt elephants continued to be taken in Cape Horn
waters off and on during the mid-nineteenth century, no record has
been found of American sealing voyages to South Georgia between
that of the *Pacific* in 1830–1 and the visit of the New London sealer
Trinity in 1870–1. The *Trinity*'s master buried his cooper on the island
and thus left a record of his visit which was visible for many years.
Similarly the graves of five men from the London sealer *Esther* who
died of typhus in 1846 long marked Grytviken (later a center of whaling
activities).[25]

These visitors were the exception; after the end of the fur seals in
commercial numbers, only the occasional elephanter came to these waters,
and even these fell off between 1840 and the end of the American Civil
War. There was no serious revival of elephanting here until the 1870s.
It continued to the end of the century and in the 1890s was combined
with a final brief flurry of fur sealing, discussed in the next chapter.
Some demonstration of commercial Norwegian interest in elephants
led to the (British) Falkland Islands Ordinance in 1899, followed by
other regulations establishing control over the elephant harvest at the
Falklands and the other dependencies such as South Georgia and the
South Shetlands.[26]

In the twentieth century, South Georgia elephant sealing (no fur
sealing was permitted) was carried on under annual license, with the
island divided into four zones, one of which was left untouched each
year. Only adult males of specified minimum size were allowed to the
licensee, the Compañia Argentina de Pesca, which was principally concerned with whaling but found that its whale catchers could obtain
useful supplemental employment in the spring bringing in elephant seals
(the whaling season was November to February). Catchers proceeding
along the coast would send in smaller prams to the beach, where the

men would herd the seals to the edge of the surf, kill them, slit the skin along the back, around the neck, tail, and flippers, then yank the whole skin off into the surf and winch it out to the catcher on a whipline, for trying out at the home base of Grytviken. The Compañia was permitted to harvest 6,000–8,000 a year and by the mid-1950s had taken some 200,000 males, averaging just under two barrels of oil each. The carcasses were left to rot, though it was recognized that another half-barrel each could be obtained if they could be conveniently brought to the steam rendering plant.[27]

Though the harvest was controlled, the result, according to R.M. Laws, staff biologist of the Falkland Islands Dependency Survey, was to reduce the average age and size of the bulls taken and the average oil content of each. Indeed, the company at times had difficulty in filling its quota, which was reduced as a result. The point is worth noting, because at the same time elephant seals of this herd were repopulating the South Shetlands and other islands. Laws's conclusion is that this expansion was the result not of population growth but of human decimation of the herd: the seals were seeking remoter and safer beaches. In any case, it was clear that average harem size was increasing and average bull size falling, potentially leading to the point at which there might be a shortage of sexually active bulls to impregnate the females.[28] Since sealing ceased at South Georgia in 1964 when the whaling station was closed due to overexploitation of whales, the future may be more promising for this seal population.

Early in the nineteenth century, however, the elephanters had moved on, at some unspecified date sealing elephants from other islands where they had earlier hunted fur seals, such as those in the Indian Ocean. According to Edouard Stackpole, Kerguelen (Desolation) lost a few elephants to fur sealers as early as 1792; Jorgen Jorgenson, however, an experienced Indian Ocean sealer writing in 1807, believed that the first successful elephanting voyage here was the ex-Indiaman *Hillsborough* in 1798 (already encountered in chapter 1).[29] British elephanters were often at Desolation between 1800 and 1810 and only occasionally afterwards as the sealers went back to the South Atlantic in the early 1820s for the South Shetlands fur seal bonanza. The Prince Edward Islands and the Crozets were heavily exploited at the same time as Desolation, mainly by Englishmen.

But Americans soon turned up. William Dane Phelps, a New England master who was to become well known as a merchant-captain on the California coast in the 1840s, spent two years of his early career on Prince Edward. Sailing from Boston in 1817 with Phelps aboard, the brig *Pickering* visited the Falklands, South Georgia, Gough, Prince Edward, the Crozets, and Kerguelen, all without notable success, before returning to deposit a crew on Prince Edward. The sealing gang was

left with enough supplies and provisions for a year and did not see their
vessel for two years, by which time utter boredom had set in. Though
meat was available to keep them alive, the gang had filled all their
barrels in the first year and then had very little to do. The *Pickering*,
which eventually retrieved the 1,600 barrels of oil and 7,000 skins which
the crew had gathered, made for Cape Town, where Phelps left the
ship (the master was eventually tried for barratry), earning little for his
two years' voyage.

Phelps, by that time, had all the experience he ever desired with
elephant seals, but his first assignment as a green hand, to kill an elephant, strip the blubber, make a fire, and then cook the tongue, heart,
and liver, as breakfast for ten men, was an adventure he never forgot:

I knew nothing of the habits of the elephant, had never seen one killed, and
there I was, with a lance two feet long on a pole-staff of four feet, and seal-
club, a butcher's knife and steel, with orders to kill, butcher, and cook one of
those enormous beasts, the smallest of which looked as if he could dispose of
me at a meal. After the boats' crews were out of sight I took a survey of the
amphibious monsters, and selecting the smallest one, commenced the battle
according to orders. When I hit him a rap on the nose he reared up on his
flippers, opened his mouth, and bellowed furiously. This gave me chance at
his breast; plunging my lance into it, in the direction of where I thought his
heart ought to be, I sent the iron in 'socket deep.' This was all right so far, but
I was not quick enough in drawing it out again, and stepping back. He grabbed
the lance by the shank with his teeth, and drawing it from the wound, gave it
a rapid whicking round; the end of the pole hit me a rap on the head, and sent
me sprawling. I picked myself up, and with a sore head took a survey of the
enemy; he had not retreated, but retained his partly erect position, bleeding
and bellowing, while his companions in the vicinity joined in the roar, but
without moving off, or attacking me …

My next resort was the seal-club. With this I managed to beat the poor
creature's eyes out, and then, fastening my knife on the pole [the other lance
having broken], I lanced him until he was dead; pounding him on the head
with a heavy club, with an iron ring on the end, produced about as much effect
on him as it would have done on the rock of Gibraltar.[30]

That, of course, was only the beginning: there was now the question
of blubber, fire, and breakfast!

In 1838, the first American voyage deliberately organized to take
elephants at Desolation – the first, that is, of which the log has survived
– sailed from New London, Connecticut, in the ship *Columbia*.[31] Within
a decade, the New London area had made a substantial investment in
the Desolation trade, which it continued to dominate. This supremacy
was gained only after some serious conflicts with Englishmen, though

that struggle was fought out more often on Prince Edward or the Crozets. Ellery Nash, master of the Stonington bark *Bolton*, arrived at Pig (or Hog) Island in the latter group in 1843 only to find that the beaches had been shared out by the English and American vessels already there. Nash was warned off by the Englishmen, who promised to defend their territory "as long as they could hold up a lance," and, when Nash attempted to seal anyway, not only stole his blubber but drove seals into the surf rather than allow the Americans to get them, telling Nash in the process that "my voyage was up."[32]

But after 1820, according to Stevenson, 94 per cent of (American) elephant seal voyages were made by the sealers of three ports: New London, Mystic, and Stonington, all of which lie along ten miles of Connecticut shore.[33] Why their domination of trade? Desolation, which produced the bulk of the oil, was no secret, and indeed other ports sent elephanters on occasion. Moreover, many New England whalers visited the island, yet elephanting was the speciality of New London. It can only be said that the trade flowed naturally from Stonington's speciality in fur sealing, although Stonington itself continued to focus upon the South Atlantic as long as the fur seals lasted, and then was active in sealing only as an appendage of New London.

Once begun, however, working in and about Desolation required great command of local navigational and weather conditions, and it was not easy for newcomers to get started. The peak came in the late 1850s, when more than a dozen vessels with several hundred men collectively could be found at Desolation's Three Island or Pot harbors. Usually these were topsail schooners or barks, often under 200 tons, needing square sails for the Atlantic but fore-and-aft rig for working quickly in and out of Desolation's fiords and for saving manpower when they had large gangs ashore. Although Desolation never witnessed the fêtes and galas of Herschel Island in the golden days of Arctic whaling, there is occasional mention of a captain's wife with a piano, crew exchanges, livestock landed on the island, general companionship, and mutual assistance – until the trade declined, and every penny became of critical importance.

The procedures at Desolation, whether in the 1850s or later, have been made reasonably clear in several works. Nathaniel Taylor's *Life on a Whaler*, written in 1859, is a full-length memoir of elephanting there. Late in the century, the standard works of George Scammon and A. Howard Clark discussed this industry, and the recent publication of Joseph Fuller's account, written near the turn of the century, adds considerable detail. Fuller spent fifty years in and about Desolation, including eleven months marooned there after being wrecked on the aptly named "Rocks of Despair," and he knew whereof he spoke. Even an expert like Fuller had few defenses against the full range of raging winds and hellish seas which could be experienced in those waters.[34]

It behoved masters to be as cautious as possible. Normally they sailed from New London in late June or early July, arriving at Desolation in early or mid-September, the start of the southern spring, just as the bulls and pregnant cows arrived. By November, the first killing was over and with the pupping and mating season ended, the surviving bulls and cows departed, and the sealers turned to the young adults arriving to moult in the wallows, followed over the next two months by the returning cows and bulls. By midsummer, February-March, there were only a few isolated animals left until the following spring, as the herd moved off to feed at sea.

On arrival at Desolation, vessels would send down and land their topmasts, which were only a handicap here. Moving from bay to bay, it was possible to collect blubber off the rougher beaches only when weather permitted; a small tender was normally used to bring in the blubber to be tryed out. Given adequate precautions, however, the fiords of Desolation, particularly the several anchorages in Royal Sound on the less exposed east coast, offered adequate shelter, at least in the austral summer. Heard Island, the one seal refuge which the sealers had overlooked in the early days, was another matter.

Heard Island lies 260 miles to the south of Desolation, a distance more important than the number of miles implies, for Heard is on the other side of the convergence and thus considerably colder. Surprisingly, it escaped potential discoverers until first sighted by Capt. James J. Heard of the bark *Oriental*, bound from Boston to Melbourne, in November 1853. Within the next four years at least nine other vessels caught sight of the island, in addition to unnamed sealers who went there on purpose – a startling fact, when the number of earlier Indian Ocean voyagers who had missed the island is considered.[35] The explanation for the sudden plethora of sightings probably lies in the race to get to Australia after the discovery of gold there in 1851, coupled with the advice of Matthew Fountaine Maury to ship captains on the significance of great circle courses for quicker voyages (specifically, his judgment that a vessel swinging wide to the south of the Cape of Good Hope would make better time to Australia).[36]

The first elephanter here, and as far as is known the first man to land on the island, was Capt. Erasmus Darwin Rogers, an experienced New London sealer at Desolation.[37] Rogers inaugurated the method of making short visits to Heard from his base at Desolation, which unlike Heard had safe harbors. The passage required three or four days downwind, but at least four or five working back (one master who misjudged the weather spent a disastrous twenty-three days on passage from Desolation to Heard).[38] Rogers and the crew of his 500-ton ship *Corinthian* found a cold and barren rock, some twenty-five by ten miles, almost entirely covered by permanent ice, and dominated by a 9,000 foot inactive volcano, "Big Ben," whose shoulders cut off the various beaches from each other – withal, "more desolate than Desolation."[39] When it became clear that Heard beaches offered elephant seal in plenty (it was not a major fur seal refuge), Rogers was followed by many of the Desolation sealers, and a fleet of perhaps half a dozen vessels would sail to Heard from the larger island in September, unload men, supplies, and empty casks, and then depart again for Desolation, returning at the end of the season for the oil.

At Heard, where a cyclone could be expected at least once a week the year round,[40] the utmost caution was essential, including massive anchor systems. Fuller carried anchors of 2,500 and 3,200 pounds for a 200-ton schooner; the 400-ton bark *Alert* put down anchors of 2,800, 4,000, and 4,750 pounds and still dragged the first night. As the years went by, anchors often had to be left, hopelessly fouled on the anchors and chains abandoned in Corinthian Bay, for example, by vessels run aground or forced to slip in a sudden gale. Wrecks were frequent; a dozen logs from Heard record half as many wrecks experienced or directly witnessed. These professionals took it in stride, though seldom as blasé as the last log entry of a schooner destroyed in a midsummer gale in 1863: "All hands on the Island employed in fishing things out of the Surf. So ends the day and the schooner *Pacific.*" The master of the bark *Trinity*, sent out in 1880, was convinced, according to Captain Fuller, that he would be wrecked at Heard, since his hard owners had sent him there for the entire season without a tender, and he *was* wrecked, within a week, and forced to live out fifteen grim months there.

When the take began to fall off, gangs were left at Heard to kill anything that came ashore; as early as 1856, men were wintering on the island.[41] If the oil was not tried out there, it was minced and put into six- or eight-barrel casks and floated out when the mother vessel came in the spring. Heavy surf made this necessary; some beaches were so bad that men would be stationed on them simply to drive the elephants back into the water to haul up on a more accessible landing spot. At Desolation, not without surf itself, it was more common to take the blubber off the beach in blanket pieces, roped together in raft lines – a practice leading to considerable waste, since voracious birds of several species would devour any blubber in sight, in or out of the water. No blubber could be left unattended on shore unless it was sunk into water-holes or buried in the sand – or more likely at Heard, in the snow. Predatory leopard seals, of which Heard Island has the greatest concentration in the world, were another danger.

Heard was colder, more sparsely gifted with vegetation, but more

plentifully endowed with winds and snow. Life was grim, above all
for men put ashore for months on end with a cask of salt meat and
some flour and molasses, to live in a shanty of rocks and canvas. The
diet could be supplemented with birds and penguin eggs in season,
however, and there was the added attraction on both islands of Kerguelen cabbage, *Pringlea antiscorbutica* (unrelated to common cabbage,
it tastes like spinach or horseradish, depending on the eater). Above
all, by the 1870s there were rabbits on Desolation, many rabbits, doing
their best to exterminate the cabbage. Fuller's wrecked crew ate rabbit,
not shot but dug out of their burrows with pointed sticks, every day
for eleven months, stewed at lunch, often fried in elephant oil for dinner,
unless there was enough young albatross or local teal to make a meal.
On Desolation they began to live rabbits and to use expressions like
"First catch your rabbit before you cook him," for "first things first."
The rabbits were probably introduced by the sealers, although the British exploring ship *Challenger* has been credited with, or accused of, the
deed. The schooner *Charles Colgate*'s log for 1883, at nearly the end of
the industry, records that it had live rabbits aboard as it moved about
Desolation, ready to plant another colony where one did not already
exist. No natural enemy existed for these animals until a German scientific party abandoned sled dogs at the turn of the century and the
two could compete in one of those interesting struggles so often pre-
cipitated by mankind.[42]

The sealers, with little else to do, gloried in the biggest bull, or the
first take of the season, after waiting through the night for the likely
dawn arrival of the first bull.[43] Yet there is little way to make the
profession appear romantic or attractive. On the contrary, elephanters
were doubtless as callous as Rallier du Baty, a French visitor to Desolation on the eve of World War I: "One need not sentimentalize over
sea elephants. Their only use to the world is to provide blubber." As
L. Harrison Matthews, who worked as an elephanter at South Georgia
long before becoming one of the world's leading mammalogists, put
it, "slaughtering seals always degrades and brutalizes those who do
it."[44]

And yet the industry lived as long as traditional whaling for several
reasons. First, elephanters had several trades: they took whales whenever they found them and if not returning home with a full cargo spent
the off-season "bay whaling" in and about Desolation's fiords for right
or humpback whales or went north to the Madagascar grounds or
Delagoa Bay. Nor were they averse to taking a rare fur seal or two.
Capt. Gurdon L. Allyn, in his memoirs, *The Old Sailor's Story*, shows
just how far elephanters were willing to go for skins, in this case on
Possession Island in the Crozets (January 1843):

Having been informed that there were some fur seal on a certain beach on the side of the island opposite to where we were; and as there was no anchorage there for our vessels, and as going around and landing in boats was both difficult and dangerous; the captain and mate of the *Emmeline* [schooner, from Mystic, with which Allyn's New London schooner *Franklin* was working in tandem], together with my mate, six men and myself, formed a party to go across lots, which proved no easy task. We provided ourselves with penguin-skin moccasins, as boots were too cumbersome and would soon cut through, and started early one morning.

First we waded through the tussock-bogs, then clambered up the sides of a mountain, over the loose, rough clampers, to an altitude of three or four thousand feet, where the snow capped the summit, and down the other side, which was much steeper and equally jagged and uneven.

The man who acted as guide made a mistake in his reckoning, and no wonder, for it snowed and blowed like fury let loose; and behold, when we descended we were about one-half of a mile from our beach, and no way had we of getting to it but by ascending and descending in another gulch, for the mountain-wall between us and our intended landing place was perpendicular and impassable. Up we clambered to the very top, where the wind blew a tornado, and down we scrambled nearly half way when the guide discovered he was again on the wrong track. Up we went again ...

The third time we reached the desired beach, where we camped for the night in a dismal den or cave, through the fissures of which water was constantly dripping. Three of our men, disgusted with the guide, were separated from our company and camped all night in the open air, and these men had the bread which we needed for provision. But although we were tired and foot-sore we secured a sea-elephant; and making a fire of blubber and some sticks which we carried for staffs, we fried his liver in an old broken camp-kettle which we found in the cave; and after boiling some sea-water for salt with which to season it, we ate it with better relish than many have for better viands. In two days' time we obtained sixteen seal-skins, and made our way back to our vessels, which we were very glad to reach.[45]

It might be expected that with such an effort for sixteen fur skins, no fur seal would ever again be seen in these waters, but it proved impossible to corner them all. There were always unreachable rocks or crevasses, especially at Desolation, where a few survived. H.N. Moseley, a naturalist with HMS *Challenger*, found a considerable number there in 1874. He killed two himself, and reported two whaling schooners taking seventy in one day. "It is a pity," he added, "that some discretion is not exercised in killing the animals ... The sealers in Kerguelen's Land take all they can find."[46]

A second reason for the survival of the industry was the fact that a man could always turn to other work; there was no need to spend a lifetime in elephanting, as Fuller did, and no reason why a sealing schooner, which did not require a whaler's built-in gear (such as multiple davits) could not be turned to other uses. Allyn, master of Lawrence and company's schooner *Betsey* for a sealing voyage to the coast of South West Africa in 1834–6, after an indifferent catch was sent with the same vessel cattle-trading in the West Indies; soon the *Betsey* was after seals once more, this time to Chile. Allyn then turned to American coastal vessels, even farming for a few years, before going sealing again in 1842.

Third, when they did go sealing, elephanters could expect to find some oil to fill at least part of their hold. The elephant seals never were exterminated in the southern Indian Ocean, though they certainly were reduced; the oil was always saleable. Allyn sold his 1843 take of 450 barrels of elephant oil at Rio de Janeiro for 29 cents a gallon, bought 350 bags of good coffee with the proceeds, and sold this cargo at home for an amount which brought the value of his oil to 50 cents a gallon.[47] Elephant oil, in other words, was a steady commodity in many markets.

Fourth, the trade continued right through the Civil War, a conflict which hardly upset Indian Ocean activities. Desolation was a 23,000-mile round trip, and no *Shenandoah*, such as destroyed the Arctic whalers, ever ventured here, though the *Alabama* did make one quick swing along Indian Ocean routes to Singapore.[48] New London sealing was of course hurt by the war, and never again did Desolation see vessels in the same numbers, but two to five ships still made the long voyage each year where eight to twelve had done so before. Finally, with its predictable return and reduced overhead (men could live off the land in part, and the ships were small, wet, and less expensive to outfit), elephanting was ready-made for sweated industry status.

The owners of course made money, otherwise they would have dropped out of the business. On average, through the years 1840–90, roughly one out of every four "whaling" vessels leaving New London went to Desolation and/or Heard for elephant seals.[49] Kenneth Bertrand has counted eighty voyages made to the two islands between 1855 and 1880, only eleven of which were not from New London (and of the eleven, five were from nearby Mystic).[50] In New London at least eight firms were involved, four of them in a major way, meaning at least half a dozen vessels each during the period. Perkins and Smith, a well-known firm, was Erasmus Roger's employer; Joseph Fuller sailed for Williams and Haven, a firm already encountered in connection with the Alaska Commercial Company's lease upon the Pribilof fur seal herd.

Williams and Haven was the leading New London shipowner and agent in the whaling and sealing industry of that town. Major Thomas W. Williams is the man generally credited with developing whaling as a permanent industry in New London; he was also one of the first owners to insist upon temperance on his ships. T.W. Williams was partner to another well-known agent, Henry P. Haven, and when Williams was elected to Congress in 1838, he became a silent partner, and Haven assumed active control of the firm. After 1848, Williams returned to active membership, and during the next two decades other partners were added, including, in 1858, Williams's son, Charles Augustus Williams, Richard Haven Chappell, and "Rattler" Morgan. After 1876, C.A. Williams, who had been the firm's agent in Hawaii at the time the *Peru* was sent to the Pribilofs, took over the company following the deaths of his grandfather and Chappell. C.A. Williams was one of New London's most noted citizens, serving as mayor from 1885 to 1888. By the time of his death in 1899 his company had been dissolved (1892), and the New London whaling and sealing industries were dead thanks to the extermination of whale, seal, and sea elephant stocks and the development of the petroleum industry. But Williams and Haven interests had sent out more vessels by far than any other New London firm; in 1845, at the industry's height, the partners owned eleven whalers. C.A. Williams under his own name had sent out twelve vessels, making twenty-three voyages between 1878 and 1892 – four under Fuller's command.

Some estimate of the possible profit can be made from records of the voyage of the 106-ton schooner *Francis Allyn* with Fuller as master in 1886–7 – nearly the end of the sealing era.[51] Her return cargo of 650 barrels of elephant oil (as usual classified as "whale oil") was valued at $6,552 as a result of a very low price of 32.5 cents a gallon, there being 31 gallons to a barrel.

return from voyage	$6,552
crew costs (average 30 per cent)	1,967
remaining to owners	4,585
outfitting costs ($50–75 a man per annum	
food and supplies, plus refitting, etc.)	1,500
profit to owners and master	3,085

The value of the *Francis Allyn* is not known but, using prices for comparable schooners of her size in the British Columbia sealing fleet, was probably close to $8,000.

Fuller's share, as captain, was 1/12 "lay" of the proceeds ($546), but he also – as was common practice in whaling and sealing – had a share (8/32) of the ownership of the vessel, which returned him $771. He would have shared in outfitting costs – a deduction – but no doubt would have managed to make up that much or more from management of the ship's stores or "slop chest." His year's income, therefore, was approximately $1,300, and no doubt was more in some earlier years when the price of oil was higher. An annual income of $1,300–1,500 was certainly acceptable in the 1880s but not nearly enough to qualify Fuller as a wealthy man: he was never able to buy into the firm in the style of "Rattler" Morgan.

At the other end of the vessel, the rest of the sixteen-man crew did not fare so well. The first mate served for 1/16, lesser officers for 1/32 or 1/52, boatsteerers for 1/75, and the hands for 1/150 or 1/160 – and in the case of two green hands, 1/195. An experienced seaman's "long lay" of 1/160 came to a mere $41 for the same voyage, but doubtless he had drawn clothing and tobacco from the slop chest, and probably a cash advance before sailing (for which he would be certain to pay interest, often 25 per cent per annum). He was likely to be assessed sundry additional charges for a share of the medicine chest, or ship-loading and unloading, justified in much the same manner as in Newfoundland. The crewman, dumped aboard shortly before sailing and leaving the ship immediately the last sail was furled, was unable to help load or unload. All in all, the seaman might well finish his voyage in debt to the owners as a result of his year's work, departing with a pat on the back and a charitable dollar or two, having cost no more than his probably skimpy rations. At that level, it was indeed a sweated industry which was "at its best, hard, and at its worst represented perhaps the lowest condition to which free American labor has ever fallen."[52]

But the men were not all "free Americans." Towards the end of the trade, practically the only men who would put up with conditions to be found at Heard Island and such dismal places, aside from crimped sailors, were Cape Verde Islanders, eager to escape the periodic famines and epidemics which visited their islands off the west coast of Africa.[53] They were willing to work, and work hard, for little pay in the hope of eventual discharge in the United States. As a result it became common to recruit at least half the oversized crews there (bigger crews, after all, were not necessary simply to sail the vessel to the sealing grounds); fresh provisions could be obtained there, and, for those planning to take fur seals, salt as well.

The lays of Cape Verdeans were very "long" indeed. The schooner *Charles Colgate*'s letter of instruction of 1877 ordered her master to pay 1/265th lay to green hands (and all Cape Verdean recruits were rated green hands), an amazingly low figure even by the standards of the later, depressed days of whaling.[54] These orders were for a typical late nineteenth-century "elephanting, whaling, and sealing voyage" to the South Shetlands and Desolation and all intermediate islands and grounds. Nor surprisingly, more and more the crews were the dregs even of the impoverished Cape Verdes, not experienced sailors (despite the fact that these were islands, the Cape Verdes had little deep-water seafaring tradition). Witness *Colgate*'s sarcastic mate in 1871: "At 9 P.M. the Boat come off [Bravo] with the men fourteen in all not a man among them that can Stear the Schooner or go aloft; good sailors to run the Indian Ocean with."[55]

They were also black, treated as second class, and often in conflict with white shipmates or officers, hardly surprising when color, language, and social standards all clashed under harsh conditions. Above all, racial conflict emerged in a crisis such as a wreck. Fuller's Cape Verdeans lived apart, as did those of the *Trinity*; racial humor or slurs became more frequent, and in Fuller's recollections "damned white negro" was as bad a term as one man could use for another.

Indiscipline, even mutiny, was the order of the day. "A bad set of Negrows," complained *Colgate*'s log for 1875, constantly recording this or that man in irons for knife fighting or general troublemaking. The next year a plot to take over the vessel and kill the officers nearly came off.[56] Fuller in 1880 had to maroon several men on an islet off Desolation so that they would not trouble his own marooned crew, while he and his officers walked around with weighted sticks in case of trouble. *Trinity*'s men refused to obey when the captain ordered them to collect blubber while they were wrecked on Heard.

Of course, even with the lowest-paid Cape Verdeans, the trade could not go on forever. Fuller was the last captain to sail regularly out of New London for Desolation, and his voyage in the *Francis Allyn* of 1887–9 was the end of an era – although he made two more voyages in the same vessel from New Bedford in the 1890s. In these last voyages he had little confidence in securing a full cargo at Desolation and normally left substantial gangs at Gough Island near Tristan de Cunha and at Possession in the Crozets on his outward passage. Fuller was not the last American elephanter at Desolation, however; that honor probably goes to Benjamin D. Cleveland, a well-known master, who took the bark *Charles W. Morgan* there in 1916.[57]

Others were still interested in Kerguelen's seals in the twentieth century, however. A Norwegian whaling company killed substantial numbers there in the years 1909–13, and in the interwar years a British firm, Irvin and Johnson, acting through a Cape Town subsidiary, leased sealing rights from the French lease-holder, René E. Boissière et Frères. The subsidiary, the Kerguelen Sealing and Whaling Company, continued to be active into the 1930s, with its four-masted brigantine, the *Sound of Jura*, making regular visits. In the aftermath of World War II France established effective control over Kerguelen for the first time, and though a private company did some sealing during the 1960s and 1970s the establishment of a national park over much of the island together with legislation to control sealing seems to have effectively ended the business for the time being; even the fur seals are recovering.

Heavy Russian and Polish commercial trawling in these waters (under agreement with France), the prospect of off-shore oil exploration, and even of nuclear testing, however, leave Kerguelen's seals with a most uncertain future.[58]

Heard Island, on the other hand, was entrusted by Great Britain after World War II to Australia, which established a temporary weather station here in 1947; this remote isle is now visited only by a rare cruise ship such as the *Lindblad Explorer*. In the Crozets, the Norwegians were sealing just before World War I and with the help of a French floating factory did still further damage to the seal population. These islands now fall under the same regulations as Kerguelen.[59] The Prince Edward Islands belong to the Union of South Africa. The last sealing here was in the 1930s from Cape Town, and they are now a wildlife sanctuary, though a base is maintained on Marion Island. Barring unforeseen developments, seal populations of all these islands may be on the way to recovery; one development which has been foreseen, unfortunately, is that overfishing by the trawlers of Russia and other nations around these islands may prove to be the most important limiting factor.[60]

Interestingly, the American sealers had not always had the great years of Kerguelen elephanting all to themselves. Nearly a century before Australia assumed control of Heard Island, Australian interests had made one attempt to share in the Desolation oil. In 1858–9, Capt. J.W. Robinson, master of the bark *Offley* (a sperm whaler), was at Heard but found the going very difficult. The voyage was a failure for her owner, Dr. Crowther of Hobart (sometime premier of Tasmania), largely because the *Offley*'s tender never arrived. Robinson left a lively narrative of his voyage, and his wife gave birth to a son at Desolation (named James Kerguelen Robinson, perhaps the only child born there, although other masters had their wives along), but the *Offley* returned to Hobart with a total cargo of only 120 tons of oil after an eighteen-month voyage, and Crowther took a substantial loss on the venture. No other Australian vessels intruded upon the American preserve of Desolation and Heard at least until the very end of the century. Australians had, however, their own elephant seal industry which, while only marginally a part of the American sealing story, deserves brief mention here because of its effect upon the southern elephant seal species.[61]

Commercial sealing in Australian waters should probably be dated to the establishment of a partnership between Henry Kable and James Underwood – both ex-convicts – with Samuel R. Chase, to undertake a voyage in mid-1800 with the sloop *Pioneer*, "for the purpose of catching Seals or Sea Lions, tanning such skins, converting or manufacturing the same into upper and sole leather for strong shoes and also for the preserving of the oils of such seals and sea lions for such market as shall be deemed most beneficial for the general interest and advantage of all the parties concerned."[62]

Kable, Underwood was the most important elephanting concern in early Sydney days. Sealing for oil was under way at King Island by 1801; by 1803 the firm was contracting for 300 tons of oil at £12 each for export – a contract which would have meant 1,500–1,800 elephants. American sealers were also present: the ship *Charles* of Boston was whaling, sealing, and elephanting in the Kent Group of Bass Strait in 1804. That same year Kable, Underwood, now a thriving concern, brought in 28,282 skins and over 180 tons of elephant oil, together with 220 gallons (not tons) of "seal oil," meaning that they were also using sea lions for this purpose, though their oil content is so low that they have seldom been bothered by oil sealers. The firm employed sixty-three men at the time. By 1810, however, the nearer Bass Strait resources were exhausted, and its islands left to small gangs of runaway convicts and loners, who were particularly adept at using aboriginal women in the hunt for the last fur seals, much as the Russians had used Aleuts in other waters. The commercial sealers of Sydney were already fanning out to exploit the shores of New Zealand and its outlying islands such as Campbell and the Antipodes.[63]

Not until the discovery of Macquarie in 1810 was it known that the basic breeding elephant stock for this region existed there, but it was the fur seals which were exploited first. As early as 1811, the New York ship *Aurora*, 180 tons, which had found only 100 skins and 140 gallons of oil at Campbell, brought back 3,000 skins (hair seals in this case) and 60 tons of oil from its first visit to Macquarie.[64] The Macquarie fur skin rush lasted only five or six years, but the production of elephant seal oil from its sizeable herd continued until the middle of the century. As with other areas, the elephant herd, though much diminished, was not exterminated; the difficulty of moving blubber to the tryworks limited the number of beaches where they could usefully be killed, and, while demand was steady, the price was not such as to bring the sort of "gold rush" which had occurred for otters or even fur seals.

In the first decades of the nineteenth century, elephant oil was bringing £20 a ton in Sydney, and exporters might sell it for £30 in India, a profitable margin. But each ton paid £2 tax in Sydney, and freight added a further £3, to which had to be added production costs. While these did remain low, they nevertheless included cost of the vessel, its outfit, and wages. Early Australian exports were likely to be a risk in any case: the English ship *Surrey* found that its 500 tons of oil taken to London in 1819 brought only £16 10s 0d a ton, and her captain, Thomas Raine (for whom an island in the Great Barrier Reef is named), lost a substantial sum of money.

Still, elephanting was continued by the gangs who were regularly

left at Macquarie. Bellingshausen, the Russian explorer, left an interesting account of his visit of 1820. He found two parties of Port Jackson
(Sydney) sealers (twenty-seven men altogether) on the island. They had
been here less than a year, but complained of boredom; having run
out of casks for their oil, they had no reason to work, and meanwhile
they were fast exhausting their provisions. Bellingshausen considered
that these men were still better off than those in South Georgia, which
he had also seen; not only did they have sea birds, penguin eggs, and
the like to eat, but also "wild cabbage" (*Stilbocarpa polaris*, in this case)
to help ward off scurvy. The men lived in huts made of the wild grass
that grows on the island, lined with sealskins, and lit by lamps burning
melted elephant blubber. "Inside," wrote Bellingshausen, "it was so
black and dark from the smoke that the smouldering light from the
lamp and from the holes in the wall over which bladders were stretched
scarcely lit the interior of the hut, and until we got accustomed to the
light the sealers had to lead us by the hand."[65]

Bellingshausen had met Kable, Underwood's men, but there were
sometimes rival gangs on the island, and battles for the beaches were
not unknown.

… the combatants, in their long beards, greasy seal-like habiliments, and grim,
fiendlike complexions, looked more like troops of demons from the infernal
regions, than baptised Christian men, as they sallied forth with brandished
clubs to the contest.[66]

The men, as might be expected, were not highly regarded by more
civilized beings, as the *Sydney Gazette* made clear in 1822:

As to the men employed in the gangs the most appalling account is given: They
appear to be the very refuse of human species, so abandoned and lost to every
sense of moral duty. Overseers are necessarily appointed by the merchants and
captains of vessels to superintend the various gangs, but their authority is too
often invariably condemned, and hence arises the failure of many a well-pro-
jected and expensive speculation.[67]

By the 1840s, with the herd over-exploited, the oil industry languished. Between 1850 and the 1870s, there was scarcely any activity,
though in the later decade it resumed and continued until 1889 when
the Tasmania government prohibited the taking of fur and elephant
seals, principally in order to assist New Zealand authorities in con-
trolling their own sealers, by reducing the temptation to proceed to
Macquarie after visiting New Zealand's outlying islands. Sealing did
continue at Macquarie, but now under license to Joseph Hatch, a man
whose entire career was associated with that island.[68]

Hatch not only took elephants. He developed a considerable establishment designed to render penguins into oil, using first King Penguins
and then the more prolific Royal Penguins which seemed to decline not
at all despite his harvest. His refinery came under much criticism, partly
because of the numbers of birds involved, partly because of his methods.
The birds were driven into an enclosure upon their seasonal arrival
(normally January), clubbed, and carried to a shed to be placed into
one of four "digesters," great bubbling pots over coal fires which consumed 4,000 penguins a day. The penguins were put in at the top, and
carried up a railed plank for the purpose. During the operations, the
plank became slippery, and at least one man lost his footing and plunged
into the bubbling cauldron, according to one observer.[69]

Hatch's enemies, particularly one Frank Hurley, claimed that the
penguins were driven up the ramp and forced to jump alive into the
digester. Hurley's account in the *Sydney Morning Herald* in 1919 is a
model of a sort:

The penguins are mustered like sheep, and driven up a narrow netted-in runway,
which terminates at the far end over the open door of a boiler digester. At this
end a man stands with a club. A most remarkable feature about these birds is
their expression of the familiar human emotions of fear, affection, curiosity,
anxiety, and solicitude. They waddle along at their quaint gait almost laughing,
with no suspicion whatever of the cruel fate in store for them. They round a
corner full of curiosity, and that is the last of them; for a knock on the head
and a kick send them into the boiler. It is one of the most pitiful sights I have
ever witnessed. It made me feel quite sick, in fact; when the boiler is full the
lid is sealed down and the steam turned on. This wanton butchery takes a toll
of some 150,000 birds annually. The industry is an unessential one, and, owing
to the primitive plant and wastage, the profit is small and the revenue to the
Tasmanian Government, which leases the Island, is negligible ... I have also
seen vast numbers of sea-elephant carcases polluting the foreshores, with the
oil yielding blubber removed only from their sides, the under portion, owing to
the trouble in handling the heavy carcase, being allowed to remain.

Hurley repeated his charge in a public address the same year, adding
another detail: "Owing to the hardiness of most of the birds this blow
only stuns them, and many of them go into the boiler alive."[70]

The charge, though untrue, was now repeated by Sir Douglas Mawson, noted polar explorer, as part of his campaign to set aside Macquarie
as a nature sanctuary, adding that the elephant seal, unless given protection, was near extermination. Such pressure was successful, and
Hatch's lease was not renewed. Hatch, in his eighties, did his best to
clear his name of the charges, and the Tasmanian government eventually
absolved him of cruelty, but by his death in 1928, aged ninety-one,

his company, the "Southern Isles Exploration Co., Ltd. (Cable Address CELEPHANT)" had long been in liquidation. In 1933, after lengthy consideration, Tasmania declared Macquarie to be a sanctuary for birds and animals, and the populations of seals and birds have been recovering since then.[71]

Thus at long last safeguards against exploitation were established in all the southern hemisphere islands. The question, as always, remains: how many elephant seals had been taken over the years? There are no certain figures, and the paucity of data makes an intelligent estimate difficult. Obviously thousands were taken off the Falklands, South Georgia, and the South Shetlands, as well as the Indian Ocean and Australia/New Zealand shores and islands. If estimates are correct which put the South Georgia herd at about 300,000 after recovery, adding 150,000 each for Kerguelen/Heard and Macquarie, and another 100,000 for the other islands frequented by the southern sea elephant, then the total population is in the order of 750,000.[72] An average of 10 per cent of that stock taken each year between 1800 and 1900 would total 7,500,000 – an impossible figure since no such volume of oil, at roughly one barrel an elephant, was ever recorded on world markets. We must try another method.

The most certain figures come from Desolation and Heard, where fairly good evidence exists of the number of voyages and their hauls, although even here the returns are largely guesswork. Logs, where they can be found, simply say "cleaned elephant off the beach" or "all day ashore elephanting"; there are no "elephant stamps" comparable to those found in whaling logbooks, and only rarely are numbers mentioned at all. Estimates, therefore, range widely. Robert Decker, in his *Whaling Industry of New London* (1973), notes that "slaughter of sea elephants became so great that 1,700,000 were killed in a single year."[73] That is a staggering number, and deserves some attention. Elephants yield roughly a barrel of oil, 31.5 gallons, a cow, and up to seven or eight for a big bull. Fuller's largest was ten barrels. At least 1,700,000 barrels would be the yield from 1,700,000 elephants, even allowing for wastage and animals in poor condition. But this leaves us with the same problem encountered in taking an arbitrary 10 per cent: 1,700,000 barrels would be at least three times the highest total combined for all sperm and whale oil ever brought into the United States in a single year, and six times the average. To put it another way, even a good crew seldom took more than a hundred elephants a day; Fuller's crew of roughly thirty men could kill, mince, try out, and stow at best sixty barrels a day. At one hundred a day, to produce 1,700,000 barrels for a year, fifty ships working 365 days a year at Kerguelen would be necessary, all quite impossible figures. The explanation for Decker's figure may well lie in a remark made by Schwatka in his *Nimrod of the*

North, discussing fur seals, not elephants, at Kerguelen: "so abundant were they at first, that as many as 1,700,000 were killed there in a single year by the crews of the vessels which flocked there from all quotas," though Schwatka, like Decker, gives no source for his information.[74]

Far better, then, to take a low estimate, that of A. Howard Clark, made in 1887.[75] Clark calculated that the take of oil from Kerguelen was 2 million gallons – not barrels – in the decade 1850–60, 1.5 million 1860–70, and 1 million 1870–80, making a total of 150,000 barrels (4.5 million divided by 31.5) over thirty years. At a barrel an elephant, this is some 5,000 elephants a year, certainly too low for good years, but probably not far off for a thirty-year average. Even quadrupled for a good year, 20,000 elephant seals, or 2,000 barrels each for ten ships, the total kill is closer to 170,000 than 1,700,000.

Assume as a base, therefore, 5,000 a year over the period 1840–80, the total for Kerguelen/Heard would be 5,000 multiplied by 40, or 200,000. The Macquarie herd seems never to have been exploited quite so intensively over the years, but since elephants were hunted in the Bass Strait and elsewhere in the general area, perhaps the same annual calculation would not be amiss for the era 1800–40, and again for the Hatch years, 1870–90, so that 5,000 multiplied by 60 yields a total of 300,000.

To this running total of a half-million, we must now add the South Georgia take. Stevenson calculated in 1904 that between 1803 and 1900, 242,000 barrels of elephant oil worth $5,420,000 had been taken from this island and its dependencies. Without knowing the precise nature of his calculation, the figure is certainly possible, allowing for a base herd of over 300,000 animals (R.M. Laws calculated the population to be 250,000 as long ago as 1953). Assuming one barrel an elephant again, 242,000 barrels (let us assume 300,000, knowing in advance that this will help simplify our calculations), is an appropriate order of magnitude.[76] The total harvest of southern elephant seals thus becomes:

Kerguelen/Heard	200,000
Macquarie	300,000
S. Georgia to 1900	300,000
S. Georgia, licensed sealing, 1900–64	200,000
Total	1,000,000

With artificially induced precision, we arrive at the neat figure of 1,000,000. This figure is no more than an estimate, and a conservative one at that. But if it does nothing more, it will at least complicate retrospective whale census calculations compiled solely upon the basis of "whale-oil" returns from South Atlantic, Indian Ocean, or Australian grounds by vessels from such sealing ports as New London or Stonington.

"Elephanting," however, was also practiced upon *Mirounga angustirostris* in California waters. There is no evidence to show that this species was ever as numerous as their southern brothers, but the breeding range is still most substantial, running from Point Reyes and the Farallons off San Francisco Bay, south for 1,000 miles into Mexican waters as far as Cedros, Cabo San Lazaro, and the more remote Guadalupe Island.

There is a considerable variety of possible habitats over such a range, but elephant seals prefer island beaches, moving to the mainland to reproduce only when traditional haunts such as Año Nuevo or the Channel Islands are crowded enough to encourage colonization. "Crowding," however, also involves the population of *Zallophus californianus*, the California sea lion familiar to so many from its zoo performances, and the Steller or northern sea lion, *Eumetopias jubatus* (which cares very little for zoos or captivity in general), the southern end of whose range overlaps the northern limit of the elephant seal's. In general, the three species dwell conveniently together, though conflicts are known to occur.[77] Seasonal patterns differ somewhat, even on the same island, as do living arrangements: Stellers like large outlying rocks sea lions sandy beaches near the water's edge, and elephants higher and drier sand.

In the late 1870s, with a population approaching 70,000, the elephants have reoccupied breeding locations once deserted, even moving onto one mainland beach at Año Nuevo. Whether the herd can expand beyond the current size – particularly given the changes which have taken place along the coast and among its inhabitants over the last two centuries – remains to be seen; a good deal depends on the level of density the species is able to tolerate, and that may well depend upon the degree to which they were hunted in the nineteenth century.

Before the coming of sealers here, elephants had few enemies to contend with in California waters. Sharks took some. California Indians, though lacking the same deep-sea traditions of the far northwest coast, took others, and not only from mainland rookeries. The Shoshonean and Chumash islanders of the Santa Barbara channel made extensive use of plank canoes, and the few harpoon heads which survive were far more likely to have been used on sea lions and elephant seals than on the grey whales which frequent this coast. Greys are quick and active, and can demonstrate considerable hostility if harassed, as many whalemen found to their cost; for this reason, they were not exploited in large numbers until after 1846 when other species were in shortage, and whalers found large stocks of greys at Magdalena Bay. Many whalers were along this coast in the 1830s, but further seaward than grey whale migration routes, searching mainly for sperm whales. Spanish mission fathers might have encouraged their hunters to take seals, but it is more likely that if their charges were directed seaward at all, it was for the more valuable sea otter. While elephant oil was useful, the missions and ranchos had at hand another source of fat for most domestic purposes, in the tallow produced by the annual cattle slaughter. The hide and tallow trade, after all, was the principal means of commercial exchange along this coast before 1850.[78]

Though the large-scale fur seal hunting expeditions had ended with the withdrawal of the Russians and the decline of the species on such islands as the Farallons, small parties of fur sealers and otter hunters still worked the coast until nearly mid-century, when other opportunities such as the California gold rush, together with the scarcity of prey, drew them away. No doubt they had taken elephants whenever they desired to do so. Henry Delano Fitch, a San Diego merchant who held a hunting license from the Mexican authorities, wrote to Abel Stearns, also an expatriate merchant, in 1840: "If you want some lamp oil, I can supply you with 30–40 gallons of good elephant oil at a dollar per gallon. I am burning it in my house and think it burns equal or better than sperm." He sent out a boat, and "in four days two men brought in blubber sufficient to make 120 gallons."[79] Illumination, it is true, was a use for which oil was well suited, but tallow candles served as well and were cheaper – and the population before the mid-1840s was never great enough to stimulate a major demand.

Whalers doubtless took elephants when whales of any sort were scarce. So did trading captains such as William Dane Phelps, master of the *Alert* (a Bryant and Sturgis vessel), who found elephant seals and sea lions still numerous along the coast in 1840–2, but not a sign of fur seal. Phelps himself shot elephant seals at the Coronados Islands, some twenty miles south of San Diego, in 1840 – an experience that may have reminded him of his breakfast assignment many years before on Prince Edward. In this case, Phelps intended selling the oil to Stearns (for six rials, or the same number of hides, a barrel). In May of the following year he took some at Santa Barbara Island, though they were very thin (it was the pupping season); when he returned in September, he "found the landing place (a small beach) so thickly covered with Eliphant & seal that we had to wait sometime for the seal to clear out before we found room to haul the boat up," though those he took were still not very fat.[80] The implication is clear: the elephant seals were there for the taking in the early 1840s, even by a hide-and-tallow trader who took a few seals only as a sideline. Similarly, the ship *Stonington* of New London found seals at the Coronados in 1846; the Stonington ship

America and the New Bedford ship *Magnolia* took elephants at both Cerros and Guadalupe the same year.[81]

Twenty years later, there were no elephants left to exploit. From 1846 onward for a decade or so, they were cleaned off by sealers, presumably both local – based on San Diego, Santa Barbara, or San Francisco – and deep-sea multi-ocean voyagers based on America's east coast or Hawaii. Hard evidence for this harvest is lacking, and can only be inferred from the easy availability of seals at the beginning of the period and their near-total dearth at the end. Several explanations are possible for this fairly sudden rush, though undoubtedly elephants had been taken in small numbers since near the beginning of the century. First, with grey whaling now a serious business, a number of whaling vessels were to be found in elephant-occupied waters, and they were hardly likely to reject such easy prey so close to the semi-permanent camps they established along the Baja coast. J. Ross Browne, in his *Sketch of the Settlement and Exploration of Lower California* (1868), reported five or six such encampments, particularly around Magdalena Bay, established in the mid-1850s. "In some years there are reported to have been not less than thirty different whaling and sealing camps below San Diego, aggregating some 2,000 men; and as seals and the affiliate families are in the greatest abundance, cargoes often prepared with great rapidity."[82]

Second, the population of California grew rapidly after 1846, and even more rapidly after the gold rush began. Demand for oil was high in the Sierra Nevada to light the miners' tents and saloons: very likely the problem then was to find sealers who had not gone off to seek their fortunes in the diggings. In the mid-fifties, sealing was still possible, as shown by the account of Prentice Mulford. Mulford arrived in San Francisco by clipper in 1856, and shipped as a cook on the schooner *Henry* (totally without experience, by the voyage's end he was up to stews but little more) on a mixed voyage of ten months down the coast which brought back among its cargo 500 barrels of elephant oil (Mulford, on the strength of his 1/50 lay, then headed for the mines).[83]

But from 1865 to 1880, only stragglers were seen over the entire range, a few each on Islas San Benito and Guadalupe. In 1800, discovery of a small herd at San Cristobal Bay, Baja California, led to a small-scale revival of sealing: 300 were killed in the next four years. But from 1884 to 1892, not a single elephant seal was seen anywhere, although several museum-organized searches looked hard enough. In 1892, a Smithsonian expedition under Charles Haskins Townsend located precisely eight seals on an exposed beach of Guadalupe Island. "Some of these elephant seals were secured," wrote Townsend, meaning that he killed seven. As Burney Le Boeuf, a leading authority on this species, has written, "This was unquestionably the nadir in the history of this species, the low point in its population history." In 1892, there may have been as few as twenty animals surviving in the entire species, and probably not more than 100. It is important to realize that the entire modern population has been reconstituted from this tiny remnant.[84]

After 1892 came some recovery, but very slowly. Charles Miller Harris saw a small herd in 1907 (he shot ten for the Rothschild museum). A joint Mexican-American expedition counted 264 seals on the whole of Guadalupe in 1922. The Mexican government, wisely concluding that the seals were likely to be exploited again, placed the species under complete protection – not simply by legislation, but also by stationing a small garrison on this waterless rock (previously used as a penal colony for fifty convicts in the 1880s). The officers' quarters of this rather unattractive billet featured a rude sign, bearing the warning, in English and Spanish, "Prohibit by law kill or capture elephant sea" [sic].[85]

Thus protected, elephant seals began to spread again into United States waters, where they were also given legal sanctuary. In the 1930s, elephants were breeding at some of the islands closer to the Baja coast (San Benito, San Miguel); in the 1940s, at Los Coronados; in the 1950s, the Channel Islands; in the next decade at Año Nuevo; and finally, in the 1970s, at the Farallons. The population was estimated at 13,000 in 1957, tripled to some 48,000 in 1976, and today is roughly 70,000.[86]

This recovery had occurred at an amazing rate, "one of the most spectacular mammalian population recoveries for which documentation is available."[87] Le Boeuf has speculated, however, that there are some ominous overtones. Though the species has returned from "extaction," a state approaching extinction, it has done so with a gene pool which is considerably smaller than that of the southern elephant seal species. The recovery has had a "bottleneck" or "hourglass" configuration ("genetically depauperate" is the proper scientific term) which occurs when a substantial herd has been reduced to a mere handful from which a large recovered herd must all descend. It is possible, of course, that northern elephants always had a smaller gene pool, but it seems more likely that this distinction comes from the fact that in a population of 20–100, there would be only one or two breeding bulls.

The result of this limited gene pool might be, as happens in such cases, a lack of adaptability to changed conditions – chemical or petroleum pollution, for example – and the species which has expanded so enormously might be in jeopardy of dwindling just as rapidly, a "flush-crash" relationship more typical of lemmings or other species in which such a phenomenon has occurred. While it is a hopeful fact that species with such a history are customarily lower on the trophic ladder (food chain), still a sudden "crash" in the elephant seal's food supply, for whatever reason, could be disastrous. Of course, genetic variability has not proven totally necessary for species survival – the case of the northern elephant seal bears testimony to that – but there is no guarantee that it might not be necessary in the next crisis.

A more tangible issue, at least at the moment, lies in the elephant seal's relationship to man. Protected by law as a marine mammal, and in addition inhabiting principally natural refuge areas, the elephant seal is not now directly threatened by human confrontation (so that chemical spillage, for instance, belongs to the category of "indirect" threat). But when *Sunset* magazine revealed, in December 1973, that elephant seals had reappeared at Año Nuevo beach, the response was not particularly encouraging.[88] Various forms of harassment and casual brutality (such as cigarette burns), required official intervention to control man/seal contacts on the beach, and to protect each species from the other (for elephants, particularly bulls, are not entirely without defenses). The California public has never resisted such control, for Año Nuevo beach is remote and most admit the need to protect the species, encouraged by the sort of anthropomorphic description popular among writers such as Jacques Cousteau. "Their globular, sunken eyes, for example, often have an unexpected look of intelligence ... the eyes which are usually brown, also convey sadness, and even supplication, when an elephant seal is disturbed on the beach and forced to fight, or to flee despite its crushing weight. In the sea, the animal's look is more alert than on land, and also more mischievous and more self-assured."[89]

Such subjective writing serves a purpose in evoking public sympathy for conservation efforts (in addition to selling books and supporting Cousteau's various projects). So long as that purpose is understood, Cousteau may be forgiven for the confusion which such attitudes bring to the serious issue of understanding – and indeed protecting – the species. But how far does such sympathy, evoked by the written word or televised science/travel accounts, or even organized tours to observe the Año Nuevo elephants, extend? Should the elephant seek to take up residence on beaches popular with summer surfers and bathers, a wider range of responses might be met. Sea otters provide something of a comparative object lesson, as will be seen in the next chapter. Perhaps even more ominous, at least to the tourist industry, is the clear indication that the recovery of California seal populations has brought with it the growth of a predatory shark population, the prime example being the great white shark, which from time to time finds a kicking surfer paddling on his modern short board indistinguishable from a tasty meal of elephant seal or sea lion. Contrary to popular mythology, most such attacks occur in clear water: the shark thinks it has found a meal.[90] It would be premature to assume that each participant, surfer, seal, shark,

will quietly accept its niche in the environment, for, after all, that
environment will never return to what it was in the days before civilization found California.

For the moment, it must suffice to estimate the take of northern
elephant seals at an arbitrary figure of 250,000. If we assume a base
population of perhaps 100,000, and heavy sealing (5,000–10,000 a year)
in at least 1840–60 (150,000), with another 100,000 for all the seals
taken in the remainder of the century, we reach a figure which is roughly
3.5 times the current population, for a century's harvest. This would
have been enough to support a minor industry – even fairly substantial
in the critical mid-century years – but not enough to make an impact
even of the extent of the New London Kerguelen elephanting. Once
again, this is guesswork at best, but until a better calculation arrives,
we will leave the total elephant seal catch at 1,250,000.

The disappearance of the elephant seal from the coast of California,
Alta and Baja, was not the complete end of the independent small-scale
coastal entrepreneur whom we may call "sealer" for lack of a better
term. After 1865 his prey was likely to be one of half a dozen possible
choices among seal species, whales, and other marine products, but at
this point, such voyages need to be considered in relationship to the
last of the fur seals.

CHAPTER SEVEN

The Last
of the Fur Seals

*The appearance of this iron-bound coast, cleft
asunder as it were into harbours by some awful
confusion of nature, presents a scene truly grand
and solitary ... The mariner has little inducement
to seek anchorage here, unless it be the whaler
who would take refuge from the coming gale
when he has not sufficient room to keep the sea,
or the sealer in the legitimate pursuit of his
calling.*
G. H. Richards and F.J. Evans, eds.,
New Zealand Pilot (1856), quoted in A.C. and
N.C. Begg, *The World of John Boultbee*

The voyage of the Boston sealing brig *General Gates* (Winthrop and
Company, 197 tons; Abimelech Riggs, master) at first sight seems
unusual only for its length and for the occasionally disastrous adventures
of the crew.[1] Sailing in October 1818, the *Gates* visited Prince Edward
(where, the sealing world being small, she gave a dog and a cat – to
help against mice – to Phelps and his crew), left her own gang on St.
Paul and Amsterdam islands, and arrived in Sydney with a reduced
crew in June 1819. Local authorities refused permission to recruit labor;
Riggs took a dozen illegally, was subsequently caught and fined, but
went sealing anyway, reaching Tahiti in November 1821, with 11,000
skins on board. From Tahiti, the *Gates* made for Canton in the spring

of 1822 to sell her skins. Two years later she was still in the Pacific, in the interim, it is to be hoped, having revisited her people left on Indian Ocean rocks. A gang left in New Zealand faced starvation and Maori hostility: a number of crewmen were captured, and several eaten, a very real risk in this otherwise attractive land (the *New Zealand Pilot* to the contrary). The crew went seventeen months without sight of Europeans, at which point the survivors were rescued by a small colonial vessel.

Altogether, the voyage of the *Gates* is an extreme example of the hardships and difficulties of tracking the last of the fur seals in the early nineteenth century. But there is another reason for taking note of the *Gates*: after leaving Hawaii for China, the brig almost certainly paused at the Leeward Islands to take whatever skins she could find in these tropical waters, thus very probably creating havoc among the small population of *Monachus schauinslandi*, the Hawaiian monk seal. [2]

Once there were three species of monk seal: Mediterranean (*M. monachus*), Caribbean or West Indian (*M. tropicalis*), and Hawaiian, all probably descended from a single Mediterranean stock, drifting westward from the Canary Islands and somehow transposed across the Isthmus of Panama. At least the Mediterranean and Caribbean stocks were once common. The former often swims through the literature of classical civilizations but today is very rare, numbering in all only a few thousand in a range which extends as far west as Madeira in the Atlantic and east into the Black Sea. The Caribbean stock found cohabitation with man most difficult and is extinct or nearly so. The last authenticated sighting was in 1952; the scientific expedition which killed forty on Triangle Island west of Yucatan in 1886 probably finished off the species forever. The Hawaiian herd is marginally better off with probably fewer than 1,000 animals; careful censuses in recent years have not found more than 800.

All monk seals have in common a preference shared with few other species for living in warm or tropical waters. To accomplish this, some body adaptation is no doubt necessary, but little is actually known of their biology. Apparently the metabolic rate is adjusted to compensate for heat; adults of the species (both sexes measure 7–8 feet and may weigh up to 500 pounds) have a lower temperature on land than in the water, despite the cooling capacity of the latter environment. When on the beach, the monk seal may dissipate heat through radiation into the moist sand into which it wriggles. If so, comparison with the thermoregulatory systems of harp seals shows just how varied adaptation can be within one animal order. [3]

But a good deal of this is speculation, for the species is most difficult to study. Although it is not known to be migratory, even on a seasonal basis, its territory is very small in total land area (about 3,400 square acres of reefs and atolls); this chain of still often uncharted and dangerous reefs is spread over 2,200 miles. The monk seals, few in number to begin with, are solitary and wary, seldom grouping in any significant numbers. They are easily disturbed by any intrusion on the beaches which they inhabit by day, feeding at night on various reef fish and squid

As might be expected, details of the numbers and timing of exploitation are very hazy, though it is certain that the population was never of a size in the sealer's day to require calculations of the sort made for other species. The *General Gates* is the best candidate for the first sealer of record here, but most accounts, basing their remarks on William A. Bryan's classic *Natural History of Hawaii* (1915),[4] refer to the brig *Aiona*, which in 1824 brought skins and oil into Hawaii and thus was better recorded than the *Gates*, which did not return to the main islands. The monk seal is not a fur seal, but the skin may be used for leather; for sealers, this species was primarily a source of oil.

Monk seals apparently were not thought worth hunting for many years, until the bark *Gambia* came into Hawaii in 1859, according to a report in the *Polynesian*, with 1,500 skins and 240 barrels of oil. This cargo record has been seriously challenged by Karl Kenyon and Dale Rice, who point out that in the 103 days the *Gambia* was at sea, it would have been most difficult – a polite way of saying impossible – to take 1,500 monk seals from those atolls, given the species' low density and wary habits. Since *Gambia*'s log is presumed to have been destroyed in the San Francisco fire of 1906, the truth will never be known, though the *Gambia*'s niche in history is secure: on that voyage her master, Capt. N.C. Brooks, formally discovered Midway Island.[5]

Whoever the sealers were, very few monk seals remained after mid-century. They had to be sought out to be seen, although a rare straggler might turn up near the main islands, such as the one seen in Hilo Bay in 1900.[6] In general, Hawaiian natives had little association with this species. A few were seen on the various reefs – Midway, Laysan, Hermes, Pearl, French Frigate Shoals – in the early twentieth century, and by 1915 the population was estimated to be about 400 and virtually extinct.

In 1909 Theodore Roosevelt declared the Leeward Islands to be a natural refuge (in 1940 it became a National Wildlife Refuge). However, in addition to the whalers, sealers, and bird-hunters who disturbed the seals, Kure was briefly in U.S. Navy hands, and Midway, a coaling station in the 1870s, became an important naval installation and locale of a critical World War II battle. The monk seal population recovered somewhat, nevertheless, hitting a high of perhaps 1,200 in the 1950s. In the 1980s, 800 is a more realistic estimate. Something, whether direct human activity, or pollution, or some undiscovered factor, is at best limiting population expansion and at worst precipitating a downward spiral towards extinction of this species, officially "endangered" since 1976.[7]

The Hawaiian monk seal was only one among the Pacific Ocean species inhabiting the central latitudes which were exploited despite a small population base and yet survived. *Arctocephalus galapagoensis*, the fur seal of the Galápagos Islands, is yet another example. As previously noted, fur seal nomenclature on the western coasts of the Americas is still confused, owing at least in part to the paucity of examples for study. Sealers did not care whether *galapagoensis* was a subspecies of *A. philippi* of Juan Fernández or *A. australis* of the South American coast, let alone related to *A. townsendi* of Guadalupe, and we need only note the issue in passing. Whatever their precise relationship, all of these island species have small populations today. Unquestionably Más Afuera once had a sizeable herd. There is less certainty about the Guadalupe population, unless it is conceded that all fur seals taken from Islas Revillagigedo to the Farallons, including the Baja and Santa Barbara Islands, were of this species. But the Galápagos herd was probably never very large.[8]

The Galápagos, like the other islands along the western coast, support a seal population because of the cold Humboldt current which passes through them. Because of the climate of this latitude, however, fur seals and sea lions both need windward rocks to facilitate cooling, and the fur seals, less widely distributed among the thirteen larger islands (and many smaller, some still being created by the volcanic process which explains the presence of the entire chain), also require caves for shade and shelter. There is enough congenial habitat among the 3,000 square miles of islands (scattered about 40,000 of water) to have preserved at least a remnant population. Ecuador, some 600 miles away, owns the chain (officially named Archipiélago de Colón) and does its best to preserve the natural life. The islands are of unique interest to nature-minded tourists, and 2,000 visit them each year.[9]

Sealers had no such protective interests and were willing to travel halfway round the world, as did the Madras-owned snow *Harrington*, which left Sydney in 1804 bound, as the master wrote the local governor, to collect skins at Más Afuera and "the Galapagus Isles."[10] Fanning, who was here in the ship *Volunteer* on her voyage of 1815–17, was simply one of many when he collected 8,000 fur and 2,000 hair skins (the latter *Zalophus californianus*, the California sea lion) on his way to complete his cargo at St. Mary's and the Falklands, pausing, as did so many visitors to the Galápagos, to capture as many tortoises as possible to augment his food stocks.[11]

Little record remains of mid-nineteenth century sealing here, but several voyages of the 1870s and 1880s are known. A Capt. C.W. Reed took 6,000 skins in four voyages between 1872 and 1878; Capt.

W.P. Noyes of San Francisco made several trips in the 1880s in his schooner *Prosper*. Noyes shot his seals both on land and in the water, and his losses in the latter case were considerable. Noyes's voyage in the *Julia E. Whalen* of 1898–9 was probably the last. The survival of the herd, even in small numbers – probably due to the presence of many volcanic caves – had at least the effect of demonstrating to David Starr Jordan and others that, regardless of ethical considerations, the proposed total extermination of the Alaska fur seal herd was a practical impossibility.[12] In the mid-twentieth century, only a remnant population of perhaps 1,000 is left, concentrated on Isla Genovesa (or Tower Island).[13]

The Guadalupe fur seal, more closely related to his brothers at Juan Fernández than the Galápagos version (the head is flatter and the snout more pointed – collie-like, in fact), had an even more dangerous passage through history, similar to that of the elephant species of the same island. Once there were many Guadalupe fur seals, proof of which lies in the rocks of their preferred rookeries, polished smooth by the wear of passing seals over countless generations.[14] At one time, perhaps 30,000 fur seals were found on Guadalupe, enough in any case for substantial hauls by a few vessels early in the nineteenth century. An English sealer, the *Port au Prince*, took over 8,300 in nineteen days in 1805; the 600-ton Boston ship *Drome* another 3,000 in two weeks in 1808, but this was from what its crew called "Shelrack's Island," perhaps Sorocco in the Revillagigedo group (and very likely the same species).[15]

The seals of course were soon savagely reduced, though sealers continued to visit Guadalupe periodically for several decades, as testified to by inscriptions dated 1834, 1835, 1839, 1849, 1851, 1869, 1873, and 1881: the declining frequency shows how seldom sealers came after mid-century. Once it was big business, to which the remains of sealers' huts and storehouses bear mute witness.[16] But even in the last quarter of the century, sealers still came. Charles H. Townsend, who at the time interviewed all the sealers he could find, calculated that 5,575 seals were taken here in the years 1876–1892, though his data for this calculation seem rather scant. Capt. George N. Chase took 217 seals here in 1878–80, and sold 114 at San Diego for $1,600, and another lot of 74 for $1,300 – hard evidence for at least that much of a catch.[17]

But seals were taken by others besides sealers. At some point in the nineteenth century, whalers (more likely culprits in this case than sealers) introduced goats to Guadalupe as a food supply, with disastrous results. What was once a paradise to the naturalist (and was so described in 1875 by Dr. Edward Palmer) became a wasteland, and many of the birds seen by Palmer were never seen here again. The goats were better controlled – after the fact – in the twentieth century, when they were killed for pelts and tallow, though generally sorry specimens even for that purpose. For the island, the result was "a most striking example of the utter ruination which man, within comparatively few years, is capable of effecting in nature's long-developed scheme."[18] Unfortunately, goat-herders on the island had every reason to kill fur seals encountered in their rounds for supplemental income, as reported by another scientific visitor in 1897.[19]

That visitor, Dr. W. Thoburn, could find no sign of living seals (but plenty of goats), and indeed none was seen in this "biological sepulcher" until the late 1920s, when fishermen from San Diego spotted a small herd and reported their find to the San Diego Zoological Society. The society's scientists quickly took up the chase, only to find that the fishermen had killed off the herd of about sixty seals and taken the skins to Panama for sale – perhaps after a falling out with the biologists over their share of the fame (or the reward). Not until after World War II were fur seals again seen on Guadalupe, and then only in very small numbers.[20]

As was the case in the Galápagos, species survival depended upon the caves which line Guadalupe's shore. Sealers had no qualms about following seals into their lairs with torches to hunt them out, but in the caves it was difficult to catch all the animals. This form of seal adaptation is not unique and in fact explains the survival of the grey seal (Halichoerus grypus) on the Welsh coast, where seals, in order to avoid the higher pup mortality from gales in such exposed cave rookeries, take to the water considerably sooner after birth than their Scots or Farne Island relatives – an interesting case of adaptation.[21] To put it more scientifically, of the Guadalupe herd "only a nucleus with secretive behavioral traits may have persisted, thereby modifying the behavior of the surviving population."[22] Such modification might have interesting side effects. What happens, for example, to visual display in the life history of the fur seals which now live only in dark caves when not at sea? The fact of becoming cave-dwellers could alter the entire behavior pattern of the species. The likelihood of this question being studied sufficiently to be answered is small, however, for, although protected, the total Guadalupe population is under 1,000; there are signs that the species is recovering, but it has far to go.[23]

These seals might have survived in larger numbers were it not for the fact that casual sealers had throughout the nineteenth century a succession of alternate cargoes which insured that at least a few small vessels were always active in the trade. There was nothing to prevent a fisherman from becoming a part-time sealer if the potential reward was great enough. The most lucrative prey was always the sea otter, though by the 1890s it was generally considered extinct, or virtually so, and the few remaining hunters had mostly turned to other pursuits.

Because otter-hunting in a way kept seal-hunting alive, and because the otters offer a story in protection which is very relevant to seals, it is worth giving a moment's attention to this trade. Otter-hunting was a serious profession at least until the California gold rush among such professionals as George Nidever and Isaac Sparks.[24] These hunters used "otter boats," double-ended 15-foot surf boats adapted to the old tradition of following the coast and beaching nightly, propelled by short-bladed paddles which were serviceable in the kelp beds frequented by otters. North of San Francisco Bay otters were more commonly shot from the coast by waiting hunters – a wasteful process – and in Washington from three-poled 20-foot towers built on the surf's edge (some moveable), on which the hunters waited, employing Indians to swim out in the surf for animals once shot. Pursued with such determination, it is little wonder that otters, whose numbers were never as great as the pinniped populations, were soon scarce indeed.

Until California joined the United States a local license, or *permiso*, was required to hunt otters, though as might be anticipated this requirement did not stop determined poachers, grouped under the general term of *contrabandistas*. Hunting privileges might also be based upon the license of another, who hired it out. Sealers also at mid-century still traded for otter skins with Indians along the Baja California coast. The decline of the otters together with the gold rush much reduced the hunt after 1849 but did not entirely end it. Some hunters soon found life tiresome in the mines and returned to the coast, perhaps to hunt elephant seals (but always with an eye out for otters). George Nidever's son followed him in the profession, as did his sons in turn. Nidever himself was famous for "rescuing" the last Indian woman on San Nicolas Island in 1853: she had lived there alone for eighteen years but died a few weeks after exposure to the wonders of civilization. Nidever continued ottering into the 1870s; his sons at the end of that decade found a "raft" of over 100 at the San Quentin Islands, worth $50 each.[25]

By the end of the century, the price for a prime otter skin in London had climbed from $40 in the 1840s (a goodly sum at the time) to $400–1,000 and double or triple that according to some sources (sales were scarce) in the first boom years of 1920s.[26] But there were few otters left to sell. As a sympathetic but fortuitous afterthought, a clause protecting otters from hunting was included in the Fur Seal Treaty of 1911. Few of the negotiators could have predicted otter recovery: the total population then from the Aleutians across a thousand-mile otter-free gap to California was probably 1,000–2,000 animals, from a base which was at least 100,000 in Alaska alone before the arrival of the Russian hunters. Protection helped, but with such exorbitant prices poachers poached and buyers bought: Harold McCracken witnessed a buyer from San Francisco openly touring Alaska outports in 1922, paying $300 as a local price.[27]

The small boom soon disappeared, and protection efforts helped restore the Alaska herd to 20–30,000 in the 1960s and 100–140,000 by 1980. A population of that size permitted experiments in transplantation to British Columbia, Washington, and Oregon, where there were no otters left at all, in 1969–70. The projects, despite considerable effort, were of questionable success, though that in British Columbia has the best record. Otters are among the most difficult of creatures to transport because of the trauma of capture and travel and the overriding need to keep their pelts free of any contamination. Transplant is also handicapped by the otter's homing instinct. But losses in these experiments were surely warranted: far more (about 1,000) were killed in the Amchitka atomic test of 1969, which was in fact one incentive to capture and move the otters. The Alaska population is thriving, so much so that 200–300 were allowed to be shot and sold in the 1960s, though that particular experiment has not been repeated. [28]

In California recovery has occurred, but not at all to the same degree. From 1916 almost to World War II, very few were seen at all. Then in 1938 a small herd of fifty or sixty was seen fourteen miles south of Carmel (near Monterey). Though verified by biologists from nearby Hopkins Marine Station, the discovery was for the most part kept secret, and the war soon provided other distractions. [29] Given that start and legal protection, in the subsequent decades the California herd has expanded to over 1,000 – but probably not the 1,800–2,000 commonly cited. Either way, it is still a small population and unlikely in the short run to grow larger. Otters are territorial, and the density along the California coast is about a dozen per square mile (it is considerably higher in the Aleutians). With the reproductive rate for otters estimated at 2.5–5 per cent per year (otters give birth every two years, so their rate is about half that of fur seals), expansion of the herd from its protected Monterey reserve, established in 1941 and expanded in 1959, is only a few miles each year at best. [30]

When the otters reached Morro Bay in the 1970s, expansion slowed, and in the early eighties it stopped. Morro Bay has the first substantial open beaches south of Monterey, and they form one obstacle: otters do not normally frequent such waters. More important to some Californians is the fact that Morro Bay – specifically the town of Pismo Beach – is a center of the speciality trade in the meat of the marine snail known as abalone. Abalone have been a popular delicacy since introduced by an innovative Monterey restaurant about 1915 (before that time, they were taken in quantity only by Chinese, mainly for export to China). [31]

Abalone are taken commercially by divers – Santa Barbara has long been a major base for this industry – working from small boats, either anchored ("dead boats") with one man to tend the diver, who may be

100 feet down, or moving above the diver ("live boats") with a third man to shift the boat in the diver's wake. The size of the industry has been substantial, with forty or more divers working from Santa Barbara alone in 1962. For Morro Bay, the peak year was 1961, when 1,597,844 pounds of red abalone (the preferred type) were landed. In 1970 only 163,306 pounds came in, and by the end of the decade there was no more commercial abalone fishing in Morro Bay, though the going price was $100–150 a dozen. The otters had arrived, and the abalone fishermen have proved determined opponents of otter range expansion. To abalone divers, the otters are responsible for declining takes; to "The Friends of the Sea Otter," overfishing is even more at fault, and the organization correctly points out that the decline of production south of the Mexican border and off the Channel Islands has nothing to do with otters, who are not found in either area. Logic has not always prevailed, however; some otter-haters have taken matters into their own hands, shooting as many as possible. Even assuming equal rights for otters and abalone hunters (an assumption many people on both sides are unwilling to make), this wastage is in the long run short-sighted and demonstrates considerable ignorance of near-shore ecology.

Otters eat abalone, it is true, and where the otters return, the abalone quickly dwindles. But the abalone is something of a parasite, whose population will "flush" – that is, suddenly expand at an unusually high reproductive rate – only when its natural enemies, otters above all, vanish. Worse, where otters decline, sea urchins also flourish, and sea urchins, because they sever kelp roots, tend to be very destructive of kelp beds. Conversely, kelp beds quickly return when otters are present. Abalone do not eat kelp but rather feed on drift benthic algae which inhabit kelp beds. In other words, if otters go, in the long run, so may the kelp, though the association is complex and may be affected by other factors such as commercial fishing for sea urchins (the roe of which is a favored Japanese delicacy) as well as harvesting of the kelp itself. Still worse, however, if the kelp goes, so do the fish and kelp-crabs which make it their home and provide food for inshore seals, sea lions, and sea birds alike (it should be added that crabs and lobsters or cray-fish also feed on urchins, and overfishing them has an intertidal effect similar to that of otter removal). One other factor to consider: the otter also fancies the delightfully edible Pismo clam, and where the otters have returned, the clams have vanished (though otters have not yet occupied the whole Pismo clam range by any means). As with abalone, however, entire beaches have had to be closed to clammers due to overexploitation of clam beds well before the return of the otters. Nothing, alas, is simple about these relationships.[32]

The overall conclusions are, first, that substantial alteration in the population of a top carnivore in the trophic chain has important effects on near-shore ecology. Second, abalone hunters might do well to think of the complex long-run balance of man, seal, otter, lobster, urchin, abalone, fish, and kelp. This is admittedly to ask abalone hunters to forget the pre-World War II days when an artificial imbalance existed, and the abalone was there for the taking. Though abalone devotees, like those belonging to "s.o.s." (Save Our Shellfish), will not soon advocate otter range expansion, there is evidence that some have come to appreciate the question's complications.

The otters are still holding on, but only just. The Marine Mammal Act of 1972 prohibited killing of otters (already protected by California law since 1916) and in 1977 the species was declared "threatened," which is not the same as "endangered." Threatened status was granted because it was felt that while the population seemed to be recovering, there was still serious danger of oil spills which could quickly jeopardize the whole herd (what was envisaged was a tanker accident or something of the sort; no attention at the time was given to offshore oil production). Considerable discussion was given to the issue of whether the California and Alaska otters were one separate species divided by a 2,000-mile gap, without resolution of the issue; there is insufficient evidence to prove distinctness, though there are differences in habits and life history. As Victor Scheffer has commented, the question is "depressing," since the issue, as recognized by federal legislation, is whether this species is or is not at risk in this particular habitat.[33]

And in this habitat it is not flourishing. A careful population count of 1,194 otters made in late 1982 over the entire 200-mile range from Pismo Beach to Santa Cruz compares unfavorably with a count of 1,561 made in 1976. There has been no northward expansion since 1977; 1982 was the first year in a decade in which there was no southward expansion. The population is not growing, and given its vulnerability to the perils not only of petroleum, but of such pollutants as cadmium, mercury, and PCBs (known from tissue studies of carcasses washed ashore), and to the growing shark predation, it can only be concluded that the verdict is not yet in.[34]

When the otters dwindled in mid-nineteenth century, there were elephant seals to turn to, until by the mid-1860s few of that species were to be found either. Then, at least by the 1870s, it was the turn of the sea lions. Sea lions, because of their fondness for near-shore rocky haunts and because they do not engage in seasonal migration – though they may move limited distances and indeed are capable of shifting rookeries altogether to a new locale with greater ease than many other seal species – have long been hunted by the Indians of both North and South America.[35] Commercial utilization, however, had to await the exhaustion of more desirable species. In general, sea lion pelts are of low value and sea lion oil of lower quality than that of elephant seals, to say nothing of requiring considerably more labor: three or four average-sized sea lions are required to make one barrel of oil, an important consideration particularly in mid-century California, "a country where normal labor was so highly valued," as Charles Scammon put it in a brief note on sea lions.[36]

In California two species of sea lion occur, overlapping between San Francisco and the Channel Islands. The larger is the Steller sea lion, *Eumetopias jubatus*, which inhabits northern Pacific waters of Japan, Russia Alaska, and the American coast as far as California. An adult Steller bull may be 10 feet long and weigh 2,000 pounds; the considerably smaller female is more likely to weigh 600 pounds. the California sea lion, *Zalophus californianus* (itself possessed of three subspecies, Japanese – now extinct – Californian, and Galápagos Islands), in both sexes is smaller and shorter. Aside from size, to the casual observer the principal distinction is the sharp bark of the "black" or "barking" seal, as the California sea lion was called, which is quite easily distinguished from the lower-pitched roar of the Steller. The term "sea lion" in the nineteenth century generally meant the Steller, not the California, species.[37]

Unfortunately for both, while the demand for seal oil fell off in the last third of the century owing to competition from petroleum oils, pelts became of value when a use was found for them in the production of glue though the "value" was only 5 cents a pound or less. A steady if minor demand also existed for trainable California sea lions from the Channel Islands for zoos and aquatic shows. Several Santa Barbara captains made a decent living at this work. It took considerable skill to retrieve live and undamaged animals from rough-water rocks, and they were as a result worth several hundred dollars each, sometimes considerably more – though half or more died on the train journey east.[38]

Even after harvesting in considerable numbers, sea lions of both species survived, though the Steller range shifted to some extent northward. By the end of the century the Steller bred in the Channel Islands only at Richardson's Rock at the northern end of San Miguel Island, where it is directly exposed to the colder southerly current (fur seals lasted longer here than anywhere else on the islands).[39] Still, whether sought for captivity or for pelts and oil, the sea lions had to be taken where they could be reached in order to get blubber and skins off to a waiting vessel, so that inevitably some were left on the least accessible rookeries. In the last third of the century, another use developed which required only that bull elephants or sea lions be killed, and that the hunter reach the animal to remove just a few pounds of "trimmings."

The trimmings trade was very destructive.[40] A set of trimmings was composed of the genitals of bull seals – penis and testicles – the gall bladder, and the whiskers (at first individually extracted, later cut off complete with cheeks or lips). The genitals were used for medicinal and aphrodisiacal purposes by the China market and the gall bladder for a sort of medicine. The whiskers, due to their sharp points and wonderful flexibility, made excellent opium pipe cleaners or even tooth-picks, sometimes elaborately mounted in gold. The whole "set," which could bring $2–5 in San Francisco, weighed only a pound or two, and unless the skin was also taken for glue the entire remainder of the animal – hundreds of pounds of meat and blubber – was simply left to rot.

By the late 1870s the trimmings trade was in full swing. Wherever possible, the sealers, operating out of Santa Barbara, Ventura, and San Pedro, took entire sea lion herds. In 1877, trimmings went for $1.25, skin for 2.5 cents a pound, and oil for 37.5–60 cents a gallon (a big 1,000-pound bull might make 18 gallons): a bull elephant therefore might be worth $8.00 altogether. Scammon, who in 1852 observed the slaughtering of a herd of seventy-five for oil on Santa Barbara Island, found the process disturbing even to his hardened senses (he was, after all, a successful whaling master) when a young sea lion was driven into the hills. "The poor creature only moved along through the prickly pears [cactus] that covered the ground when compelled by his cruel pursuers; and, at last, with an imploring look and writhing in pain, it would hold out its fin-like arms, which were pierced with thorns, in such a manner as to touch the sympathy of the barbarous sealers, who instantly put the sufferer out of its misery by the stroke of a heavy club."[41] Scammon was emulating the romantic and evocative journal-ism of his era (this was published in 1872), but the scene had remained in his memory in the twenty years between the experience and the recording of it.

Where the whole herd could not be shot, the trimmings were still worth taking. The method was for the vessel, with a crew of three or four men, to approach the lee side of the rookery rock or island and drop off a hunter (who had to be a good shot) to creep round to the seals with a .45-caliber Winchester repeating rifle and quickly kill three or four bulls before the herd took fright: simply wound one and they would all make for the water. Then, working his way to the carcasses, he removed what was desired. If the skin was to be taken, it was hauled out to a small boat rowed by a second man who approached as close as he dared to the windward side and floated in a light line which the hunter tied to the skin. Meanwhile the master waited offshore with his schooner, often with only the cook to give him a hand.[42] It was meth-odical, reasonably lucrative, and incredibly wasteful. Yet in the 1890s half a dozen sloops and schooners out of San Diego continued to find such employment in American and Mexican waters, regardless of legal prohibitions.

The effect upon rookeries was disastrous and not simply because every possible bull of every species was killed. As John Rowley, whose efforts in part were responsible for protective legislation, explained in 1929: "when the killing continues at close intervals during the entire breeding season, as it usually does by professional seal hunters, the domestic activities of the animals are so upset that a protracted series of these persecutions extending over consecutive seasons is sufficient to drive the remaining animals from the rookery and break up the breeding station for all time to come." Trimmings-men made an effort to clean up their work only when poaching in federal or state reserves, for example at Año Nuevo (a lighthouse station). As one hunter wrote to Rowley in 1906, " 'We have been towing out a load of floaters [bloated carcasses] around the island [Año Nuevo] and today we sent 30 adrift to Santa Cruz and Del Monte [mainland beaches] to perfume the air for the bathing beauties.' "[43]

The extent of the kill in 1908 so reduced the Channel Island population that Rowley could find only one bull at the height of the season – when he also found 200 salted skins in a railway warehouse in Santa Barbara consigned to a San Francisco dealer. Fortunately, the next year under his urging California passed legislation protecting sea lions in the Channel; Santa Catalina Island to the south already had its herd protected by Los Angeles County legislation, not out of conservationist concern but because the Banning Company, which owned the island and its resort town of Avalon, realized that the local sea lion rookery was a popular tourist attraction.[44]

By the early decades of the twentieth century the sea lion population, predictably, had fallen off badly. A census of 1938 counted only 5,841 Steller and 2,020 California sea lions on the entire California coast. Just at this point a San Diego company was beginning to harvest unprotected Mexican sea lions to be canned for dog and cat food – but, perhaps assisted by a mild public outcry, the enterprise failed. Protection north of the border was never permanently guaranteed, however, even after the end of the trimmings trade in the 1930s (the Sino-Japanese War had at least one beneficial side effect of destroying that market). Several battles were fought – notably one in 1927 – to remove legal defenses protecting sea lions on the grounds that these animals consume large quantities of commercially useful fish. They do, but according to biologists who have studied the problem the amounts consumed are probably offset by the number of parasitic fish – lampreys, especially – which they also eat.[45]

The best protection was the sea lion's own defenses: his habitual wariness; his willingness to migrate easily, often for long distances (witness the Steller sea lion seen in 1977 at Bonneville Dam, 150 miles from the sea, calmly selecting choice victims from the top of a fish ladder);[46] his rapid reproductive rate; and the fact that breeding stocks often are located in generally protected (and observed) wildlife refuges. North of California, sea lions, far from being endangered, have often been regarded as nuisances. Serious attempts have been made to reduce the herds of British Columbia, Washington, and Oregon, but somewhat surprisingly the practical problem of eliminating unwanted sea lions remains unsolved. "Persuasion" is ineffective: seals appear to be unimpressed by sound recordings of killer whale noises, even when accompanied by models of killer whale dorsal fins. Paying a bounty requires proof of kill, such as a flipper, snout, or jaw. Shooting is inefficient in any case and leaves unwanted carcasses to pollute beaches. Poisoning, assuming a successful method could be found, could have dangerous side effects. Depth charges – a serious proposal at one point – are not particularly helpful to fish stocks. Land mines on beaches, also seriously considered, might injure unintended victims. Patrol vessels with machine guns have been used in British Columbia, dynamite on favored sandbars in Oregon – yet the sea lions, with perverse persistence, survive.[47] In California, however, the Steller stock has suffered an ominous decline from low pregnancy rates and too many stillborn births, probably due to an epizootic disease. Species competition, in this case with *Zalophus*, has also been blamed. The world Steller population is perhaps 300,000, but the count in California of 6,100 in 1927 had risen only by 1,000 in 1958 and in the sixties began to decline: the current population is only a few thousand. Perhaps the Steller is less capable of adapting to changed circumstances at the outer limit of its range; at any rate, its niche is rapidly being filled by the rising California sea lion population, currently 156,000, with about 55 per cent in California and the rest in Mexico.[48] The struggle now is against fishermen, who are convinced that so large a population is a serious threat to an already limited commercial fishing industry.

The relationship of a sea lion stock of this size to limited fish populations is not at all clear. Sea lions are undoubtedly lazy, and they will pick prime specimens from a baited line in order to chew off choice morsels. But the larger question of marine mammal interaction with fish species needs vastly more study than it has received before such questions will have clear answers. For the time being, the tendency has been to focus upon gill-netting and the all-too-frequent tendency of seals to become enmeshed, often fatally, in synthetic monofilament nooses (in 1982 gill-net fishing was forbidden in Monterey Bay, since otters were also being thus entrapped).[49]

Detailed consideration of that subject must be left to others. It is sufficient here to note that sea lions do not seem to have been hunted commercially since World War II, angry fishermen aside, and the story of man's pursuit of the species, while important to its survival, is a small one in the larger history of commercial sealing. For the future, sea lions are protected by both Federal marine mammal legislation and the limits of the Channel Islands National Marine Sanctuary (N.M.S.) designated in 1980 by President Carter (six nautical miles around the islands of San Miguel, Santa Rosa, Santa Cruz, Anacapa, and Santa Barbara) as well as the Point Reyes-Farallon Islands N.M.S. which includes both the Farallons and the adjoining coast, important as much for seabirds as for seals. Both preserves – indeed, the entire sanctuary program – remain under concentrated attack by the oil and gas industry, and it is safe to assume that they will continue to be for some time.[50]

To return to the trimmings trade and the utility of blubber and skins, it should be clear why sealers were generally available to chase down rumors of rediscovered fur seals. Small-scale enterprise continued at least until World War I, seeking not only elephant seals, fur seals, and sea lions, but also green turtle for meat and shells, grey whales, and abalone for the Chinese market (dried meat) and Europe (the tough outer shell provided a sort of mother-of-pearl favored in decorative inlay work). Always there was a Nidever, or a Sparks, or a Captain Noyes ready to sail for the Galápagos. Prentice Mulford, already met elephanting out of San Francisco in 1856, actually had signed aboard for a "whaling, sealing, abalone curing, and general 'pick-up' voyage," with a crew of Kanakas (Hawaiians). Though it lasted for some years, that life, save for the capture of zoo animals, was gone for good by World War I.[51]

Well before the war, the sealers, expelled from the Bering Sea and finding little to hunt in California or the Indian Ocean, had returned for the last time to the southern hemisphere. Fur sealing in the higher southern latitudes had not been abandoned simply because the rush of the 1820s was over, any more than gold mining in the Yukon was ended when the river beds were cleaned out: there were always a few hardy individuals prepared to stay on for the gleanings, and, with luck, an overlooked horde. The same was true of seals.

While none might be found on the obvious beaches of South Georgia, a lucky sealer would fill his hold from the desolate sands of South West Africa or the dangerous but extensive coastline of Patagonia and Tierra del Fuego (though pelts from that coast, being coarser than Alaska or Juan Fernández skins, commanded a lower price). Though rarely now, vessels came from the old ports. The schooner *Chile* of Stonington searched the Strait of Magellan, and then Mocha, in 1825; the schooner *Penguin*, from the same port, managed in 1828–31 to collect 1,400 skins (400 from Statenland), while a gang left on the rocks called the "Diegos" took another 1,300. In the same years the *Rob Roy* of New London visited not only Patagonia for whales and seals but also the Crozets, Australia, Chatham Island, and the Antipodes, eventually circumnavigating the globe in search of seals. Meanwhile, the *General Putnam*, with Gurdon Allyn aboard, took 2,500 seals from the coast of South West Africa in 1830, returning for similar cargo in 1834–5. In 1831 the Stonington schooner *Hancox* found at least eleven vessels sealing in either Patagonia or Falkland Islands waters.[52]

But these visitors of the 1830s had far more difficulty than the Stonington fleets of the early twenties in filling their holds. By necessity they gathered in catchable whales as well as elephant seals or, from time to time, even sea lions ("hair seals"). They traded for otter with the Indians; they fished; they landed to hunt beaver or guanaco. In a typical sealing contract, it made little difference what the prey was. In the schooner *Betsy* of Stonington, 1842: "each & every seaman & Mariner for him Self agrees to discharge his duty faithfully ... & to obey all lawful commands of sd. master or other officers of Said vessel at all times & place both when on shore during Said voyage employed in taking Oil & Skins or doing any other duty, and when on board sd Vessel or her Shallops in her boats when engaged in any & all their duty appertaining to sd voyage."[53]

By the late 1830s even these voyages had become rarer: there simply were very few seals. When the *Ann Howard* reached the desolate rocks known as "the Evangelists" in January 1837, she found two of her own men and two from the schooner *Pacific*, all four of whom had taken a total for the season of one yearling seal: "their was no seal this season at the Islands."[54] The *Ann Howard* had "mated," share and share alike, with the schooner *Montgomery* that year; the total take of the two vessels was 216 prime skins. The partnership was dissolved, and the *Ann Howard* went off to leave eight men on "Devil's Rocks" and "the Judges" (every small rock or ledge frequented by seals and sealers was known and named), to stay over through the next season. But like the New York brig *Athenian*, "bound on a voyage of adventure & discovery to the south seas," these voyagers had to be prepared to go anywhere (the *Athenian*, in fact, devoted most of her 1836–9 voyage to Patagonian hair seals and elephant seals from the Crozets).[55]

With such small harvests there was little cause to remain, and the sealers moved on, seldom to return for many years to these shores, if the negative evidence of the absence of logbook records is accepted. Occasional exceptions prove the rule, but usually the evidence is inferential, such as that concerning the schooner *General Jackson*. The *Jackson* visited St. Paul in the Indian Ocean in 1883, a fact known only from her rescue of twenty-one survivors there from the English ship *Lady Munro* wrecked while outward bound from Calcutta.[56] Few if any expected to find fur seals in the 1840s or 1850s, and then came the Civil War. The total lack of logbook traces indicates that American sealers did not return to far South American waters until the 1870s.

When they did return, at first it was probably on simple speculation.[57] Certainly this was the motive for the course set by Williams, Haven of New London for its bark *Peru* (already met at the Pribilofs) and attendant schooner *Franklin* in 1871. Both vessels were to make for the Falklands, there to meet with the schooner *Francis Allyn*. Skins, elephant oil, whales: any cargo would do. Capt. G.W. Gilderdale of the *Peru* was equipped with vessels, stores, and crews for two years, and he was not expected to return empty. South Georgia was not recommended, for the owners expected little to be obtained from that now barren island, "but we have considerable confidence of getting a good cut at the Shetlands & are disposed to try it if we can arrange to do so to our satisfaction." If there was nothing to be had at the Shetlands after all, Gilderdale was to go for sperm or right whales off Patagonia "or some other good ground."[58]

Sealing these waters was no easier than it had ever been. On the contrary, because there were fewer sealing ships to lend a hand in case of disaster, and because the most remote and dangerous rocks had to be approached, sealing on the "Coast of Padegony"[59] at the end of the century was probably the most dangerous in the entire history of the industry. The weather was typically horrendous in the best of seasons, and a vessel might spend weeks waiting about, "gales when we want fair and when fair not wind enough to raise a curl from a ladies cheek," as Gilderdale put it. "It requires the patience of a saint and a *most* amiable temper to withstand the harassing and laborious work we have to content with to earn our daily bread."[60]

Bad news, more often than good, revealed the presence of sealers. The Stonington schooner *Charles Shearer* did bloody battle with some Tierra del Fuegans in 1875, costing the lives of two crewmen and four Indians before it was ended.[61] The same vessel left a boat's crew on Diego Ramirez in October 1877 with three months' provisions; six months later, a wasted, scurvy-ridden gang was rescued just in time by the clipper *Jabez Howes* which fortuitously passed close by on her rush to San Francisco.[62] A starving boat's crew from the wrecked schooner *Surprise* of New Bedford was plucked off outlying rocks by the *Thomas Hunt* and taken to Ushawia Mission (Tierra del Fuego) in 1882,[63] and so on.

And this was still at Cape Horn. Approaching the southern islands was no less hazardous. The *Sarah W. Hunt* took thirteen days in 1871 to cover the 280 miles from South Georgia to the South Shetlands and had four of her five boats stove in during the passage.[64] Ice was a particularly serious obstacle, as the *Charles Colgate* found in 1877–8, apparently a particularly bad season, "the worst one that I ever saw the ice is such that we cannot work through it and their is no way to get around it," wrote her logkeeper. The Shetlands were thickly sur-

rounded in early December; the Orkneys were packed in twenty miles
north of the group, and bergs were everywhere: "God only knows
what we shall do." By mid-January the sails were worn out, the crew
likewise; the mainmast had sprung; "Capt [Simon] Church Sick he
cannot sleep or Eat he has worried himself so that he cannot get on
deck."

Turning back, the *Colgate* managed to make the Falklands, send down
and lash her topmast to the mainmast to provide support, and then
crawl home to New London, though her state on arrival was not en-
viable: "about one half of the crew is off duty the most of the time
with the scurvy. The forecastle is worse than a hog pen the watter
runes through the deck and ceeps it wet all of the time and it is not a
fit place for a man to Live."[65] She was lucky. So too, in a different
sense, was the *Sarah W. Hunt* which went round the world in 1883–4,
replacing three full crews in the process, but returned as empty as the
day she left New Bedford.[66]

The schooner *Flying Fish*, a veteran New London sealer, had a different experience. Making directly for South Georgia in 1878, she found
only trouble. Caught in a thick late-August snowstorm (still winter in
southern latitudes), she soon was so iced-over that she was well down
by the head, her pumps unable to cope. The entire crew was put to
chopping ice out of the rigging for days, but despite numerous cases
of frostbitten hands and feet could make no progress: every block was
choked with ice, every sail solid from spray which froze the instant it
hit. By 9 September, with yet another gale in progress, fear had joined
the crew. "If gail continues long schooner will founder. She mosions
[motions] but a very lite [little] to the Sea. The Sea Brakes on her the
same as a Berg of Ice." Two days later, still headed south, she hit pack
ice; the few hands able to work were still chopping ice or manning the
pumps. By the 21st, they had managed to set a close-reefed foresail
and reefed trysail, but she was leaking badly, with oakum hanging out
of her working seams, and her bulwarks stove in in many places.

On the 28th the mates "refused to go enny further South in the
Schooner and requested the Capt to git her to the Nearest land or in
the track of Ships." They handed Captain Dunbar a petition: "Sir We
the undersigned Officers of Schooner Flying Fish do consider the Schooner
leaking to much to Prosad on the Voyag and considder it your Duty
to Prosead to the nearest Port." With the vessel "working like a basket,"
Dunbar at last turned northward. On 3 October he decided to abandon
the vessel; with great good luck the *Flying Fish* had reached the traffic
lanes (42.20° S. by 46° W.) and encountered the Rostock bark *O.
Kohan*, whose master stood by long enough to take the men off, though
one boatload capsized in the heavy seas, "and the last we see of the

little Flying Fish she was Roaling very hevy as tho She was unwilling to give up the Gost."[67]

The journal of William Henry Appleman aboard the schooner *Thomas Hunt* of Stonington provides a last sample of the typical sealer's workday. In January 1875, Appleman tried for nearly two weeks to reach a rock off Tierra del Fuego on which he could see a thousand seals, but each day he could do no more than lay off in a ship's boat and watch his prey, as unobtainable as if they had been on the Pribilofs. On the eleventh day he wrote, "god knows how long it may last returned on board and sat down didn't swear a damned bit."[68]

In November 1873, Appleman had attempted to land two boats with crews and a month's provisions on "Seal Rock" in the South Shetlands. The captain landed with the first boat and ordered the second to lay off:

in attempt to launch again was swamped by an unfortunate Breaker But all finally reached the shore in an exasted state. Saved our boat but lost the oars. Ordered the other boat with third officer to proceed immediately on board and leave us and supposed he had done so But after leaving us and starting for the schooner in fear of not reaching he attempted a landing and the result was the loss of all there cloths most of the privisions all the boats tackling. And nearly there lives. But succeeded in climbing a rock and roosting with the Penguins till the morning and arrived on board in the boat with two oars the following morning in an exhausted and played out state having had no Fire and no means saved of getting one. The first boat myself in command having matches entered a cave killed a leopard, sea dog [probably a fur seal] and two elephants one young and one old. Pealed them [i.e., of blubber] built a rousing fire sent the boys up the hills brought home a half bushell or more of eggs which we fraid boiled and roasted together with variety of liver & tongues we managed to live on the fat of the land through the day and night comfortably.

The next day a boat from the ship supplied the necessary oars. Only once did Appleman show anger in the log: "Patience nearly gone Whiskey having done its work in my absence with Third mate."[69]

Appleman's adaptability was a necessary survival trait, but it was not always enough. Two months later, two boats were smashed in a similar episode and four men lost: "May Lord receive them in his kingdom. Day ends gloomy." The next day Appleman searched unsuccessfully for the men, "and we got underway with saddened feelings to proceed westward on our cruisings. Pity the poor hearts made desolate and May God sustain them in their greaf," adding in the margin, then or later, "This was a sad day."[70]

Complete stoicism was perhaps too much to ask, but a certain fatalism helped. In November of the same year, on a return voyage, Appleman was at Diego Ramierez:

Found on Boat Island sother Diego two graves The first in memory of a young man aged 19 years of Portland Maine name Thompson Buried from schooner Amelia in year 1832 with marble Headstone in good preservation the other with sone of the country No name save the square and compas to commemorate him. Such is life and such is Death. How little it matters them whether buried in a quiet New England church Yard or here amidst the ceaseless roar of South Atlantic and Pacific waves with timultuous cry of legions of Birds and the quiet tread of Seals. And so it is ...

A month later Appleman was building a cairn of shingle rock above the grave of one of his own seamen who had fallen from aloft and died of a broken skull. The place was Low Island, most southerly and least visited of the South Shetlands, where there were truly only birds and seals. "It of course made little difference save for the satisfaction of his friends whether the body was cast into this stormy icy sea or placed on this lonely isle, its only the body, his spirit has gone to the god who gave it."[71]

When the *Thomas Hunt* reached Stonington in May 1875, with 1,600 skins, she had truly worked for her cargo. Her crew would doubtless have agreed with a later logkeeper on the same schooner (1879–80), who recorded on turning north, "We have all had a full dose of South Shetlands and are going home to Cape Horn."[72] But by that time, once again, the seals had been exhausted. Sealers continued to come during the 1880s, but in declining numbers, and hardly at all after 1890, though a few poachers visited South Georgia in the first decade of the twentieth century.

The southern tip of South America, on the other hand, continued to see considerable activity though the 1890s and well into the next century. In part this was due to the growth of small local sealing industries at the Falklands and at the sealers' "Sandy Point" (Chile's Punta Arenas) on the Strait of Magellan. Using local labor, a dozen or so small (20–60 ton) schooners kept up an industry that was on its last legs. Because it really cannot be considered North American, Sandy Point's sealing enterprise need not concern us in detail: the numbers involved were small, the season short, and the profits absorbed by local merchant middlemen who shipped the skins (10,000 in a good year) to Europe by the first available Royal Mail steamer. At the Falklands there were another dozen such vessels; their habits were understandably secretive and their takes unknown, but they did continue sealing in sufficient numbers to prevent any recovery of remnant fur seal stocks for many years.[73]

A second source of Cape Horn sealing in the same era did involve North Americans, this time from Canadian ports on both sides of the continent. Nova Scotia vessels sent around the Horn in the mid-nineties for the halcyon days of northern Pacific pelagic sealing could, and did, pause to take seals from the Falklands and Cape Horn on their way out, as did the *Director*, which took 610 skins from the Falklands in early 1895, though the Japan coast was her destination.[74] Only in the next decade, with the Bering Sea increasingly closed to pelagic sealers, was a serious effort directed specifically at the "Cape Horn Grounds." A. Alfred Mattsson, who himself was wrecked on the West Falklands in December 1903 in a Halifax sealer, explained forty years later how it began: "When word got around that a new sealing ground had been found some of these men [Victoria sealers] returned east and got capital interested in Halifax, and the result was that one or two vessels were fitted out to go sealing in South America in 1900. They must have done well, because next year one or two more were sent out and some of them wintered in Port Stanley ... In 1903 a fleet of eleven schooners left Halifax for the sealing grounds."[75]

The best hauls were made on vessels sent from Victoria, not Halifax. The *Florence M. Smith*, ordered south by the Victoria Sealing Company in October 1902, took 2,300 skins by the following March – a catch equal to more than half the total taken by a dozen schooners prowling the North Pacific. The *E.B. Marvin*, which had departed Victoria ten days later, had 1,100 by the same date. At least another four vessels were dispatched from Halifax, but by the Victoria interests of Capt. Sprott Balcom and associates, not by local Halifax interests. By no means all of these trips were successful, and the rush was soon over, taking in all perhaps 35,000 fur seals, mainly at South Georgia. Some, like Mattsson's second vessel, attempted to fill their holds by raiding Uruguayan rookeries, a procedure not generally recommended: Mattsson spent seven and a half months under guard in Montevideo when the vessel on which he was boatsteerer was seized by a Uruguayan patrol boat with a persuasive 3-inch gun.[76]

Sporadic sealing voyages were made to these grounds until World War 1 but records are very scanty. In the Falklands, a tax on seals – which could only be taken under license – was only another obstacle to be avoided by poaching sealers. Others had similar objections to South Georgia regulations concerning elephant seals. One notable voyage, however, was that made in 1912–13 by the substantial New Bedford whaling brig *Daisy* (383 tons), under the command of a grizzled veteran whaler and sealer, Benjamin D. Cleveland, who had visited Georgia more than once before. The voyage was less remarkable for Cleveland's disregard for the conditions of the license he obtained at Grytviken (it entitled him to take only bulls; he took every elephant seal he could find without regard to size and sex to a total of 1,107) than for the presence on board of a young naturalist, Robert Cushman Murphy, whose several accounts of this voyage began a remarkably successful career. One, written for his wife, is still a classic of its kind: *Logbook for Grace: Whaling Brig "Daisy," 1912–1913*. Cleveland continued sealing as long as possible in his sixties, taking – as noted in the last chapter – the *Charles W. Morgan* to Kerguelen in the midst of war.[77]

Following World War I there was little sealing activity of any kind in these waters. Declining world demand and local regulations combined to permit the southern fur seal stocks to recover. Very scarce everywhere in the 1920s and 1930s, by the 1950s *A. australis* had reached 10–15,000. The next two decades brought considerable growth of the basic South Georgia herd, which today is again established in all its old haunts, with colonies on the South Shetlands, South Orkneys, South Sandwich Islands, and even on the Palmer Peninsula of Antarctica. Estimates of the total population reached 350,000 in the mid-1970s, and projections carried into the mid-1980s reach one million. This rapid growth – nearly 17 per cent per annum – is another case of population "flush," due primarily to the increased availability of the small shrimp-like planktonic crustaceans, in several species, known collectively as "krill" (Norwegian for "small fry").[78] That availability was purchased at the substantial cost of the very widespread reduction of stocks of baleen whales for which krill is also the principal food.

Such discussion touches the frontier of scientific study of ecological systems, in this case that of whales, seals, and krill, though penguins and sea birds are also involved. Nor is it simply a matter of fur and elephant seals: four other species are involved in these southern waters.[79] The widespread but solitary leopard seal, *Hydrurga leptonyx*, has already been met lurking off Heard Island, awaiting the first penguin or seal pup into the water; though useful for oil (females, bigger than males in this species, may be 14 feet and weigh 1,000 pounds), thin distribution of a world population which may total 200,000 or more, and determined ferocity, detract from the leopard's desirability. The Ross seal, *Ommatophoca rossii*, with a population estimated at 20,000, is the Antarctic species least seen and studied; it is a shy type which lives among oceanic pack ice.

The Weddell seal, *Leptonychotes weddelli*, named after sealer/explorer James Weddell who captured half a dozen in 1823, exists in larger numbers, perhaps half a million. This seal is larger, averaging 10 feet and up to 1,000 pounds and, though sluggish, is one of the deepest diving of seals, capable of staying under water for over an hour at maximum effort. Its habitat, the fast ice surrounding the Antarctic mainland (it lives under the ice in winter, using its teeth to cut air holes), has so far protected it from any serious exploitation.

It is the fourth species, the crabeater seal, *Lobodon carcinophagus*, which deserves the greatest attention, if only for its numbers: estimates go as high as 75 million, but even at the generally accepted 15 to 30 million, it is the largest pinniped population by far.[80] The crabeater is similar to the harp seal in that it prefers floating pack ice, though crabeaters are not known to be migratory in the same degree. Long, slender, and about half the size of Weddell seals, crabeaters are much quicker in the water They feed exclusively upon krill, thus competing with whales, penguins, and some fur seals, though several species of krill are involved. The Kerguelen fur seal, *A. gazella*, is a krill-feeder, which may explain why its reproductive rate is nearly 16 per cent, compared to under 11 per cent for *A. australis*, though both are found together at a few islands like Prince Edward. *A. australis* eats a wide variety of foods, and, with a longer lactation period, is able to offer more "maternal care." In theory, its pups should thus survive at a higher rate; in practice, it is *A. gazella*'s quicker feeding on the krill, not care, which matters, and upon such differences are based the replacement of one species by another.[81]

None of these four species has been commercially exploited by American sealers to any great degree, though leopards were commonly taken at Heard and Kerguelen if available. Even had the vast herds of crabeaters been known and reachable, the very common scarring of their pelts (they are not fur seals in the sense used in this book, in any case), principally by killer whale attacks, makes them of value only for oil. Only once in the nineteenth century did sealers take crabeaters in any numbers. The experiment was carried out by six Dundee whalers in the 1890s. Though the voyage was not a failure, when weighed against expenses the returns were inadequate to justify repetition.[82] Not until 1964 was another feasibility study attempted, this time by the Norwegian vessel *Polarhav*; the 1,100 seals taken did not prove an economic cargo. A similar Russian harvest of 1,000 in 1972 had the same result, and crabeaters remain an unexploited population.[83] No doubt some scientists would welcome a small sealing industry involving this species if only because of the probable resulting growth in scientific knowledge.[84]

The international thirty-year Antarctic Treaty, first signed in 1959 (it came into force in 1961), laid the basis for the Convention for the Conservation of Antarctic Seals, signed in 1972 and activated in 1978 after ratification by the majority of treaty states. The treaty bars sealing for Ross, elephant, and fur seals south of latitude 60° south and sets maximum catch quotas (to a total of 192,000) for crabeater, Weddell, and leopard seals in the same area, divided into six separate zones. The number is hypothetical; whether such a figure will prove to be a maximum sustainable yield remains a moot point. Numbers alone would seem to argue that it would be sustainable, but numbers alone are not the entire story where disruption of a habitat or larger ecosystem is concerned. No nation has attempted commercial sealing in these waters since the Russian test of 1972, and should one desire to do so, at least legal limits are on the books – if they can be enforced, and if the several governments claiming Antarctic territory will freely exchange information and responsibility. The commercial harvest scientist/managers of krill may actually be the greater danger: no international agreement of any kind exists to limit use of an obviously rich food resource for a hungry world "closing in for the krill." The only safe conclusion is that there is an Antarctic "problem," and that seals, like krill, are significant ingredients.[85]

Since North American sealers have not hunted them, these Antarctic species have been given only summary consideration here. Similarly scant attention must be directed towards one last important marine mammal, the walrus. Walruses, like harp or fur seals, were hunted commercially in large numbers, but principally by whalers as an adjunct to their whaling trade rather than by professional sealers – though as in the case of whaling and elephanting where oil was concerned overlap was inevitable. But elephanters were "Elephanters," and not whalers, though they might take a whale or two; hunters of walrus (aside from native peoples) were simply whalers filling in their time, and their holds, until whales turned up.

The association of man and seal is most ancient, and Norsemen can claim, as a part of that story, long familiarity with the "hvalross," or "whale horse," from which is derived the popular name given to *Odobenus rosmarus*. It is difficult for any one to gaze upon this creature without some astonishment, for it is not only big (only slightly smaller than elephant seal bulls, a large walrus may weigh 3,000 pounds) but also equipped with an impressive set of tusks and mustachio-like bristles. It is no wonder that when Captain Hudson exhibited a captured walrus at the court of James I in 1608, "the king and many honourable personages beheld it with admiration for the strangeness of the same, the like whereof had never before been seen alive in England."[86] If one literary work can mark a creature in perpetuity, Lewis Carroll must be regarded as the walrus's Melville, though the walrus of *Through the Looking Glass* was hardly as malevolent as Moby Dick. But the walrus has never been the same since 1872, when

The Walrus and the Carpenter
Were walking close at hand,

discussing shoes and ships and sealing-wax – and, very properly, oysters most of all.

Though found in vast numbers at Spitsbergen in the late sixteenth century, the walrus was marveled at more than hunted for over two centuries. A typical response was that of Stephen Bennet of the Muscovy Company: "It seemed very strange to us to see such a multitude of monsters of the sea lye like hogges upon heapes: in the end we shot at them, not knowing whether they could run swiftly or seize upon us."[87] Serious commercial hunting seems to have come first at the hands of whalemen-fishermen reaching the New World who found walrus in plenty off the Nova Scotia coast (especially at Sable Island), and in the Gulf of St. Lawrence. Extensive hauls of walrus for "trayne oyle" were made throughout the seventeenth century, hauls so large that by its end there were few walrus left in areas commonly visited by the Europeans. Remaining Atlantic walrus stocks were pushed back towards Greenland and the Davis Strait, and by the mid-nineteenth century, Spitsbergen and Franz Josef Land witnessed large-scale walrus kills; few survived by the 1860s.[88] Further walrus exploitation, therefore, could only come in the North Pacific Ocean.

Whether the Pacific walrus deserves the status of a separate species is arguable, but at least some mammalogists find enough differences in the larger size, longer tusks, and broader head of the Pacific genus to assign it the name *Odobenus rosmarus divergens*, calling his Atlantic relative *O. rosmarus rosmarus*. Neither the Inuit (Eskimos) who had long hunted walrus nor the whalers who took up the chase in the mid-nineteenth century cared at all. At least it may be said that the Pacific walrus, which does survive in much of the habitat area it occupied when walrusing began, is better known, comparatively speaking, than the vanished Nova Scotian walrus, but as with so many marine mammals much of its life history remains unexplained.[89]

The overall range of the Pacific walrus is based upon the Bering and Chukchi seas and extends from the Eastern Aleutians and the Russian coast in maximum winter extension northward to the edge of the permanent ice pack in summer.[90] Migrations of 2,000 miles are possible, yet some walrus herds appear to move hardly at all. At any rate, the habitat is bound by three factors: ice must be present, for the walrus hauls out on ice floes but does not live in permanent ice; it must have at hand the bivalves (clams) upon which it feeds, and within its diving range (to 300 feet for the most part); and the monthly mean air temperature should range from −15° to +5°C.[91] Though land haulout beaches (*ugli*, or *uglit* in the plural) also exist at such places as Walrus Island in Bristol Bay, generally the walrus hauls out to rest on ice floes, apparently in a change of habit resulting from the onset of heavy hunting on the preferred gravel beaches of the mainland (like the elephant seal, the walrus is no rock climber). Though capable of maneuvering easily in water, the walrus does not have a seal's stamina and needs those nearby floes.

Mating and pupping habits are little known (the *uglit* are places for hauling out to rest, not for reproduction), but it is at least clear that the walrus is probably the most "thignotactic" of mammals – meaning that when out of the water the animals are in nearly constant physical contact with one another.[92] Doubtless a life-saving means of conserving heat in winter, the practice is continued even on a warm summer day, when an entire group may be observed belly-up, flippers waving, tusks raised to the sky in an effort to dissipate heat. Though its skin is nearly white in the water, in the sun blood rushes to the surface, turning it a bright-reddish color. Equally unforgettable is the sight of a brilliant pink herd of walruses each resting its tusks upon the other (another possible reason for such "clumping"; the tusks are awkward attachments for a lone animal trying to rest on its stomach).

But what is the exact purpose of the tusks? There is little question that they are used for defense, not only to strike downward, but also sideways or upwards, always substantial implements which, while not particularly sharp, are backed by heavy weight and muscle. They also help the walrus climb aboard an ice floe ("Odobenoe" means "those who walk with teeth") and pile onto those already present until the floe sinks slowly into the water and, with the addition of one too many, eventually tips over, at which point the whole clump scrambles aboard another single floe to repeat the process. As some hunters have found to their cost, tusks are also good for hooking onto the bulwarks of a small boat and serving out considerable damage to its contents.

It was long assumed that the tusks were used above all for rooting up the clams upon which the walrus feeds, until it was realized that a submerged walrus is like a weightless human swimmer under water, who would find swinging a pickaxe, particularly if attached in tandem to his nose, no simple task. Moreover, tusks so used would probably be much more worn than they ordinarily are. The most likely explanation is that the walrus uses its extensive whiskers to sense the presence of bivalves, the tusks serving rather like runners on a sled to guide his search of the bottom (the front and tips do get considerable wear), and then sucks the meat out of its prey: a captive walrus can suck the meat off an entire mackerel, leaving its keeper with a handful of fish bones.[93] Those tusks, however, provide a desirable form of ivory, which whalers were unlikely to overlook.

It is not only its protective blubber layer and ivory which makes the walrus a huntable object: its lifestyle also facilitates the process. Walruses are both thignotactic and extremely herd-minded and will attempt to rescue a wounded fellow or calf (according to some authorities, seals have pups, but walruses have calves), much in the manner of land-bound elephants and some small toothed whale species. The walrus is possessed of curiosity, like a dairy cow quite apt to investigate some new creature visiting her pasture. Finally, the walrus is much given to frequent vocalization out of water, doubtless a means of contact in areas where visibility is often limited by fog conditions.[94] These several characteristics – grouping, protectiveness, curiosity, vocalization – all make life easier for hunters.

The Inuit, who traditionally hunted the walrus with harpoons, knew these weaknesses, but they also respected an intelligent and potentially serious adversary. A ritual was observed in the killing of the walrus to propitiate the walrus spirit – a ceremony normally not awarded to common varieties of seal. Nor was the walrus killed at the uglit. Inuit women butcher seals, but men carve up walrus; to the Inuit, there clearly is a difference in kind.[95]

That special relationship was profoundly altered by the coming of the affordable repeating rifle to Alaska (gunshots seldom in themselves alarm walruses, who are accustomed to the constant noise of cracking ice). Well before that time, however, whalers were taking walruses. At the end of the 1840s, American whalers entered the Bering Strait; though some walrus encountered at the edge of the ice may have been killed, most authorities are agreed that it was not until approximately fifteen years later, when whale stocks were becoming depleted and whalers had little recourse but to await the melting of the ice and hope for a quick passage through the Bering Strait to the whales beyond that walrus were taken in large numbers, thus killing time and meanwhile filling casks. The traditional date of 1863 is often used, but it is too late; the Honolulu brig *Victoria* was trading whalebone and ivory with whalers in the Bering Sea in 1858, and taking some walrus herself, and hers is only the earliest walrusing log so far found.[96]

While it cannot be claimed that the demand for walrus oil kept whaling alive in the second half of the nineteenth century, it certainly helped keep it alive at the level of sixty to seventy American vessels sailing each year rather than the thirty to forty which might otherwise have been the case.[97] Walrus oil was not as fine as seal oil and commonly was simply mixed with the oil of whales. When kept separate, walrus oil could occasionally command a few cents more a gallon. The similarity to whale oil was another reason why whalers, rather than specialized sealers, brought home walrus products.

Extensive harvesting from the mid-sixties over some twenty years killed some 10,000 a year, or the equivalent of the estimated base herd size of 200,000. Between 1869 and 1874 alone, some 50,000 barrels of walrus oil were returned, the equivalent of 85,000 walruses.[98] Though almost the size of the elephant seal, the walrus has more meat and less blubber and on average makes less than a barrel: a 1,500-pound walrus will make only as much oil as a 600-pound elephant. This was the sort of statistic in which whalemen were interested: it was useful to know, for example, that the ship *Progress* (in 1869) made 585 barrels of oil from 700 walrus, or about 26 gallons from each animal (a barrel containing 31.5 gallons).[99]

The meat was of little value. Though it is certainly edible, the principal Inuit use was as dog food, save for the delicacies of flipper and tongue. According to one experienced epicure, the meat resembles "tough Texan beef, marbled with fat, and soaked in clam-juice." When sled dogs were replaced by snowmobiles, the walrus lost its importance for meat alone. The hide, on the other hand, is valuable, since that of bulls is the toughest and most resistant known among sea mammals. It has long been used for shoe and harness leather, especially the latter in the age of horse-drawn transport. At the end of the nineteenth century, with the coming of the bicycle age, there was a rush for walrus leather for bicycle seats. In the modern world, aside from the many uses by natives, the main utility of walrus hide is for billiard-cue tips (mostly from Greenland) and abrasive wheels for polishing silver, for which purpose there is still no fully adequate synthetic substitute. The hide can also be eaten in case of dire necessity, but, as the same epicure has recorded, even cooked for several days, the result is "simply equivalent to rubber belting of the same thickness," and, if it happens still to have attached the short bristly hairs which serve as outer layer, the erstwhile gourmet will find that "he is cutting up and eating a wire brush."[100]

The tusks, above all, have kept their value. Walrus ivory was an important part of Russia's China trade, and the Alaska colony exported tusks from at least half a million animals between 1800 and 1867. Domestically, walrus ivory was long preferred by dentists for false teeth, since walrus is hard, wears longer, and is slower to yellow than elephant ivory. Particularly in the twentieth century, when walrus ivory has become a scarce commodity, and the oil, meat, and hide have been worth very little, walruses have been taken for the ivory alone, a practice known as "headhunting." The result could be spectacles such as that witnessed by Karl Kenyon in 1958: thirty-nine walruses killed for tusks weighing three or four pounds each (up to twelve pounds for a good pair) and perhaps 100 pounds of edible delicacies, with 117,000 pounds of meat and hide abandoned – not unlike the practice of African elephant poachers. When the trade was at its height, about 1890, 10,000 tusks a year went to San Francisco alone; twenty years later, perhaps 100 arrived at the same port.[101]

Obviously the herd was not flourishing. In 1891, the last walrus was killed on the Pribilof Islands, and the herd was nearly extinct in the entire Bering Sea. Fortunately, the population, then reduced to 20,000–40,000, survived at the ice edge in the Chukchi Sea, reprieved by the general if temporary extinction of the whaling industry.[102] Though the whalers were gone, walruses were still hunted by the Inuit and "pick-up" voyages which traded in native goods of various kinds, such as that on which Albert Tucker sailed from Seattle in 1925: a diesel-powered schooner which at Petropavlosk and along the Siberian and Alaskan coasts dealt in land furs taken by natives, and at the same time shot walruses from her own deck. The Coast Guard killed a goodly number for natives in need of food, as photographs in the Coast Guard archives testify.[103]

Commercial operations for walrus were not yet over. In the Atlantic, Dundee whalers took walrus when they found them. The steamer *Diana*, after harp sealing in 1899, managed in a northward swing to collect ten whales, seventy walrus, three narwhales, twenty-two seals, and fourteen polar bears.[104] Spitsbergen and the Kara Sea stocks were heavily hunted in the 1930s. In Canada the main exploiter was the Hudson's Bay Company, which shipped some 175,000 hides in the period 1925–31. A salted hide at that point was worth $150, hence the demand, but the walrus was now under heavy assault everywhere. Fortunately, protective legislation was passed in Canada in 1931 which blocked the export of walrus products and limited their killing to Inuit and some indigenous whites. By this date, perhaps 2 million walrus had been taken in the Bering Sea area alone since the beginning of exploitation, and the total world population of both species was under 100,000.[105]

The Pacific walrus has begun to climb back, and current estimates are usually in the 170,000 range. But walruses are very difficult to count and these estimates are just that – estimates.[106] Though protection has helped in both Canada and Alaska, where the walrus comes under the umbrella of the Marine Mammal Act, the take of walruses each year is nevertheless roughly 10 per cent of the population. To put it another way, the margin between normal annual increase in the population of 170,000 (to which perhaps 25,000 Atlantic walrus should be added to reach a total for the species) and the annual harvest is very slight. Some developments, such as the abandonment by natives of King Island near the Bering Strait in the 1960s for mainland settlements (King Island was an important indigenous walrus-hunting center), have helped. On the other hand, the growth of commercial clam-harvesting means that the walrus population is perhaps at its maximum limit or even beyond it, and indeed could soon be in serious jeopardy should those clam beds essential to its survival be overexploited, though in the definition of that word "overexploitation" lies much controversy.[107]

The problem of walrus conservation or management is not simple. Once again a large ecological system is involved, which includes man, walrus, clams, and even polar bears – the main enemy of the walrus aside from man – in a relationship that needs far more study. The walrus issue is further complicated by the international locale of its habitat and such indirect dangers to the species as pollution from new oil production in walrus-inhabited waters.

Even protective legislation can be complicated. In this case, the state of Alaska had legislation governing walrus kills and establishing a sanctuary at Bristol Bay's Walrus Island before the passage of the Marine Mammal Act of 1972. After that act it was found that in effect, by lifting any limits on kills by native peoples, it had loosened, not tightened, protection of the walrus. On the other hand, natives also complained: while they now could kill as many walrus as they wished for their own purposes, they were denied an economically useful trade in guiding hunters eager for a walrus trophy, for such exploitation was forbidden by federal legislation. The 1972 Act prohibited trophy hunting; the Inuit responded by increasing their walrus take for ivory. In 1976, in part to answer Inuit complaints, control over the walrus was returned to the state of Alaska, but the species remains, like the otter and the polar bear, under the ultimate jurisdiction of the Department of the Interior's Fish and Wildlife Division.[108] Seals, to add to the confusion, are rather the responsibility of the Department of Commerce's National Marine and Fisheries Services and the Marine Mammal Commission, though, to enlarge the bureaucratic jungle, northern fur seals also fall under the aegis of the (international) Fur Seal Commission.

For the walrus, the future is uncertain. Its restricted range and limited number of food species appear to argue that its survival, in the long run, may be more at risk than that of some other marine mammals considered in this volume.[109] And yet, like all the other species harvested commercially by North Americans, the walrus hangs on – at least as well as, and perhaps better than, the Hawaiian monk seal or the Galápagos fur seal. Two species of seals, however, have continually been harvested in large numbers in the twentieth century and for that reason have been the subject of considerable public controversy: the harp seal of Newfoundland and the fur seal of the Pribilofs thus deserve a final chapter to themselves.

PART FIVE

Epilogue

Taking three as the subject to reason about —
 A convenient number to state —
We add Seven, and Ten, and then multiply out
 By One Thousand diminished by Eight.

The result we proceed to divide, as you see,
 By Nine Hundred and Ninety and Two:
Then subtract Seventeen, and the answer must be
 Exactly and perfectly true.

The method employed I would gladly explain,
 While I have it so clear in my head,
If I had but the time and you had but the brain —
 But much yet remains to be said.

In one moment I've seen what has hitherto been
 Enveloped in absolute mystery.
And without extra charge I will give you at large
 A lesson in Natural History
  Lewis Carroll, *The Hunting of the Snark*

CHAPTER EIGHT

Northern Climes, Modern Times

*Now to conclude and finish, of my people I will
  tell
A thousand men have frozen still out on the
  Arctic hell
Some say they are barbarians, our dignity they
  slander
While I am proud to tell you, I'm a native
  Newfoundlander.*
> Gary O'Driscoll, Bay Bulls, "A Sealing
> Song" (c. 1978) in Cynthia Lamson,
> *"Bloody Decks and a Bumper Crop": The
> Rhetoric of Sealing Counter-Protest*

By the outbreak of World War I the only large-scale commercial North American sealing was of the two substantial herds remaining in the subarctic waters on either side of the continent: the northern fur seals of the Pribilofs, and the harp and hood seals of the Northwest Atlantic. Both industries have continued to the present; both deserve supplementary discussion.

To some extent the two industries share problems, such as the concern of conservationists and, more recently, preservationists opposed to all killing. The initial difficulty faced by United States officials administering the Pribilofs after 1910, however, was less a matter of conservationist activity – though Elliott and Hornaday were early representatives of a movement which was still a novelty in the early years of the century – than of the obvious shortage of seals. For that reason a moratorium upon sealing, except those animals needed for Aleut food, was declared for five years. The ban went into effect for the 1913 season, but Pribilof officials had restricted the 1912 take in anticipation of the prohibition passed in August of that year. The herd thus had six seasons to recover before sealing was fully resumed in 1918.[1] The population did rise from what is generally agreed to be a nadir in 1911 (some 300,000, in contrast to the 2–3 million of 1867), but it did not reach a peak until the late 1930s. Meanwhile, scientists were only gradually coming to understand the herd's population dynamics, as demonstrated by their recommendation to *reduce* herd size in the 1950s.

The tale is complicated but basically stems from pup counts made in the period 1912–22; only a few counts of individual rookeries were made after that date, since growth of the herd made such censusing impossible by traditional methods. Subsequent figures were always estimates based upon some logical assumptions. Alas for accuracy, these assumptions proved erroneous. The biologists thought that females started breeding at age two and that the pregnancy rate was 100 per cent. Though breeding age varies, it normally is a year or two later. Similarly, only 60–80 per cent of females carry a fetus to full term. Compounded over time, these two errors produced a very warped calculation. As table 6 demonstrates, a cautious kill rate was maintained through the 1920s, at which point it was increased through the thirties to approach – and in the forties to pass – the 60,000 per annum which at one time had been the anticipated yield of the North American Commercial Company's lease. By the end of the 1930s the herd had approached 2–2.5 million and seemed capable of sustaining such a yield indefinitely.[2]

Not everyone concerned was happy with this recovery. Japan, already embarrassed by having yielded pelagic sealing for a share of the harvest, only to find the harvest ended for six years, claimed in 1926 that the recovered herd was now eating its way into important fish stocks. The Japanese government requested an international meeting of the signatories of the 1911 accord. The United States declined to cooperate, first on the grounds that the seals off the Japan coast came from the Commanders and Robben Island and that there was no mingling of the herds, and second because the United States had not recognized the Soviet Union and had no intention of doing so even implicitly through a meeting on fur seals.[3]

Japanese authorities remained dissatisfied, facing internal pressures from both the Japanese fishing industry and sealers eager to take once more to sea. Paradoxically, Japan subsidized pelagic sealing vessels while at the same time remaining pledged to conservation at the international

TABLE 6
Pribilof Seal Harvest, 1912–1980

		1946	64,523
1912	3,764	1947	61,447
1913	2,406	1948	70,142
1914	2,735	1949	70,990
1915	3,947	1950	60,204
1916	6,468	1951	50,771
1917	8,170	1952	63,922
1918	34,890	1953	66,669
1919	27,821	1954	63,882
1920	26,648	1955	65,453
1921	23,681	1956	122,826
1922	31,156	1957	93,618
1923	15,920	1958	78,919
1924	17,219	1959	57,810
1925	19,860	1960	40,616
1926	22,131	1961	95,974
1927	24,942	1962	77,915
1928	31,099	1963	85,254
1929	40,068	1964	64,206
1930	42,500	1965	51,020
1931	49,524	1966	52,472
1932	49,336	1967	50,229
1933	54,550	1968	46,893
1934	53,468	1969	32,817
1935	57,296	1970	36,307
1936	52,446	1971	27,338
1937	55,180	1972	33,173
1938	58,364	1973	28,482
1939	60,473	1974	33,027
1940	65,263	1975	29,148
1941	95,013	1976	23,096
1942	150	1977	28,444
1943	117,184	1978	24,885
1944	47,652	1979	25,762
1945	76,964	1980	[25,000]
Total			3,139,221

Sources: 1912–65: Francis Riley, *Fur Seal Industry of the Pribilof Islands, 1786–1965*;
1965–80: North Pacific Fur Seal Commission, Annual Reports (Washington, D.C.,
1966–80)

level. Japan made another unsuccessful attempt at renegotiation in 1936,
insisting that pups branded or sheared in Pribilof tagging experiments
were being taken in Japanese waters. Finally, in October 1940 Japan
gave notice of withdrawal from the treaty, as she was permitted to do
by its terms. The treaty therefore expired a year later, leaving Japan
legally entitled to take seals pelagically, and but for the ensuing war,

doubtless she would have done so in considerable numbers (some, indeed, were taken in wartime). In the international relations of those tense years, seals were a minor issue; on the other hand, the way to solution of seal problems had not been smoothed by the general tension resulting from the Sino-Japanese conflict and the approach of war between the Pacific powers.[4]

On the Pribilofs, meanwhile, confidence stemmed from the herd's recovery. In anticipation of Japanese dislocation of the harvest, just over 95,000 seals were taken in 1941, the highest total by far since the 1911 treaty. World War II and fears of a Japanese assault on the islands prevented any harvest in 1942, but in 1943 a contingent of sealers landed briefly on the islands and took 117,000 seals. In the early postwar years the catch approached the 70,000 level but fell off slightly in the early fifties, bringing biologists to conclude that the herd was too big to produce the maximum number of pelts and therefore should be reduced.[5]

The decision to cut herd size was defended on the grounds that the 1911 Convention required the population to be "maintained at the levels which will provide the greatest harvest year after year," even though by Japanese action the treaty had fallen into abeyance.[6] In 1942 American and Canadian authorities had reached a temporary agreement (Canada was to receive 20 per cent of the raw skins), and when the war was over American occupation authorities obtained a Japanese pledge not to resume pelagic sealing. With the first aerial census of 1948 in hand, showing 500,000 pups (but the numerical predictions had called for 1,100,000),[7] the scientists argued that the rookeries were overcrowded, bringing a fall in the average pregnancy rate and a rise in the average maturity age, together with an increase in pup mortality from the effects of crowding. Logic dictated that the herd should be cropped to reverse these trends, thus producing a maximum sustainable pup productivity of some 400,000, and a harvest of 55,000 to 60,000 male and 10,000 to 30,000 female seals annually. Since over half a million pups were then being born, the formula concluded, herd cropping sufficient to produce 20 per cent fewer pups, would in the end produce more pelts. Needs of the harvest appeared to harmonize with benefits to each surviving seal, now likely to live a longer and healthier life. Later arguments from those opposing management altogether were not in evidence; it would be some years before Victor Scheffer would write, for example, that management "quiets the very struggle for existence that has shaped these magnificent creatures over millions of years."[8]

Management specialists would reply that nothing is so simple, but in this case their voice was muted for the cropping program of 1956–

68 (during which a total of 321,000 females were killed) proved to be a failure. There was, to begin with, considerable opposition from those directly involved: the Aleuts, in whom it was ingrained not to kill females, refused at first to do so. The company which processed the pelts opposed the plan, fearing damage to its carefully cultivated reputation for prime male pelts. Also it proved awkward to take females; unlike bachelors they were in the main herd, in harems, and it was difficult, sometimes dangerous, to separate them out until the main weeks of the breeding season were concluded. More important, while pup mortality did indeed fall – from over 13 per cent to about 3 per cent per year – any alteration in pregnancy rates by age or impregnation did not produce the predicted increase in pups per adult female. Put another way, the MSY of 60,000 and 30,000 could not be reached, in part because after cropping the number of available bachelors fell.[9]

The reasons for this decline are not clearly known, because it is not simply a matter of pups born vs. seals killed. One important consideration is the development in the 1960s of a commercial fishery for pollock in the North Pacific Ocean and Bering Sea. Pollock is an important staple food for the northern fur seal, but by 1972 nearly 2 million metric tons were being taken by the trawlers.[10] Seals may have begun to lose to human competition for fish resources, but as always it is difficult to prove this connection. One clear link, however, is seal mortality caused by discarded trawler netting: studies conducted since 1967 show roughly 0.5 per cent of harvested seals are entangled in net fragments or wrapping bands – and these seals are the temporary survivors, the number dying at sea being unknown. Yet another cause of the decline may be contaminants such as heavy metals and organochlorine pesticides now found in fur seal tissues, though here scientists disagree on "normal" percentages for uncontaminated seals. Finally, there may have been "systematic bias in the estimating procedure."[11]

Once again, only large ecosystem studies are likely to provide the answers. For the time being, the harvest is held at about 25,000, the proceeds of which are distributed according to an interim international agreement of 1957 between the United States, Japan, Canada, and the Soviet Union. Regularly prolonged since that year, the convention is now supervised by a permanent Fur Seal Commission on which all signatories are represented and which produces occasional research papers of value along with notably uninformative annual reports. The treaty commits the signatories to share the proceeds (the United States and the USSR each yield up 30 per cent of their catch, to be divided equally in each case between Canada and Japan) as well as to continue to strive for the maximum yield. This international pledge is worth noting, for it complicates the efforts of those who would stop the harvest altogether.

It is no accident that conservationist organizations such as Friends of Animals and Greenpeace have directed their attention to the Pribilofs. Greenpeace, by no means the first group involved with these islands, serves as an interesting example of the multiple reasons for focusing upon sealing here.[12] In the first place, Greenpeace – headquartered in Vancouver, though it has several semi-autonomous branches in the United States and Canada – has been strenuously engaged in the anti-harpsealing movement and wished not to give the impression of being a western Canadian movement directing its influence solely against Newfoundland and Newfoundlanders. Second, Pribilof action widened Greenpeace interests, appeals, and educational impact. Third, the feeling was strong that in the Pribilofs the problem was one of educating the Aleuts to their responsibilities; although the U.S. government was involved, it was not a question of large capitalist interests similar to those (mainly Norwegian) active in Newfoundland. In this, as will be explained, Greenpeace was not entirely correct, but there was a warm nativist feel to this effort. As Patrick Moore, Greenpeace director, wrote in an internal memo, "we are asking them [the Aleuts] to return to the traditional ways in which the seal is respected as a giver of life and not as a commercial product to be exploited for economic gain."[13] Finally, for American Greenpeacers, there was the further advantage of adding U.S. government sealing as a target to assist in the struggle against international whaling: whaling advocates had always been able to counterattack with references not only to American support for bowhead whaling by native peoples but also to the Pribilof seal harvest.

Not all within Greenpeace agreed with this analysis. Some feared a dispersal of effort when neither whaling nor harp sealing yet were battles won. Others were less sure about the stance toward the Aleuts: perhaps to stop sealing would be to *disrupt* a traditional native lifestyle. Were Aleuts sealers because they had always been sealers or because they were paid government laborers? The answer was not at all clear. Either way, Greenpeace received little sympathy or cooperation from the Aleuts.

Nevertheless, the organization mounted a plan in 1979 to land protesters on St. Paul. Alaska state officials had a counter-plan to fly in a detachment of state troopers. Moore told his followers, "This is a pilgrimage, in the spirit of the Kwakiutl nation, the Oglala Sioux, the Warriors of the Rainbow, and the 5,000-year history of the Aleut people. The Whole Earth Church will preside!"[14] Alas for such inspiration, internal Greenpeace disagreement on how the $20,000 costs were to be met led to the abandonment of the entire mission. A future confrontation is unlikely, since conservationist efforts have increasingly been directed to killing the market for seal products, not their production. It is worth noting in all this that Greenpeace was arguing against commercial exploitation of the seals and for "traditional" Aleut ways, and not against the method of killing, or even waste, for since World War I, a by-products plant of one form or another has operated on the islands to make use of the carcasses for meat or oil or fertilizer.

The question of profit once the federal government assumed full control warrants more discussion. Until 1912 all the pelts were shipped to London for the regular auctions of C.M. Lampson and Company, but in 1913 the small harvest of that year (for Aleut food) was entrusted for sale to a St. Louis firm, Funsten Bros. and Company. The guiding spirit of Funsten, one "Colonel" Philip B. Fouke, had since cancellation of the NACC lease been working to persuade Washington that there were financial advantages to be gained by transferring the fur sealskin industry from London to the United States – and incidentally to his own St. Louis firm. In 1915 Funsten sold the combined small catches of 1914 and 1915, and in the same year, after a forty-minute conversation between Fouke and President Wilson, Funsten was given a contract to process as well as auction the skins (the 1913 and 1915 sales had been of salted skins, not fully prepared furs).[15]

That contract revolutionized the industry. Since the purchase of Alaska, London had been the center of Alaska skin processing. Fouke had been successful in persuading – after considerable effort – W.G. Gibbins and J.C. Lohn, both members of the firm of Rice and Bros., a principal London fur dresser, to move to the United States and to establish, under his aegis, the Gibbins and Lohn Fur Seal Dressing and Dying Company. Rice and Company now brought suit against the use of their own secret dye process, but Fouke argued – successfully enough to satisfy United States patent officials – that his altered alkaline formula was different enough from their vegetable dye to justify protection at law.

Though early sales were successful enough for Funsten, the firm went bankrupt in 1920 in the wake of a brief but severe depression in the fur industry. Soon it emerged in a new guise, now joined with Gibbins and Lohn into the Fouke Fur Company.[16] Fouke's firm not only is still in existence but still processes the Pribilof seal harvest, having proved successful since its first ten-year contract of 1921.

The actual process has undergone some changes since the early days, particularly in the early 1920s. "Blubbering," removal of fat from the skin, is now done on the island, not after shipping, and by machine rather than by hand. The skin is removed from the animal by "stripping" rather than skinning. In this process, "clubbers" or "stunners" kill the seals, "stickers" bleed them, "slitters" make a cut along the stomach and around flippers, neck, and tail. A "barman" then pins the carcass to the ground with a sort of heavy two-pronged pitchfork, and the entire skin is yanked off by three "strippers" using special tongs resembling those used to manipulate blocks of ice. In this way, a skin can be stripped off in a few seconds, as opposed to the minutes an old-style skinner needed. More important, there is no more danger of a knife mark which would reduce the skin's value. Refined processes were also developed in other areas, such as dyes: two standard colors emerged, black and bois de campêche or mantara (brown), though other shades have been tried through the years.[17]

Fouke's profits were directly related to both price and catch, and both increased as the years passed. The contract was renewed in 1931, and by the mid-1930s, with harvests in the 55,000 range and auction prices averaging nearly $27, fur seal skins were witnessing a revival in popularity. Even in 1930, in the wake of general depression, fur seal coats were the preferred style for those who could afford them. Eight or nine skins were necessary for a long coat, which sold that year for $600, though a top-of-the-line model cost $1,000. By the time of this final manufacturing the skins had moved from 75 cents worth of labor to $70–140 each. When Mrs. Herbert Hoover, proud owner of a bois de campêche coat, sent ten skins for a similar garment as official presidential wedding present to a Belgian princess who was marrying Prince Umberto of Italy, Fouke, reaping excellent free publicity, could be confident of surviving the depression.[18]

Fouke's control of processing was not unchallenged. From the beginning, antimonopolists asked why one firm should be given such a powerful and profitable position. Henry Wood Elliott, in his eighties but still leading the fight for seals, as he would continue to do until his death in 1930, now entered the lists against Fouke. Antisemitic political groups associated Fouke's company with the Jewish elements which, they claimed, dominated the retail trade; Henry Ford's vitriolic *Dearborn Independent* was notable in this regard.[19]

But the principal basis of criticism was the American government's continued acquiescence in Fouke's high level of profitability, though there was considerable disagreement over what that level was – not least since, as a condition of the contracts, profit and loss figures have never been made public. In 1953 the company argued for a bigger slice of the profits. The Department of the Interior, now supervising the Fur Seal Service, disagreed, and a study which it commissioned, while not revealing details, charged that Fouke set an artificially high net worth on its assets, thus minimizing its profitability rate as a percentage of assets. In 1955 contract changes were accordingly imposed upon the company, which now would be paid a flat $15 a skin for processing, together with an additional 17.5 per cent of the average selling price (as of 1961, $45–50 for male pelts, and about $40 for female skins).[20]

The Interior Department, meanwhile, considering such fees to be unnecessarily high and realizing that without competitors there was little choice but to meet Fouke's terms, undertook what proved to be an unproductive search for alternate means of processing the skins. Only one other firm (in London) actually processed skins in the same manner, and interlopers were reluctant to challenge a company with such a dominant position. (By 1960 Fouke was processing 90 per cent of the world's fur seal harvest, only 50 per cent of which was American; the rest came from South America and South Africa.)

Fouke's gross income for 1961 was $4,100,000. Profitability, however, cannot be calculated without access to company records. The company was compensated for its services, and the measure of profit in part depends upon how such "costs" of production are determined. The 1954 report, for example, noted that the company had a payroll of $1 million, 20 per cent of which was paid to seven officers of the company, a figure which was considered high at the time. Public government records show that on average over the period 1947–52 net profits were 17.9 per cent of book value of assets, and for 1956–61, 13.09 per cent, or some $200,000 a year.[21]

These figures do not mean a great deal; it was difficult for Fouke's opponents, as it is for the historian, to be specific about the company's finances. A different sort of opening was provided, however, by Fouke's shift of operations from St. Louis to Greenville, South Carolina, in 1961. When Alaska was given statehood in 1958, the company was asked to consider processing at least some of the furs in the new state but replied that this was not possible owing to the need for the skilled labor of the St. Louis area. Shortly thereafter Fouke constructed a new plant in Greenville to process the female skins – now to be taken for the first time – and then in 1962 moved its entire operation to Greenville, apparently no longer requiring the same skilled labor force. (Little incentive was offered workers to relocate; while St. Louis staff were offered jobs, they were at Greenville wage rates – about 50 per cent lower – and no relocation expenses were provided.) A St. Louis union filed charges with the National Labor Relations Board, and the government was now in an embarrassing position as principal client of a firm with labor difficulties and which, moreover, was considered to require a "much more positive approach" to the question of employing black workers before it would be in compliance with equal opportunity guidelines as of 1962.[22] The Controller General's Office refused to renew the Fouke contract, basing its decision "upon a record, extending a number of years into the past, of attitudes, actions, or inactions on the part of Fouke which were considered undesirable by your Department [of Interior], but which were not necessarily violations of Fouke's obligations under the terms of its contract."[23]

For a brief period the contract was awarded to a new firm, "Supara," actually a combination of the National Superior Fur Dressing and Dying Company of Chicago and the allied Superior Fur Dying Company of New York. Despite Interior's support for Supara, Fouke's product was simply the best, and that company's protests against the injustice of cancellation of its own long-time contract, promises of higher wages in the future, and a good deal of hard negotiation brought renewal of its contract in 1965, and subsequent years into the 1980s. That the General Accounting Office had invalidated the Supara contract was a considerable incentive to resuming the relationship with Fouke.[24]

Supara would have received $14 a skin preparation costs, and 16 per cent of the average selling price. At this rate it was calculated that profits would be $177,400 on sales of $4,400,000, or 4.22 per cent; the United States would receive $1,400,000, of which expenses would consume $1,200,000. The 1965 contract with Fouke, on the other hand, was to pay the company $11.13 a skin and 16–18 per cent of the selling price, returning a profit margin roughly equal to that of the late 1950s, or 14–17 per cent.

Having weathered the move to Greenville, Fouke now went forward in a new marketing campaign, stressing "the emergence of the multi-fur woman, the acceptance of 'fun' furs and 'young' furs and the increasing usage of furs among men," as one suggested advertising program put it. Fouke knew its market, worth in the mid-sixties over $350 million a year: white, urban, northeastern, college-educated consumers who made over $10,000 a year and read *Vogue* and *Harpers*. For the trade, an even better approach was to display Elizabeth Taylor, draped in seal, in professional organs the world around – *Fur Age Weekly, Fur Review, Run um den Pelz, Pellicce Moda,* and *L'Officiel de la Fourrure.*[25] Two decades later, while the aggressive advertising approach has muted, and Fouke remains alive and well, the company must fear the sort of boycott suffered by the harp seal industry.

For the United States government, the picture has been less rosy. By the end of World War II, in its dealing with Fouke since 1919:[26]

Sales (1919–46) returned	$37,446,717
Deduct payments to Fouke of	
transportation and commission	1,728,914
dressing, dying, machinery	13,019,107

liquidation of Funsten contract	200,000
fiscal year 1946	503,065
refund to Fouke, 50 per cent of net proceeds in	
excess of guaranteed minimum sale of seal skins	1,266,358

	$16,717,444
likewise payments to Canada and Japan	2,894,154
leaving a net of	17,834,119
which combined with other sales (by-products)	
comes to a grand total of	19,012,919
from which is deducted overhead, Bering Sea Patrol, etc.	16,037,825
leaving a net surplus of 7.7 per cent of gross receipts	$2,975,094

One variable in such calculations is the estimated cost of the Revenue Cutter Service, known after World War I as the Bering Sea Patrol – $7 million. The patrol had little trouble after the war with pelagic sealing, though it had to examine native canoes exercising traditional rights and to deal with an occasional Japanese poacher; on the whole, it was basically an escort for the annual seal migration. It did valuable work, however, in preserving communications and insuring survival when coastal settlements ran low on food; life on the Bering Sea was never a sinecure. Though the responsibilities of the patrol were several and some service would have been required along the coast in any case, without seals and Aleuts on the Pribilofs it is possible to conclude that the service might not have been paid for by the federal government – hence the rationalization for charging it all to the seal revenues. There were some complaints that none of the expense of maintaining three or four vessels in the April-July harvesting season was absorbed by Fouke, or Canada, or Japan, each of whom shared in the take.[27]

These complaints could be expected to grow, for as the harvest diminished, so did government receipts, while expenses predictably declined not at all. In 1979, total administrative costs were $4,500,000, not including any operations by Navy or other vessels at sea in connection with the islands or $300,000 for research:[28]

St. Paul administration		$2,000,000
St. George administration		1,500,000
Schools		300,000
Seattle office costs		300,000
Seal processing		
plant operation	114,697	
harvesting and skinning	118,231	
blubbering	22,503	
curing and packing	30,287	

transport and storage	32,960
data collection	5,112
	323,800
Total	$4,423,800

The harvest for 1979 was 25,752 seals. That total expense figure, divided by the number of pelts, means that each cost $171.75, but this is hardly a fair figure; with or without sealing, so long as there was a dependent Aleut population there, the Pribilofs would have cost the government money. No sealing has in fact been done on St. George since 1972 as a control experiment in herd management, to the disadvantage of that island's population (who were not consulted when the decision was made).[29] But direct processing costs (the $323,800) did amount to more than $12.50 a skin. With little likelihood of a major growth in the herd and subsequent harvest but every probability of an increase in the number of Aleuts, that share of processing costs which is actual wage-payments will have to be spread ever further; that the Pribilofs will remain a net liability seems a safe estimate. The end of sealing would at least have the effect of increasing that liability, as well as increasing the dependency of the population, if the experience at St. George is any guide.

The position of the Aleuts has undergone a substantial metamorphosis in the three-quarters of a century since the government assumed total control. At first, through the 1930s, there was no visible change in their dependent status. On the contrary, their economic level declined relative to mainland averages, while at the same time psychological dependence deepened.[30]

PER CAPITA PERSONAL CONSUMPTION

Year	St. Paul	Mainland U.S.
1911	$254	$281
1916	183	358
1921	308	537
1926	249	621
1931	286	487
1936	218	483
1941	292	604

The cause for both was the moratorium upon sealing after 1911, for it meant that there was no cash money at all for the Aleuts from that source, and very little from any other. Though at first a small amount of cash was doled out, it, like the food and clothing which were also distributed, far more resembled charitable largesse than a meaningful wage from a government bound, under the 1910 act, to provide "fair compensation" to the population.

The wages which were paid soon after the government took over were roughly 50 per cent of what they had been – according to an official government logkeeper – "under the parsimonious policy of the [former] lessee dictated by shareholders keen for profits only"; the government, in other words, was even more parsimonious.[31] The weekly "reward" in addition to food distributions (which were made to heads of households only), ranged from $5 for a seal clubber or first-class skinner, through various grades of workers to a low of 50 cents for a third-class boy. Even for the skilled, it was small recompense, but just as important was the agent's role in assigning an individual's rank. After 1918, wages were again paid for actual work, but since the quotas were small and the base pay still 75 cents a skin, the monetary situation did not improve.[32]

The basic subsistence, however, was from the distribution of stores. It was not long before the natives were given no choice; the bags of staple groceries were simply made up at the store and handed out. The food included little in the way of milk or fresh meat. Though livestock, including dairy cows, was present on the island, the produce was used exclusively for the white population. "One day I picked up my bag of groceries," an Aleut recalled. "We got only two cans of tuna fish, number 2 size, that was our whole supply of meat for the week. It was my wife's birthday. So I thought, what the heck, we'll have a party and use both cans. We ate potatoes and rice for the rest of the week." Clothing was similarly rationed: "We weren't allowed to pick colors, just take what they gave us. They just threw the clothes at you. Sometimes they didn't fit. They carried only two sizes – small and large."[33]

Dorothy Jones's study, *A Century of Servitude*, from which these remarks (from her field notes) are taken, is not an impartial source, but these Aleut recollections are confirmed by the official logs, which show that even a sympathetic agent was caught in the middle. The log for 31 January 1917 provides an example.[34] John Merculieff had asked permission to go shooting, which was refused. Merculieff persisted.

He said he had not enough to eat; that he could not get it unless he went shooting; and that he could not be expected to work when he was hungry. He intimated that he expected to go shooting whenever he could find anything edible to shoot at whether the Agent gave permission or not, and stated that he should not be asked to work regularly when the return for work was insufficient to keep him and his family in proper food.

There having been mutterings of similar character from the usual members of the dissatisfied element for some time, and as many staple foodstuffs are either exhausted or so nearly so that the allowances of them are practically negligible, and as there is a great deal of work which the natives will be called upon to perform whenever weather conditions permit ...

the agent called a meeting of the men to explain the situation, adding
that their petition of the previous August and October to allow general
hunting had apparently been disapproved by higher authorities, since
no positive answer had been received. Meanwhile, all would be expected
to do their duty, and if there were still complaints, "they should petition
Washington in a concrete and explicit statement." For his part, the
agent could do no more than demonstrate that the food distributions
were the most that his current supplies allowed. On 1 February, however, the men presented another petition, "which the Agent regarded
as vague and rather impertinent in tone. Further consideration of the
matter was advised." The petition was withdrawn the next day; H.C.
Fassett, the agent, soon ceased recording such detailed notes.

If anything, the government had increased its interference in the lives
of the Aleuts since the old leasehold days. To the old game of quass-
brewing were added other issues, such as the agent's refusal to coun-
tenance any mingling whatsoever of white and Aleut; this was a larger
problem when a government radio station was established on St. Paul.
Although logkeepers had been told to keep their personal reflections to
themselves and out of the logbooks, the dry record of daily activities
indicates the Orwellian tone of life. 28 March 1929:

At about 5:00 P.M. it was reported to the Agent that one of the radio boys was
out walking with one of the school girls. The Agent immediately sent men
out to follow them and send the girl back to the village. The couple was followed
over by the By-Products building when they turned off the road towards
Kukanin, the girl coming back to the village via the cemetery and the radio
man returning via the lagoon. The radio man was called to the Agent's office
and told that such action on the part of a single man was looked on with
suspicion by this office and any reoccurrence would be reported to higher
authority and if any civil offence could be lodged against him he would be
prosecuted.[35]

Agents believed that their role was to improve the morals of the
population, and promiscuous mingling was not in order. For that reason, strict orders were issued to preserve segregation: "no white person
should visit native houses after dark and we hope that white persons
will find it advisable to refrain from entering native houses at any time,
except on business," wrote the agent to the radioman-in-charge in
1927.[36] But whites and Aleuts were not equal. When a movie house
was constructed, whites sat downstairs, Aleuts in the balcony. The
recreation hall, according to an official notice by the agent dated 9 June
1938, "has been set aside by the Bureau of Fisheries for the benefit of
white employees." Running water, indoor plumbing, refrigeration,
furnaces, and other modern amenities were restricted to white housing
until after World War II. When a new appliance was received – a kitchen

range for example – it would be put in white housing, and the old mode passed on to an Aleut house.[37] And, as always, private lives were carefully regulated:

August 25, 1938

ALL NATIVES TAKE NOTICE

Beginning today all card playing will terminate promptly at 10:45 P.M. There will be no card playing, or the playing of any other games, in native houses or on any other place after that hour ...[38]

Health standards, on the other hand, improved, following the provision of more medical care, though it is doubtful that the card-playing order, ostensibly for health reasons, had much effect in that direction. Education made few strides. The Aleuts continued to resist the government's denigration of Aleut language use and Russian religious instruction. In those cases where Aleuts did pass through the local government school, they found that no education was available at all beyond the age of sixteen. For a short time a scheme was attempted to send the best students to a special high school, the Chemawa Indian School in Salem, Oregon, with the intention of employing Aleuts with clerical skills, for example, as replacements for higher-paid whites and thus reducing overhead. What was overlooked was the inclination of the school's graduates not to return at all, and the fact that, even if they did, Aleuts used as clerks or other permanent government employees would require a regular wage, in money – thus creating a double Aleut standard which might bring a demand from the bulk of the population for more than a handout of groceries. The plan did not last long.[39]

Social improvements of the interwar years in the continental United States simply bypassed the Aleuts. No Aleut was aware of the Indian Citizenship Act of 1924 which declared Indians to be citizens of the United States, but then agents on the islands were in a similar state of blissful ignorance (they could vote by absentee ballot; Aleuts simply could not vote). Legislation of the 1930s – social security, workman's compensation, civil service regulations – had no effect whatsoever in the Pribilofs. The servile status of these "wards" was unchanged; indeed, in comparison with the situation in earlier years, when the company could occasionally be balanced against the government, it had worsened. Since the material conditions of these people had also fallen, in relation to conditions on the mainland, they were worse off in every category except health as a result of nearly three decades of government rule.[40]

World War II, however, brought considerable change. In June of 1942, fearful of a sudden Japanese attack upon the islands, the authorities ordered the removal of the entire population at a few hours' notice. The natives were taken by ship to Funter Bay on northwest

Admiralty Island, sixty miles north of Juneau and 1,500 miles from their home. Though the removal is understandable as a wartime necessity, the subsequent treatment resembled that afforded Japanese-Americans sent to "relocation camps," despite the fact that in the case of the Aleuts there was no question of potential collaboration with the enemy.

The 290 men, women, and children from St. Paul were placed in an abandoned cannery of the P.E. Harris Company (the St. George people were sent to a disused mine across the bay), which after a dozen years of neglect was "so old and rotten it is impossible to do any repairing whatsoever," wrote the frustrated agent on the scene. There was no hot water except what could be heated on two old camp stoves (the only cooking equipment for the entire group); effective sanitation measures were impossible; no bunks or mattresses were provided. The St. Paul agent, L. Macmillin, resigned in protest in October, unable to see his people suffer – and indeed, the official death rate among the Aleuts exceeded the birth rate for the first time in decades.[41]

A year after arrival, a visiting doctor reported the situation: "As we entered the first bunkhouse, the odor of human excreta and waste was so pungent that I could hardly make the grade ... The buildings were in total darkness except for a few candles here and there which I considered distinct fire hazards since the partitions between rooms were made mostly by hangings of wool blankets. The overcrowded housing condition is really beyond description."[42] No doubt one could ask that the Aleuts demonstrate some initiative and fend for themselves, but independence was not to be expected from a people who during their entire lives had relied upon higher authority for permission to hunt, or marry, or play cards after 10:45. Indeed, much credit is owed to the eight whites who voluntarily stayed with the Aleuts to do what they could (the original plan called for them to continue on to Seattle and leave the Aleuts to their own devices).

The Aleuts did complain, and the women presented a petition: "We the people of this place wants a better place than this to live. This place is no place for a living creature," to be told that it was wartime, and people were worse off elsewhere.[43] It was not an effective response. As one agent put it, "It is impossible to make them see in their minds conditions that might exist elsewhere that are worse than they have. They want to see for themselves. I don't know where they would look myself unless it would be soldiers in actual trenches."[44]

Wartime shortages and restrictions handicapped attempts to rectify matters, but by mid-1944 substantial improvements had been made at the camps – just in time for the Aleuts to be returned to their islands. Aside from some individuals, the attitude of the government was not overly sympathetic. Ward T. Bower, chief of the Division of Alaska

Fisheries, explained in a letter to Dr. Bernata Block, of the Alaska Territorial Department of Health, who had complained of conditions after a visit to Funter Bay: "It may well be that the natives of the Pribilof Islands have been coddled too much and the time has come to bring home to them forcefully the need to look after themselves in more decent ways than seems to have been the case, with notable exceptions, at Funter Bay. If they do not respond to ordinary instructions and suggestions along this line, more drastic measures will be necessary."[45]

In 1943, gangs were returned to the islands to conduct the summer sealing, since there was military need for the oil and no reason not to take the usual number of skins; the following year it was deemed safe to send the entire population home. The Aleuts found an island laid waste by the army unit which had been stationed there as a "caretaker group." As the general manager of the islands reported to Bower, "Excepting actual destruction of the village the enemy could not have done much worse." Houses had been ransacked, taps turned on and pipes frozen, the water tank destroyed.[46] The Aleuts accepted these conditions less quietly; times were changing. A number had left Funter Bay to work in Juneau defense industries, despite the attempts of Fish and Wildlife to persuade the wartime Employment Service to keep the Aleuts together. As the agent's annual report for 1944 explained: "They have accumulated many ideas and thoughts, many of which are not the best for themselves or for the Bureau. Their attitudes, in many respects, have changed for the worse, but it is believed this will all be rectified in short order, upon return to the Island."[47]

It was not rectified in short order, in part due to the intervention of a dedicated nurse turned publicist, Fredericka Martin, who published in 1946 an indictment of the government's treatment of the Aleuts, *The Hunting of the Silver Fleece*.[48] In this and other publications, and in testimony at congressional hearings, Martin, who had lived on the islands, labored to arouse public attention (the fact that she had served in the International Brigade in the Spanish Civil War did not endear her to Washington officials newly caught up in the cold war). She did not work entirely alone, and such protests resulted in a congressional investigation of the Pribilof administration in 1949. Although generally uncritical, the study, together with a sealers' strike, brought a significant wage increase in 1950, leading to the introduction of payment fully in money rather than goods. In 1951, clothing was no longer distributed, nor, in 1954, food. In each case a cash dispensation replaced the actual item, thus allowing far more discretionary use of income. An even bigger change accompanied Alaskan statehood in 1958. Alaska ceded any claims to control sealing to the federal government, in return for 70 per cent of the net proceeds after deduction of expenses, which meant that sealing was even less likely to produce a profit. The result was a totally new policy for the sixties.[49]

In the decade 1961–71, official intentions were first to concentrate the population of both islands on St. Paul, and then, over a longer period, resettle all the Aleuts on the mainland, returning only with an annual summer workforce. Seals were declining in number, harvests dwindling, costs rising, and the Aleut birth rate had climbed above the death rate once again. As the number of jobs declined, the number of mouths to feed would rise, leaving a vision of constantly spiraling costs to the government. The only safe plan from an economic standpoint was to persuade the population to leave – but there was no reason to leave so long as jobs were available. R.G. Baker, chief of the Division of Resources Management, saw the dangers as early as 1959: "Gradually the lack of incentive leads to retrogression ... Whether a limitation is placed on jobs or home, or both, the political consequences are great. To do the right thing for these people will take courage."[50]

Fish and Wildlife was caught in a dilemma. While once such a policy would simply have been ordered and enforced, in the 1960s the Aleuts had considerable autonomy – perhaps most significantly marked by the end of alcohol prohibition on the islands in 1962 – as well as potential access to the media. Compulsion would not work, but without compulsion, they would not leave. Meanwhile federal authorities were losing their capacity to control events, as the 1966 Fur Seal Act allowed the transfer of government land and houses to the Aleuts. Further concessions were made in the 1971 Native Claims Settlement Act. St. Paul became an incorporated town; health was removed from the Department of Interior's control and given to the U.S. Public Health Service; education was contracted out to the state of Alaska. Tanadqusix and Tanaq, village corporations respectively for St. Paul and St. George, have provided some alternative employment. The Aleuts now have their own lawyers, and have originated cases against the federal government for past injury.[51]

In 1978 the U.S. government agreed to pay $8,500,000 to Pribilof representatives (a judicial ruling had set a figure of $11,200,000, but under threat of appeal the Aleuts accepted the lower figure). In late 1983, the federal government by agreement completely withdrew from previous obligations and established a trust fund of some $20,000,000. Such payments have not proved to be the end of the issue: much of the 1978 award was to be divided among the population as individuals, and for the remainder investment opportunities are limited in a situation where sealing has been the only industry. Tourism and commercial fishing offer opportunities, but it is not clear what the future has in store. Nor is it at all certain that the United States has the power, unilaterally, to stop sealing, so long as the herd is managed according to international treaty. That the harvest is no longer profitable in the mid-1980s is clear from the stocks of unsold skins as protectionist influence at the marketing level becomes more and more effective. Meanwhile the Aleuts cling to those islands to which, ironically, they were first forcibly imported as "slaves of the harvest."[52] For the seals, however, the final answer probably lies in the status and control of the commercial fishing industry and the resource upon which both it and the seal herd depends.

Much the same conclusion might be reached on the other side of the continent regarding the harp seal, but in this case, recent attention has focused far more upon stopping the killing than the condition of the killers. In fact, by 1983, large-scale commercial exploitation, except as "landsmen" may be included in that category, had come to a virtual halt owing to an effective European boycott of seal products, above all the leather goods which are the principal use of harp seal skin. But to reach this stage the Newfoundland sealing industry has undergone considerable change since the days of the "wooden walls" and – just before World War I – the steel-hulled steamer. As always, it is an involved story which can only be followed here in summary.

As table 7 demonstrates, the World War I years were good for those sealers who still went to the ice. Vessels were lost, some like the *Erik* and the *Stephano*, to submarines, others, like the *Beothic, Bellaventure,* and *Bonaventure*, sold to Imperial Russia and never seen again. Twenty sailed in 1914 but only thirteen the next year, though the number of men involved remained high. Wartime need brought the price for sculps to its highest level ever in 1918 ($12), and an all-time record individual share (before World War II) of $140.[53]

In the interwar years, the price stabilized at $4–5 until the depression. The form of the hunt stabilized as well, with about ten ships carrying some 1,500 men to the ice for an average annual catch of 153,955 (1921–40). Experts agree this harvest might have gone on at that level forever assuming no new extraneous factors (pollution, overfishing).[54] The terms of trade changed little from prewar years: if anything, financial control was all the more concentrated in St. John's, especially after the depression. In 1937, to take a sample year, though the price was back to pre-depression levels ($5), only seven steamers sailed. All were substantial; the smallest were the old wooden walls *Terra Nova* (458 tons) and *Eagle* (457 tons), the largest the steel *Ungava* (1,115 tons). But only two firms owned all the ships, and Bowring Bros. controlled five (*Ranger, Imogene, Eagle, Terra Nova,* and a new *Beothic*). The catch was still worth several hundred thousand dollars, but the industry was really marginal (though not to the sealers), as demonstrated by the general disinclination to add men and ships except a few elderly steel steamers to replace those sold or lost.

TABLE 7
Newfoundland Harp Seal Catch, 1914–1946

Year	Vessels	Men	Catch	Value	Average Share	Young Harp Price
1914	20	3959	233,719	$497,979	$42	$4.75
1915	13	2932	47,004	93,479	11	4.75
1916	11	2028	241,302	642,464	106	6.00
1917	12	4390	196,228	516,717	81	8.00
1918	12	4250	151,434	803,553	140	12.00
1919	10	3599	81,293	278,445	54	8.00
1920	9	1583	33,985	159,950	34	11.00
1921	9	1264	101,452	171,243	45	4.00
1922	8	1196	126,031	197,838	55	4.00
1923	8	1224	101,770	209,136	57	4.50
1924	8	1227	129,561	241,127	66	4.75
1925	10	1505	127,882	269,467	60	5.00
1926	12	1648	211,531	395,511	86	4.75
1927	9	1634	180,459	297,423	61	4.50
1928	11	2190	227,022	394,164	60	5.00
1929	12	2154	201,856	359,920	56	5.50
1930	11	2129	218,644	405,810	64	5.00
1931	8	1268	83,077	121,115	32	3.50
1932	4	731	48,613	58,367	27	3.00
1933	6	1122	176,046	191,912	57	3.00
1934	8	1459	223,600	324,792	74	3.50
1935	9	1784	140,751	169,816	32	4.00
1936	8	1460	183,689	224,495	51	3.75
1937	7	1305	113,340	205,022	52	5.00
1938	8	1459	226,747	490,664	112	5.50
1939	7	1291	97,345	149,399	39	4.00
1940	7	1307	159,687	205,030	52	4.75
1941	4	606	42,666	67,179	37	6.00
1942	2	235	4,698	11,686	17	8.00
1943	0	–	–	–	–	–
1944	1	121	6,697	17,683	49	8.00
1945	5	114	11,543	33,621	98	8.00
1946	12	367	34,605	121,080	110	10.00
Total			4,164,277			

Source: Chafe's Sealing Book (1924 and supplement).

It was not simply that Water Street merchants were conservative. The industry itself, like many associated with the sea, was reluctant to try unproven methods, and they were unlikely to become proven without trial. A striking example is the use of aircraft. In 1923, an aviator was brought from England, and a sealing vessel specially fitted with a deck aft; the plane was to be lowered onto the ice, and then flown off to search for seals. The vessel's master refused even one flight. The next year, Capt. B. Grandy, both a sealer and a pilot, had more luck in persuading the master of his vessel, and made the first successful flight for seals. Though it was hard to admit that there could be any improvement in their methods, C.S. Caldwell, an intrepid Canadian airman, proved in succeeding years the value of using his "Baby Avro" or "Avro Avian" to find the patch – though his profession was a dangerous one: his plane had only limited range, and if forced down on the ice out of sight of sealers his chances were small.[55]

Beyond the useful innovation of aerial spotting, the years between the wars were not years of change. Sculps for leather and oil were still the goal; unlike the Pribilofs, where seal carcasses were near at hand and experiments in their utilization reasonably easy, full processing of harp seal carcasses would have required a revolution in the industry, with more men, larger ships (or more trips, with meat displacing pelts), and the construction of sufficient plant capacity to handle the more than 150,000 carcasses. Such a development was impossible. There was little demand for seal meat, and the cost of production would have surpassed any anticipated return, not least because the average young harp seal is composed of much bone and sinew as well as pelt and fat, leaving only 16 per cent of body weight as usable meat.[56] Men on the first few vessels to return to port have always found a ready market for fresh flippers, but the demand is finite and quickly exhausted. Recent times have produced canned seal meat, and appropriate recipes for seal stew or pie in loyal Newfoundland publications, but no meaningful demand has been generated.[57] There is another problem, as well: careful sanitation, of the sort untraditional in the industry, is necessary to keep the meat fresh. Contamination is all too easy through contact with the deck or hold, or antioxidants used on pelts, or inadequate gutting or washing. Only modern factory ships which process the entire seal at once, much as in whaling, seem capable of proper meat processing on any large scale. This method was in fact tried by Russia in 1962, with a 7,000-ton helicopter-carrying vessel in Newfoundland waters, but apparently the experiment was not successful enough to be continued.[58]

Indeed, the industry might simply have remained an adjunct of Bowring operations, with only the odd spotter plane to show that half a century had passed since the introduction of the steel steamer, had it not been for the coming of World War II. While seven ships went to the ice in 1940 – a standard number for the 1930s – only four went in 1941, two in 1942 (*Eagle* and *Terra Nova*), none in 1943, and only the *Eagle* in 1944. Wartime losses, desperate needs for manpower and cargo space accounted for the decline of the industry, despite an average price of $8 in most war years. What little was left was now entirely Bowring's.

But in 1945 came the revolution.[59] In that year, five diesel vessels, the total tonnage of which was less than the *Eagle*'s, went to the ice.

Not one was from St. John's; they came rather from Twillingate and other outport towns. The day had arrived of the "long-liner" (so-called from the long baited lines used to catch seals in the water) – diesel-powered, 65 feet or under, with perhaps forty to fifty men instead of the hundreds of earlier days. Some were small indeed. The *Little White-coat* of Twillingate (Ashbourne's) was a mere nine tons, but she brought back 417 seals, earning $73.71 for each of her eight men. In 1946, a dozen similar ships were sealing, eager to bring in sculps at the near-record price of $10. St. John's, and Bowring's, were back, but the *Eagle*, now the largest vessel involved by far (none of the others was over 175 tons), was a relic. But Bowring added the *Ice Hunter* (174 tons, thirty-nine men), showing that it too could adapt to changing times, and the firm continued to send a ship to the ice until the mid-sixties.

The thirteen ships of 1946 represented twelve different companies sailing from seven different ports. To the history-minded, it appeared that a new day had dawned for Newfoundland if the outports once again could compete successfully in the sealing business. In 1947, al-though another steamer (*Sable Island*, 341 tons) joined the *Eagle*, fifteen motor vessels went sealing, several making two or even three trips, the first for whitecoats, the later for bedlamers, adult harps and hoods. Crews totalling 628 men brought back 97,535 seals worth $186 a man – the price was good, but the man/seal ratio (155 seals a man) was better, having been beaten only once (1933; 157 a man).

In 1948, the fleet increased to twenty-five, from eight ports and nine firms. It all looked very promising. J.S. Colman, an authority on in-terwar sealing, wrote in 1949: "St. John's remains the most important single port, but the breaking of its monopoly has served to reintroduce the excitements and profits of the seal hunt to the more distant outports; this is likely to be a good thing for Newfoundland both financially and in morale,"[60] particularly since the seals did not appear to be in any danger of disappearing, and the long-liners, unlike the giant steel ships, could find profitable off-season use in cod-fishing.

Even as he wrote, that rosy dawn was proving false, and the New-foundland sealing industry was about to enter dark days. One cause was perhaps the very spirit of enterprise in the outports which Colman saluted. St. John's merchants on the whole had not entered into the sealing revival, but rather remained uninterested, blaming expensive safety requirements or the effects of the war. The outport long-liners thus had to face alone what proved the most damaging factor: the rise of serious Norwegian competition for harp/hood resources.

Newfoundland's development strategy in the postwar era falls be-yond the bounds of this study though it may be noted here that the course it took was largely determined by the aftermath of federation with Canada (a union achieved, not without difficulty, in 1949), and the domination of Newfoundland politics by Joseph R. ("Joey") Smallwood Smallwood was premier from 1949 to 1972, and he did his best in that lengthy era "to bring Newfoundlanders kicking and screaming into the twentieth century."[61] Much was achieved in the Smallwood years, not least a near doubling of the population, despite out-migration, to over half a million in Newfoundland and Labrador. But Smallwood's focus, rightly or wrongly, was on resettlement of the population away from the outports to the urban centers – to provide a cheap labor base for industry, said his critics. The goal, responded Smallwood, was to bring to Newfoundland the industry to develop its forest and mineral reserves and upon which, in his view, its future depended, and to develop the transportation, health, and welfare systems necessary before Newfoundland's standards could approach Canadian averages.

This is not the place to analyse what Ralph Matthews has termed the "development of underdevelopment," meaning that the strategy was designed, however unconsciously, in the long run to keep Newfoundland economically dependent upon mainland Canada.[62] Nevertheless, the focus of Smallwood's government was not the sea and traditional Newfoundland maritime enterprise. It is not that Smallwood rejected this heritage, and indeed fisher-folk helped keep him in power. Rather he recognized the bitter fact that despite fairly substantial government support for both inshore and offshore fisheries, fish stocks had declined since the war and were continuing downward as a result of overexploitation, and economic survival for his province meant finding other sources of employment and income. But without some incentive, either appealing profit levels or attractive subsidies, Water Street was unlikely in such an environment to go sealing with any special vigor.

This was all the more true in the face of an assertive new Norwegian presence. Norwegians had long hunted seals off Jan Mayen, making a particularly big harvest in the late 1920s. When the herd was badly depleted, the Norwegian vessels moved to the White Sea. By the late twenties, that herd also was down, and thoughts turned to Newfoundland. In 1937 the Norwegian ship *Ora* took hoods in Newfoundland waters, an ominous forerunner of a considerable influx. World War II prevented any immediate Norwegian participation, but when peace was restored, so were the Norwegians – propelled all the more by a 1946 Soviet order excluding the efficient Norwegian steel ice-breakers from the White Sea.[63]

At the same moment there was a sudden new interest in sealing from eastern Canadian ports – above all, Halifax – since after 1949 the seal hunt was governed by Canadian, not Newfoundland, regulations. There was no need to observe season limits, for example, until they became Canadian law through act of Parliament or international treaty. In 1949, the Canadian catch was 170,412, biggest in over a decade. Though the bulk of the Canadian vessels actually sailed from St. John's, and often with Newfoundland crews aboard (the same was often true of Norwegian ships), still the profits did not fill St. John's coffers.

The catch of 1949 was only once bettered (228,014 in 1951). In the succeeding years, Canadian involvement fell steadily as table 8 shows, although these figures are somewhat misleading, since Canadian sealing was usually under the auspices of companies of Norwegian origin established in Canada, regardless of their employment of Newfoundland and Nova Scotia officers and crews. The demand for pelts – in the 1950s, more important than that for oil – remained high through the 1960s, but the required technology was more elaborate than ever before. As in the cod-fishery struggle years earlier, Newfoundland was unable to compete. The goal now was to secure fine furs, not simply oil and low-grade leather (though most pelts still went for shoes, belts, and furniture covering). Quality as much as quantity governed the hunt, with the highest price going for the young hood seal, or "blueback," meaning that hoods were now hunted as intensively as harp seals, though they remained a small proportion of the overall catch (about 2 per cent of the Canadian catch in the 1970s). The fluctuations in price and value of such luxury pelts could be disenchantingly startling.[64]

Pelts now had to be washed on deck, treated with antioxidants to prevent yellowing, and then refrigerated for the remainder of the voyage. Such new methods, coupled with Norwegian tax benefits, gave the competitors an apparently invincible edge. Crosbie and Company of St. John's, which had first entered sealing just prior to World War I, was the last Newfoundland concern to keep on, and in the late 1960s its lone vessel, the 2,000-ton *Chesley A. Crosbie*, was commonly the single representative of an old tradition. In the 1970s, the sealers, if they worked at all, found jobs with the Norwegian-owned and Halifax-based Karlsen Shipping Company, and landsmen who hunted for pelts sold them to the one plant in Newfoundland prepared to process them before shipment to Norway, the Norwegian-owned Carino Company, of Dildo, Trinity Bay (a similar plant is located at Blandford, Nova Scotia). By the time Canadian government quotas on seals were established in 1971, the industry may safely be said to have come into Norwegian hands.[65]

At this point the present study, intended only to survey North American sealing, might fairly be ended. To stop here, however, would do scant justice to Newfoundland interests, for the taking of pelts by landsmen for sale is arguably a commercial activity. More importantly, Newfoundland sealing has, rather ironically, aroused international attention just when it could be said to have become foreign-owned. True, 500–600 Canadians, mainly from Newfoundland, kill the seals, and

TABLE 8
Harp Seal Catch, 1947–1970

Year	Newfoundland Catch	Canada Catch	Norway Catch	Total (Canada & Norway)
1947	130,128			
1948	171,982			
1949	170,412			
1950	121,908			
1951	228,014			
1952	105,245			
1953	106,336			
1954	67,357			
	(1,101,382)			
1955	55,561	168,275	165,094	333,369
1956	77,586	199,467	189,943	389,410
1957	46,182	122,483	123,057	245,540
1958	55,427	165,895	131,891	297,786
1959	32,029	102,419	217,715	320,134
1960	37,459	147,993	129,357	277,350
1961	41,450	64,526	111,940	176,466
1962	59,753	165,482	154,507	319,989
1963	77,767	176,090	138,159	314,249
1964	45,720	152,996	188,667	341,249
1965	79,954	167,096	67,157	234,253
1966	51,515	173,062	150,677	323,739
1967	42,070	147,044	187,312	334,356
1968	39,747	104,436	88,260	192,696
1969	118,072	171,398	117,414	288,812
1970	93,286	142,295	115,200	257,495
Total	2,132,495	2,370,957	2,176,350	4,547,307
Add Newfoundland 1947–54		1,101,382		1,101,382
		3,472,339		5,648,689

Source: J.R. Beddington and H.A Williams, *The Status and Management of the Harp Seal in the North-West Atlantic: A Review and Evaluation.*

thousands of pelts are landed in Canada. But a pelt for which, in the mid-1960's a sealer might be paid $2–3, was worth 20–25 when processed in Europe, $50–60 to a furrier, and $100–125 to the consumer in the form of a coat – and only a very small share of that money ever found its way back to Canada.[66]

But such high value for specialty leather goods, coupled with an almost complete absence of controls on sealing, halved the herds, and consequently aroused public hostility. Newfoundlanders who take seals, and many who do not, speak scathingly of preservationists or conservationists who malign their vocation without grasping its significance, economic or psychological. In fact, those conservationists may have prevented the virtual extermination of the herd by arousing public concern, and thus, in a further irony, continued to give work to the sealers and lay the basis for further sealer-conservationist confrontations.

Concerned scientists, led by David Sergeant, voiced their fears of the potential effects of a harvest averaging more than 300,000 each year between 1949 and 1961, when the number of pups born annually was estimated to be 350,000 (down from 750,000 in 1950). But no limitations were placed upon sealing outside Canada's three-mile limit (the 200-mile limit was declared only in 1977), except for a vague gentleman's agreement of 1952 between Canadian and Norwegian sealers not to begin sealing at the Front until 10 March (5 March in the Gulf). In 1961, a closing date of 5 May was established. But that year twenty-eight ships were on the ice, and the next season helicopters were introduced to locate seals and transfer men to areas of seal concentration (in unscrupulous hands, helicopters had other uses, as tales of stolen pelts testify). In response to mounting public pressure, Canada suggested that the seal hunt be regulated by the International Commission for the North Atlantic Fisheries (ICNAF), but not until 1966 did the requisite number of members agree, and not until 1971 did it establish international quotas.[67]

In the meantime, Canada had acted to restrict the hunt in several ways. In 1963 the closing date was moved up to 30 April (25 April in the Gulf). In 1964, a limit of 50,000 young harps was established in the Gulf; inspectors were assigned to sealing vessels to insure commercial limits; every sealer, vessel, and aircraft was required to take out a license; the baited longlines were prohibited. In a move of 1967, particularly unpopular among sealers, the gaff was outlawed as a killing tool, and all sealers required to use bats of 24–30 inches. While sealers had indeed used any handy instrument before, and unnecessary cruelty certainly occurred, still the bat was less effective than a properly wielded gaff. A blow from the latter required two hands to deliver, and, all other things being equal, was likely to be effective more often than a single-handed strike with a bat. The bat was also useless for any other purpose, while the gaff served as walking stick, vaulting pole, and rescue device. Captains had to forbid their men from taking gaffs along with bats onto the ice, knowing they would use the former if they had them. To the sealers, it seemed as if the government wanted to make sealing so difficult and dangerous that men would be discouraged from going to the ice at all. (Norwegians, incidentally, generally use a *hakapik*, a club with an iron hook on the end.)[68]

More than any other measure, the bat/gaff regulation demonstrated the mounting attention which the hunt was receiving. A landmark in that regard was the 1964 movie made by Artek films – still highly controversial – which showed scenes from the Magdalen Islands, including a seal which moves and screams *after* skinning. Subsequent charges of savagery, inhumanity, and cruelty aroused far more public attention than any discussion of numbers or endangered resources in the abstract.[69] Books such as Peter Lust's *Last Seal Pup: The Story of Canada's Seal Hunt* (1967), and Brian Davies's *Savage Luxury: The Slaughter of the Baby Seals* (1970),[70] spread the message. The response was enormous, channeled through such organizations as the International Union for the Conservation of Nature, the International Fund for Animal Welfare (itself an outgrowth of Brian Davies's work in the New Brunswick SPCA), and Greenpeace. Public people of many sorts lent their efforts, from California Congressman Leo Ryan to French actress Brigitte Bardot.[71]

By the late 1960s a boycott of seal products was beginning to be felt by Canadian sealers, but the first real victims were Inuit sealers who had taken up commercial sealing in northern Canada for the pelts of the very different ringed seal (*Phoca hispida*), and now found that the Hudson's Bay Company could no longer pay or command high prices for coat-type skins. Coats are the most obvious focus of a boycott; the average consumer is not able to tell from what sort of leather his belt or wallet is made. The growth of the protest movement continued, however, and is a remarkable story on an international scale.[72]

Much of the concern was over the "slaughter of baby seals," and the methods by which all seals were killed. More knowledgeable conservationists, aware of the fairly restrictive Seal Protection Regulations which, as enforced in 1982 and 1983, permitted only experienced sealers to obtain sealing licenses, were more worried about overexploitation of the harp seal herd.[73] Table 8 shows the large harvests between 1956 and 1971, the year in which quotas were introduced. In 1967 scientific advisers to the ICNAF reported that the Front herd had sustained a marked decline in the last fifteen years, though they were unable to provide precise quantification. As a result, the next year the closing date was moved to 25 April, taking of adults on whelping patch areas was prohibited, and aircraft were restricted to spotting only at the Front (and prohibited for hunting purposes at the Gulf in 1970).

The establishment of a quota, introduced by the ICNAF in 1971, was a more important measure, but many argued that the "total allowable catch" (TAC) was unrealistically large, reminiscent of quotas allowed to whalers by the International Whaling Commission (see table 9). It is important to note that kill totals do not necessarily fall, over the short run, merely because the population does; much depends upon the circumstances of any particular year. The TAC for 1971 of 245,000 (100,000 each for Canadian and Norwegian vessels, and 45,000 for landsmen)

TABLE 9
Harp Seal Quotas and Catches, 1971–1984

	Quotas[1]			
	CANADA		NORWAY	
Year	Large Vessels	Landsmen	Large Vessels	Grand Total
1971	100,000	45,000	100,000	245,000
1972	60,000	30,000	60,000	150,000
1973	60,000	30,000	60,000	150,000
1974	60,000	30,000	60,000	150,000
1975	60,000	30,000	60,000	150,000
1976	52,333	30,000	44,667	127,000
1977	72,000	53,000	35,000	160,000
1978	72,000	63,000	35,000	170,000
1979	77,000	73,000	20,000	170,000
1980	76,150	72,050	20,000	168,200
1981	74,900	70,800	22,500	168,200
1982	77,000	74,000	24,000	175,000
1983	77,000	75,000	24,000	176,000
1984	77,000	75,000	24,000[4]	176,000

	Catches[2]			
	CANADA		NORWAY	
Year	Large Vessels	Landsmen	Large Vessels	Grand Total
1971	86,000	47,000	98,000	231,000
1972	53,000	24,000	53,000	130,000
1973	20,000	45,000	59,000	124,000
1974	52,000	40,000	55,000	147,000
1975	61,000	53,000	60,000	174,000
1976	53,000	66,000	46,000	165,000
1977	59,000	60,000	36,000	155,000
1978	51,000	94,000	16,000	161,000
1979	77,000	63,000	20,000	160,000
1980	73,000	76,000[3]	20,000	169,000
1981	68,000	107,000[3]	22,000	197,000
1982	78,000	68,000[3]	24,000	170,000
1983	9,000	47,000	NIL	56,000
1984	NIL	30,000	NIL	30,000

Source: Government of Canada, Department of Fisheries and Oceans, Economic Branch (Statistics Group).

1 Excluding unregulated Northern Allowances introduced since 1977.

2 Rounded to the nearest thousand and not including the native catches under the Northern Allowances.

3 In addition, there were reported catches of "unspecified" seals in the Gulf Québec: 3,220, 6,824, 3,984, for 1980, 1981, 1982 respectively, which were believed to be mainly harp seals.

4 This amount was not allocated to Norway in 1984 but held in "reserve".

was not even reached, and in any case was very little less than the unlimted 1970 catch of just over 250,000. The figure for landsmen represents an estimate of the expected catch (at least until 1977), rather than a catch regulated by specifically licensed number.

In subsequent years, however, with the herd showing typical signs of decline such as a fall in the age of sexual maturity, the TAC was reduced substantially to a figure which government scientists believed could be supported by the herd and at the same time permit it to increase slowly. It was hoped that the herd might climb from a low of 1,000,000 estimated for 1972 (compare 1951, when 3,300,000 had been the estimate) to perhaps 1,500,000–1,600,000; the presealing base, it will be recalled, was perhaps 10,000,000. Officials are quick to stress that a herd of 1,500,000 is not necessarily specifically selected to produce a maximum sustainable yield (MSY), but should be able to support an annual kill of 237,000. In fact, the Canadian government estimates the population to be about two million as of 1983. (The MSY concept may be past its time in any case, since it is based upon a simple S-curve in population numbers, and pays scant regard to complex ecosystem relationships.)[74]

Unfortunately, the very numbers themselves are the subject of contention, in part because of the difficulty of taking a census of this population. Many methods have been tried on pups and adults alike. For some years in the 1970s the most promising appeared to be aerial ultra-violet photography of pups, though even this method is complicated by the need to compensate for pups hidden under ice cliffs and other problems. Today, the favored method is capture-recapture tagging, using durable plastic markers. Studies done in 1978–80 and 1983 estimate the population to be perhaps two million, derived from an estimate of about half a million pups; the TAC is 186,000 (1983 and 1984; up from 170,000 in 1981) harp seals and 15,000 hoods.[75] Population considerations, as with the Pribilof seals, are not simply the result of man/seal interaction, but involve as well the exploitation of fish stocks, particularly the capelin which makes up about a quarter of the harp seal's diet. (One and a half million seals eat approximately 1.5 tons of fish each per annum, and a quarter of the total would be some 623,500 tons of capelin, not far off the ICNAF capelin quota for the Northwest Atlantic for 1975.)[76] The ICNAF and international fishery agreements thus have and will have much to do with seals in more ways than simply setting a seal harvest quota.

Yet another controversial issue is the value of the hunt to Canada, although it is agreed that the greatest profit accrues to processors and retailers elsewhere. The standard figure used through the mid-1970s was $5,000,000.[77] The price for pelts has in fact fluctuated widely in recent years, falling from $15 in 1965 to $2 in 1968, and then rising again in the 1970s. By 1982–3, however, the standard value figure had increased through inflation to $13,000,000.[78] An idea of how the $5,000,000 figure was reached is given by D.L. Dunn's detailed study of the industry in 1976, using "value-added" calculations, that is, in each case arriving at a value of each sector and subsector by deducting the cost of raw material from total receipts:[79]

PRIMARY SECTOR		SECONDARY SECTOR	
number engaged	*value*	*number engaged*	*value*
3,045 landsmen	$ 702,000	36 subagents (who buy	
796 in vessels 35-65′	1,202,474	skins for the agents	
189 in vessels over 65′	1,060,907	of processing firms)	26,892
other	86,410	67 agents	252,960
		fresh meat sales	302,635
		61 meat processors	110,800
		55 pelt processors	1,752,000
subtotal	$3,051,791	subtotal	$2,445,287
Total value added by 4,249 employed in seal fishery			$5,497,078

Dunn has also surveyed the average receipts of small and large vessels, concluding that on average, the former (35–65 feet; anything smaller is consigned to the category of "landsman") could expect a return of $3,960 on expenses (maintance, repairs, operating costs, insurance) of $2,380, earning therefore roughly $1,500 before deduction of fixed overhead (vessel depreciation). Larger ships might receive $150,000 for pelt sales on $50,000 of expenses – a considerably better return.

Value added is only one method of calculation. The nearly 1,000 men who sailed on 203 small vessels shared with their owners $753,000 after expenses. The 3,045 landsmen received another $705,045 in receipts, meaning that the bulk of those with jobs in sealing shared $1,500,000 in wages, a considerably firmer statistic. Landsmen each averaged $231 in cash returns, not a large sum, though the value of this cash in the individual's finances should not be underestimated; the $375 which each of the 4,000 total sealers averaged, Dunn has calculated, had benefits directly affecting 18,000 people.[80]

But it is especially the large-scale, commercial sealing vessels, and their owners, with which the conservationists take issue. It has been a Greenpeace tactic, since that organization entered the sealing struggle in 1976, to try to win over the lesser sealers, the landsmen, to cooperate against the larger capitalist enterprises. This approach has not been markedly successful. Newfoundlanders have seen in Greenpeace and the others not only a threat to incomes, but also an attack upon a traditional cultural trait by wealthy outsiders. To them, a sealing moratorium had simply become the cause of the moment among the idle rich. In this impoverished province, the charge that conservation was a lucrative money-making proposition was particularly telling. But little quarter has been given in this struggle: conservationists counter that sealing defenders are simply the pawns of oilmen who wish to exploit offshore resources and would prefer to see the seals exterminated and thus present no future problem.[81] Pro-sealers in turn rally to anti-Greenpeace campaigns and the slogans of the satirical counter-group "Codpeace," dedicated to preserving the cod fish from extermination by the seals. The permutations are nearly limitless for this "Honest to Cod" movement working for "Codservation." Canadian lobbyists at the European parliament, however, presented too direct an approach to win many converts with their buttons reading, "SAVE OUR COD, EAT A SEAL."[82]

The standard defenses of sealing, issued by the Canadian government in a multiplicity of forms, deserve consideration. Outsiders, run such arguments, cannot understand the true nature of the hunt, whether it is a question of its role in the traditional life of Newfoundland, or of its economic importance, however small the return to the individual sealers. Seals, like fish, are simply a valuable and – with proper management – a permanent resource to be utilized by mankind. Nor, for that matter, is there really severe suffering on the part of the seals; the issue of *how* the killing and skinning are done, in other words, is magnified out of all proportion.[83]

Despite the merit of such arguments, they are easily undermined by just one publicist ready to cry that precisely those defenses – economic importance, traditional lifestyle, exaggerated suffering – had been put forward in the case of slavery, and that some aspects of human society do not deserve perpetuation. As Richard Adams, the author of *Watership Down*, put it in a New Zealand newspaper, the only argument used by sealers that was not used by slavers, was, "they eat too much fish," exactly the point which Codpeace and the lobbyists in Europe were attempting to make.[84]

The struggle has had many phases, from saturation of the media to the direct confrontations with sealers and the Royal Canadian Mounted Police utilized by Cleveland Amory of the Fund for Animal Welfare, Paul Watson (former Vancouver Greenpeace director), and the crew of the *Sea Shepherd* and *Sea Shepherd II* in the late 1970s and early 1980s. Their particular goal was to spray such seals as they could reach with a harmless but permanent green or red dye and thus to make the pelt worthless. To authorities, their actions were illegal, according to the 1977 Canadian Seal Protection Act: since they were not licensed sealers, they were prohibited from approaching within a half-mile of the herd.[85] Spraying of seals was one way to attract attention; talk shows are another.

David Letterman (31 July 1980): Now, when you deal with the guys actually committing the crime [i.e., the sealers], what is their response? Do they like the work, I mean, have they been doing it for generations and generations?

Brian Davies: Well, their response is a direct contradiction to the Canadian government's position, which is that it is an industry, that people go out, they enjoy doing it. In fact the men that I've talked to on the ice, and I've been on the ice for about twelve years, consecutively, is that if there was some other form of employment they could be doing they wouldn't be there. I mean, nobody in their right mind wants to spend four or five weeks of the year beating baby seals over the head.[86]

In the end, the victory seems to have gone to the conservationists, through the very effective device of persuading the United States and European countries to ban seal products. As *Maclean's* explained it, describing a 1982 European Economic Community vote to ban seal pelt imports: "The hunters and the Canadian government have history, logic, and economics on their side. The protestors have Brigitte Bardot, full-page newspaper ads, petitions from schoolchildren and yards of lurid film footage of grim-faced sealers crushing the skulls of cuddly seal pups."[87]

In the 1983 season, Norway announced that its ships would not go to the Front, and the Norwegian-controlled processing monopoly, which had purchased almost the entire Canadian catch of 114,000 pelts in 1982, added that it would take no more than 60,000 for lower-grade leather.[88] The boycott had approached the total victory it desired by an all-out campaign: "ARRETÉZ LE MASSACRE: DANS QUELQUES JOURS 120,000 BÉBÉS PHOQUES VONT ÊTRE ASSASSINÉS," announced the Comité d'Action "Sauvez Les Bébés Phoques" in a sizeable ad in *Le Monde*, in a typical example, asking its supporters to write President Mitterand: "Vous avez promis au Français le socialisme à visage humain! C'est le moment d'en apporter le preuve. Invitez les Ministres européens à sauver les bébés phoques, à la initiative de la France généreuse."[89]

The harp seal hunt is not over, but in the near future is likely to be limited to such harvest as can be used locally in Newfoundland, unless, and until, some new market is located by the Canadian Sealers' Association with the help of the Department of Fisheries and Oceans. On a higher level, the Newfoundland and Canadian governments have attempted to block a European Economic Community ban in seal prod-

ucts and, in late 1982, proposed an international convention on sealing.[90] Once again, and for the last time, the question presents itself,
how many harp seals?

1914–46	4,164,277
1947–70	1,992,356
1971–78 (excluding Norwegian)	819,133
	6,975,766
Total catch, 1817–1913	25,663,696
Grand total	32,639,462

To reach a grand total for all species discussed in this book, however,
such precision is hardly appropriate, and even rounding to the nearest
100,000, the figures are still misleadingly exact:

Harp seals	32,600,000
Northern fur seals (to 1910)	3,900,000
Northern fur seals (1911–80)	3,100,000
Elephant seals	1,000,000
Southern fur seals	4,000,000
Guadalupe, Galápagos, Hawaiian monk	200,000
Walrus	200,000
Grand total	43,300,000

The total of over 43 million seals taken commercially by North Americans between 1790 and 1980 for their fur, fat, and leather excludes
seals taken in American waters by Russian, Japanese, and Norwegian
sealers, and makes no allowance for seals killed but not retrieved, an
important variable at least in Pacific pelagic sealing. These omissions,
if added, bring the total to over 50 million:

Northern fur seal, pre-1867	4,000,000
Pelagic losses (2:1)	2,000,000
Norwegian	1,000,000
Total	7,000,000

Considering that this catch was taken over nearly 200 years by the
people of an entire continent, the number does not appear great, particularly when it is compared to the countless herds of sheep and cattle
and flocks of chickens consumed by the same continent in the same era.
But even to farmers, seals are not the same as sheep or cattle or chickens.
John Lister-Kaye, in his study of the grey seal cull in Great Britain,
notes that farmers, farm workers, foresters, and villagers – people who

deal with animal control every day of their lives – eagerly signed petitions against sealing in Britain, without being able to explain their
actions any more clearly than their fellow protesters could.[91]

Perhaps, as Lister-Kaye has suggested, it is the age-old association
of man and seal which is ingrained in the Scottish-Welsh-Irish culture
of which he wrote. This would not necessarily affect people of a different heritage, of course. Perhaps it is the seal's dog-like nature, and
indeed the seal shares common ancestry with man's best friend, and
man by a sort of Freudian transference takes the seal similarly to heart.
Perhaps it is, for those familiar with the seal hunt of Newfoundland,
the idea of so much waste for what is really a luxury trade (and indeed,
though fresh meat and flippers are still sold, there has been no harp seal
canning since 1977),[92] and a luxury trade, moreover, which requires
the sacrifice of the young.

Major William Howe Greene, grandson of Benjamin Bowring, and
himself the author of a valuable if uncritical study of Newfoundland
sealing, *Wooden Walls Among the Ice Floes*, offered in a 1934 letter a
defense which was then possible of the use of seal oil and leather:
"mankind *needs* these things, & the Nflder is his instrument – brave,
simple, white-souled, fervidly religious, uncomplaining."[93] Half a century later, Brian Davies was more persuasive:

David Letterman: Now, you mentioned that it's about 200 or 300 guys working
at it [sealing on the ice]. What are the pelts used for, I mean is this some kind
of valuable material we can't do without?

Brian Davies: Oh boy, you could certainly do without what you get from the
harp seals; oh, little harp seal bow ties, harp seal slippers, fur trim for jackets,
and so on, nothing of any consequence and nothing we couldn't replace from
man-made materials.[94]

Above all, however, it is the killing of defenseless pups, totally unprotected by burrow or nest or natural camouflage, which has aroused
such fury; as Lister-Kaye has put it, "our Western culture loves all
things young, furry, and helpless."[95]

No whitecoats were taken commercially in 1983 or 1984, though
that is no guarantee that the kill might not resume in the future under
altered circumstances. Widespread interest may dwindle, for example,
if only adults are taken. Certainly articulate sealers such as fifty-year-
old Alan Richards, of St. Anthony, a sealer for the last thirty-five years,
will continue to defend their lifestyle: "I want my freedom. I don't
want to ever have to sit down in the month of April or May and just
look out at the ice drifting by, knowing that the seals are there and not
be able to go out and hunt seals. It would be just as well to take me

and lock me up in jail. I don't want to be a drawback on my government. I don't want unemployment or handouts."[96] Similarly, some management-scientists will continue to argue, in justification of the hunt, that uncontrolled expansion of seal herds carries its own dangers, and that such populations should be as much managed as those of deer or other game hunted by license. To many such, too many protectionists fail to recognize the need to balance the interests of mankind and seal.

Yet each time they will be answered by those who would prefer to see the seals and whales left alone as wild creatures of the sea, whatever the "natural" results of a lack of management. They were more likely to agree with Victor Scheffer, one of the most eminent biologists to write, both popularly and scientifically, on marine mammals, and above all on northern fur seals. Scheffer was involved early on in the issue of the Pribilof census, and was for many years a staunch defender of the harvest, so much so that he was commended by the Fouke Company – until in 1973 he argued publicly that killing seals was not the best way to use them, and Fouke asked that Scheffer be removed from the position on the Marine Mammal Commission to which President Nixon had just appointed him.[97]

"I was simply beginning to shake off the influence of a narrow, zoological-technological training which had dictated that wild animals exist to be killed for subsistence, sport, or vanity," he explained in his autobiography.[98] Scheffer's personal metamorphosis led him to disagree with those who, like Jacques Cousteau, feel that only irrational emotion separates the use of the seal for his pelt from that of the pig for his bacon – though as we have seen Cousteau is not above anthropomorphization. In the end, our attitude toward the treatment of seals reflects society's values, values which we all know would soon come into question should a dog-breeder suggest that his Dalmatians would make up into quite attractive leisure wear. "Our mental image of a seal pup," wrote Scheffer, "is quite unlike that of a fattened pig and we see no reason to confuse the two. We gently agree with T.S. Eliot who wrote, 'Human kind cannot bear very much reality.' We know that both seal and pig are animals but we reserve the right to spare the one and to kill the other. We make a deliberate choice between two values that cannot be compared – one that can be framed in 'rational' words and one that must remain unframed in wordless feelings."[99] His own conclusion, after a lifetime spent with seals and other creatures: "the highest use of a wild animal is to let it be."[100]

Scheffer's argument is clear proof that not all biologists are of one mind upon sealing, any more than they agree on the uses of nuclear energy. Scheffer is not alone, though many who share his position might not go as far as Harry R. Lillie, who became much attached to the penguins he studied in the South Atlantic, and wrote in the preface to his 1934 study, *The Path Through Penguin City:*[101] "In the end I am content to have the same fate as the Penguins, and Whales, and Seals, and the little wild things who cry out in the night when the traps set by humans close on their paws; and all the other creatures sacrificed for this that is called humanity. If these people have no souls, then I am grateful to the Creator that he has not given me one either, and I shall not have any churchman taking up his time making arrangements for me. Perhaps I can carry a small feathered person along ... and, with the other arm round some furry neck, we can set off together into the great beyond."

Sealer, manager/conservationist, preservationist: the distance which separates these diverse positions is vast, at times unbridgeably so, as well it might be between Alan Richards of St. Anthony and Harry Lillie. Yet, as Scheffer has shown by his own career, the boundaries are not always clear-cut, nor individual positions graven in stone. For the future, one encouraging fact is outstanding: despite the enormous harvests of the seal species discussed in this book, most have managed to survive, some – after barely escaping extermination – to do very well indeed. If for the monk seal there is little hope, that seal's story is fortunately not typical.

For myself, I believe that it is in no man's interest to thus jeopardize, let alone exterminate, any seal population. I hope this position between sealer and anti-sealer has enabled me to do adequate justice to both – and to the seals as well. As for the future, it is unlikely that any such species will be deliberately attacked, if only because of the considerable public attention which the seals have received since the late 1960s.[102] A movement is unlikely to be ignored which can mobilize such a successful international boycott, nor, if the recent history of the antiwhaling movement be any criterion, is the interest in seals likely to be a passing fad, so long as the world has time for consideration of the survival of species other than man – a rather important qualification.

But the story is not concluded. Mankind's predilection to despoil the earth's lands and seas with dangerous pollutants has reached no end; similarly, current and future needs for food (and profit) make it probable that issues such as fish/seal or clam/walrus relationships will be heard of again. For the seals of the world, as for natural history in general, the issue of krill is in some ways typical, forever linking and locking together the food chain, the many species dependent upon krill in that chain, and the attitudes and policies of the human beings whose actions are, in the modern world, going to affect it for better or worse. Battles have been won and lost on both sides, but it cannot be said that we have seen the final episode of "the war against the seals."

Maps

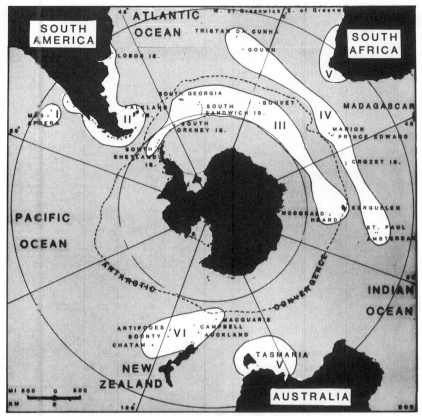

MAP I · Distribution of Southern Fur Seals

PRINCIPAL SOUTHERN SPECIES OF ARCTOCEPHALUS

I.   A. philippi: Juan Fernández fur seal
II. A. australis: South American fur seal
III. A. gazella: Kerguelen fur seal
                       Antarctic fur seal
IV. A. Tropicalis: Amsterdam I. fur seal
                       Subantarctic fur seal
V.   A. pusillus: Cape fur seal
                       South African fur seal
                       Tasmanian fur seal
                       Victorian fur seal
VI. A. forsteri: New Zealand fur seal

MAP 2 · Drake Passage and the Scotia Sea

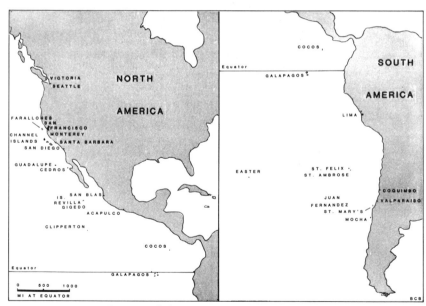

MAP 3 · Western Coasts of the Americas

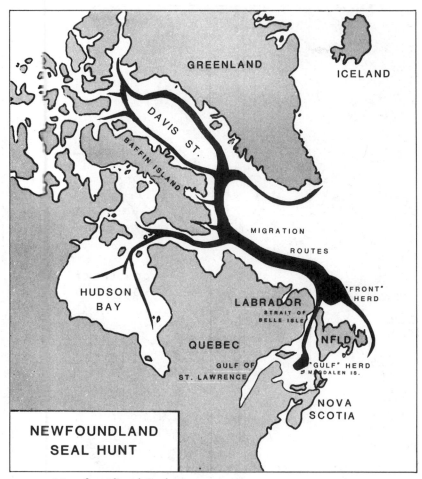

MAP 4 · Newfoundland Seal Hunt

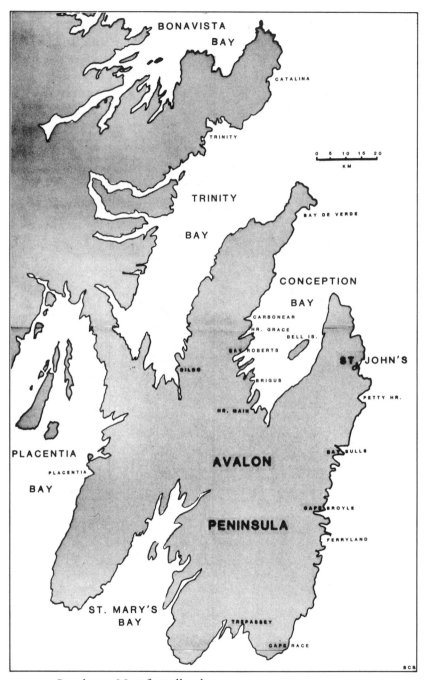

MAP 5 · Southeast Newfoundland

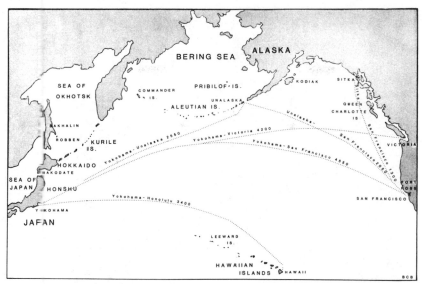

MAP 6 · North Pacific Ocean

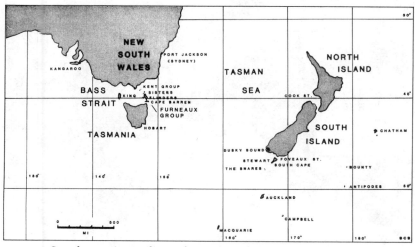

MAP 7 · Southeast Australia and New Zealand

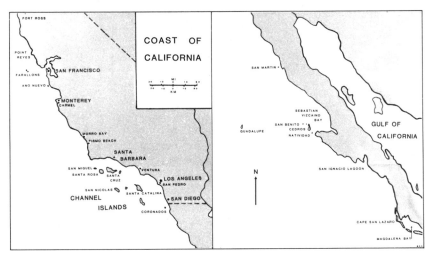

MAP 8 · Coast of California

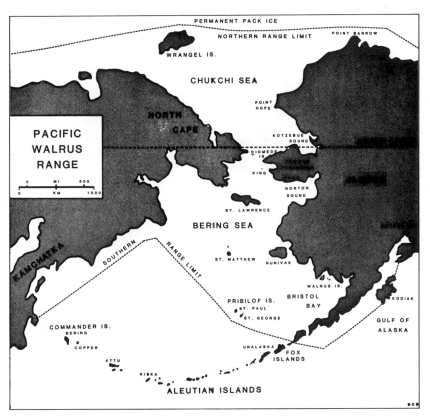

MAP 9 · Pacific Walrus Range

Notes

CHAPTER ONE

1 Remarks on the life history of the otter are based primarily upon Karl W. Kenyon, *The Sea Otter in the Eastern Pacific Ocean*. Other recent, more popular works are Jane H. Bailey, *Sea Otter: Core of Conflict, Loved or Loathed*, and John Woolfenden, *The California Sea Otter: Saved or Doomed?*.

2 Elliot C. Cowdin, "The Northwest Fur Trade," *Hunt's Merchants' Magazine* 14 (June 1846): 534.

3 John Ledyard, *Journal of Captain Cook's Last Voyage*, ed. J.K. Munford. The best account of the role of sealskins in the early China trade is James Kirker's authoritative study, *Adventures to China: Americans in the Southern Oceans, 1792–1812*. See also Frank E. Ross, "American Adventures in the Early Marine Fur Trade with China," *Chinese Social and Political Review* (Peking) 31 (1937): 221–67. On Cook's role, see Barry M. Gough, "James Cook and the Origins of the Maritime Fur Trade," *American Neptune* 38 (1978): 217–24.

4 Causes for American entrance into the Pacific have been explored by Magdalen Coughlin: "The Entrance of the Massachusetts Merchant into the Pacific," *Historical Society of Southern California Quarterly* 48 (1965): 327–52, and "Commercial Foundations of Political Interest in the Opening Pacific, 1789–1829," *California Historical Quarterly* 50 (1971): 15–33; see also Herman J. Deutsch, "Economic Imperialism in the Early Pacific Northwest," *Pacific Historical Review* 9 (1940): 377–88, and F.W. Howay, "Early Days of the Maritime Fur-Trade on the Northwest Coast," *Canadian Historical Review* 4 (1923): 26–44, and other studies by the same author listed in the bibliography.

5 Kirker, *Adventurers*, 6–7.

6 Howay, "Early Days," 27.

7 Kirker, *Adventurers*, chap. 1; Howay, "Early Days," and "Capt. Simon Metcalfe and the Brig *Eleanora*," *Washington Historical Quarterly* 16 (1924): 114–21. I have followed Kirker in spelling the name "Metcalf."

8 Howard I. Kushner's study, *Conflict on the Northwest Coast: American-Russian Rivalry in the Pacific Northwest, 1790–1867*, provides thorough detail on this conflict, and discusses Adam's role.

9 Raymond A. Rydell, *Cape Horn to the Pacific: The Rise and Decline of An Ocean Highway*. Details on the vessels may be found in F.W. Howay, "A List of Trading vessels in the Maritime Fur Trade, 1795–1824," *Transactions of the Royal Society of Canada*, 3rd ser., vols, 25–8, seriatim.

10 This point is stressed in Terrence J. Barragy, "The Trading Age, 1792–1844," *Oregon Historical Quarterly* 76 (1975): 196–224; see also Warren L. Cook, *Flood Tide of Empire: Spain and the Pacific Northwest, 1543–1819*.

11 Edouard A. Stackpole, *The Sea-Hunters: The New England Whalemen during Two Centuries, 1635–1835*, pp. xiii, 183–7, and *Whales and Destiny: The Rivalry between America, France and Britain for Control of the Southern Whale Fishery, 1785–1825*, chap. 1, discuss the role of Rotch and whaling in general. See also Kenneth J. Bertrand, *Americans in Antarctica, 1775–1948*, chap. 3.

12 Kirker, *Adventurers*, 10–12, 148.

13 General remarks on Russian activities are based upon Kushner, *Conflict*: Glynn Barratt, *Russia in Pacific Waters, 1715–1825*; Clifford M. Foust, *Muscovite and Mandarin: Russia's Trade with China and its Setting, 1725–1805*; James R. Gibson, *Imperial Russia in Frontier America: The Changing Geography of Supply of Russian America, 1784–1867*, and *Feeding the Russian Fur Trade: Provisionment of the Okhotsk Seaboard and the Kamchatka Peninsula, 1639–1856*; S.B. Okun, *The Russian-American Company*; Hector Chevigny, *Russian America: The Great Alaskan Venture, 1741–1867*, is a more popular account, as are his biographies of Baranov and Rezanov (see bibliography).

14 As viewed by Peter Freuchen, *Vagrant Viking*, 340.

15 The hunt for the otter is best studied in Adele Ogden's classic work, *The California Sea Otter Trade, 1784–1848*. See also T.A. Rickard, "The Sea-Otter in History," *British Columbia Historical Quarterly* 11 (1947): 15–32. A convenient discussion of numbers and prices which I have used here is found in Mary E. Wheeler, "Empires in Conflict and Cooperation: The 'Bostonians' and the Russian-American Company," *Pacific Historical Review* 40 (1972): 419–41.

16 The phrase is Chevigny's, *Russian America*, 3.

17 The history of early Russian fur-sealing is discussed in considerable detail in Fredericka Martin, *The Hunting of the Silver Fleece: Epic of the Fur Seal,* and a shortened and revised version, *The Sea Bears.* Martin herself is discussed below (p. 239).

18 On South African sealing, see Peter B. Best, "Seals and Sealing in South and South West Africa," *South African Shipping News* 28, no. 12 (December 1973): 55; and with P.D. Shaughnessy, "An Independent Account of Capt. Benjamin Morrell's Sealing Voyage to the South West-Coast of Africa in the Antarctic 1828/29," *Fishery Bulletin of South Africa* 12 (1979): 1–19; W. Wilfried Schuhmacher, "South African Light on American Fur Trade Vessels," *American Neptune* 40 (1980): 46–9, and W.H. Zur Strassen, *The Fur Seal of Southern Africa.*

19 On Uruguay, Hugh M. Smith, "The Uruguayan Fur-Seal Islands," *Zoologica* 9 (1927): 271–94.

20 Stackpole, *Sea-Hunters,* 208.

21 Hugh R. Mill, *The Siege of the South Pole,* 39.

22 Quoted in M.B.R. Cawkell, D.H. Maling, and E.M. Cawkell, *The Falkland Islands,* 34.

23 Ralph S. Kuykendall, "A Northwest Trader at the Hawaiian Islands," *Oregon Historical Quarterly* 24 (1923): 111–31 discusses the *Butterworth.* On this early sealing, see Gordon Jackson, *The British Whaling Trade*; 103; Mill, *Siege,* 87–8, 102; Bertrand, *Americans,* chap. 2; Kirker, *Adventurers,* chap. 4 (*Hancock,* 79).

24 In addition to Kirker, *Adventurers,* and Bertrand, *Americans,* see Ralph Lee Woodward, Jr, *Robinson Crusoe's Island: A History of the Juan Fernandez Islands.*

25 Standard references which indicate the taxonomic difficulties are Judith E King, *Seals of the World,* and Victor B. Scheffer, *Seals, Sea Lions and Walruses: A Review of the Pinnipedia.* The U.S. Marine Mammal Commission has for its own uses settled upon *Arctocephalus philippii* (Juan Fernández) and *A. townsendi* (Guadalupe): *Marine Mammal Names* (Washington, D.C.: Marine Mammal Commission, July 1976).

26 Kirker, *Adventurers,* 84–5.

27 Quoted in Woodward, *Crusoe's Island,* 39.

28 Edmund Fanning, *Voyages and Discoveries in the South Seas, 1792–1832,* 74.

29 Kirker, *Adventurers,* 66, 162. For other candidates, see Stackpole, *Sea-Hunters,* 207–8, 214 (Nantucket whaler, the *Olive Branch,* in 1795–6), and Bertrand, *Americans,* 24–5, who votes for the *Hancock.*

30 Amasa Delano, *A Narrative of Voyages and Travels, in the Northern and Southern Hemispheres, Comprising Three Voyages Round the World.* 306.

31 Fanning, *Voyages and Discoveries,* 79–80.

32 Delano, *Narrative,* 306.

33 Ibid., 255. Kirker, *Adventurers*, 34–6, discusses crews.

34 Simmons's contracts with crew members, letters of instruction to Simmons, and accounts and correspondence with Sullivan Dorr showing profits are all in the Sylvester Simmons Collection, Rhode Island Historical Society, Providence. The *Enterprise* and Dorr are discussed in Kirker, *Adventurers*, 163–5.

35 Kirker, *Adventurers*, chap. 3.

36 Joel Root, "Narrative of a Sealing and Trading Voyage in the Ship *Huron*, 1802–1806," *Papers of the New Haven Colony Historical Society* 5 (1894): 144–71. See also Edouard A. Stackpole, *The Voyage of the 'Huron' and the 'Huntress': The American Sealers and the Discovery of the Continent of Antarctica*, 227–31.

37 Barnard; Kirker, *Adventurers*, 81; 1806 date is used by W. Nigel Bonner, "The Fur Seal of South Georgia," *British Antarctic Survey Scientific Reports*, no. 56 (London 1968), 21. The story of John Wright was still common when gold rush-bound vessels paused here in the 1850s: see log of Schooner *Spray*, 12 March 1850, G.W. Blunt White Library, Mystic, Conn., Log 510.

38 Log of Brig *Rebecca*, 1797–9, microfilm roll no. 1, G.W. Blunt White Library; Kirker, *Adventurers*, 92.

39 Log of ship *Minerva*, 1802–5, microfilm reel no. 32, G.W. Blunt White Library.

40 Ibid., 20 July 1803; also quoted in Kirker, *Adventurers*, 84.

41 *Minerva* log, 1 April 1804.

42 Ibid., 29 October 1804.

43 No better account of the difficulties of dealing with a *baidarka* is to be found than Harold McCracken, *Hunters of the Stormy Sea*, especially 56,· 189; not the least problem is sitting in an "L"-shape, with legs straight out, for days at a time.

44 See sources listed in n. 13 above, and William Dane Phelps, "Solid Men of Boston in the Northwest," Bancroft Library, University of California, Berkeley, P-C 31. On Phelps, see Briton Cooper Busch, ed., *Alta California, 1840–1842: The Journal and Observations of William Dane Phelps, Master of the Ship "Alert"*, Introduction.

45 David A. Henderson, *Men and Whales at Scammon's Lagoon*, 102.

46 Kirker, *Adventurers*, 96, 153–4.

47 In addition to Kirker, *Adventurers*, 155, and Phelps, see Robin W. Doughty, "The Farallones and the Boston Men," *California Historical Quarterly* 53 (1974): 309–16, and E.Q. Essig, "The Russian Settlement at Ross," ibid. (1933): 191–209.

48 Philip L. Walker and Steven Craig, "Archaeological Evidence Concerning the Prehistoric Occurrence of Sea Mammals at Point Bennett, San Miguel Island," *California Fish & Game* 65 (1979): 50–5.

49 [Zakahar Chichinov], "Adventurers of Zakahar Tchitchinoff, an Employee of the Russian-American Company, 1802–1878," Kodiak, 1878, Bancroft Library manuscript, 1878 P-K 25.

50 Camille de Roquefeuil, *A Voyage Round the World between the Years 1816–1819*, quoted in Howay, "Early Days," 40.

51 Typescript log of ship *Amethyst*, 1806–11. Phillips Library, Peabody Museum, Salem, 656/1806R.

52 A basic source is L. Harrison Matthews, *South Georgia*; see also Bonner, "Fur Seal of South Georgia," and "Exploitation and Conservation of Seals in South Georgia," *Oryx* 4 (1958): 373–80.

53 Farning, *Voyages*, 216.

54 Thomas W. Smith, *A Narrative of the Life, Travels and Sufferings of …*, 145.

55 Stackpole, *Sea-Hunters*, 211.

56 Kirker, *Adventurers*, 35–6.

57 R. Gerard Ward, ed., *American Activities in the Central Pacific, 1790–1870* 8:35, citing a report in the *Salem Gazette*, 8 August 1797.

58 Farning, *Voyages and Discoveries*, 217–8.

59 Quoted in Bertrand, *Americans*, 28.

60 Farning, *Voyages and Discoveries*, 223.

61 Bertrand, *Americans*, 31–2.

62 Ibid., chap. 3, gives a detailed account of the *Hersilia*'s voyage; see also Stackpole, *Sea-Hunters*, chap. 25.

63 Bertrand, *Americans*, 32–7.

64 James Eights, "Description of the New South Shetland Isles," in Edmund Fanning, *Voyages to the South Seas, Indian and Pacific Oceans, China Sea, Northwest Coast, Feejee Islands, South Shetlands …*, 198–9.

65 In addition to the works of Stackpole, Bertrand, and Mill, see also Philip I. Mitterling, *America in the Antarctic to 1840*; Edwin Swift Balch, "Stonington Antarctic Explorers," *Bulletin of the American Geographic Society* 41 (1909): 473–92, and "Antarctic Addenda," *Journal of the Franklin Institute* 157 (1904): 81–8. On Palmer, John R. Spears, *Captain Nathaniel Brown Palmer*, and George L. Campbell, "Nathaniel Brown Palmer and the Discovery Antarctica," *Historical Footnotes* (Bulletin of the Stonington Historical Society) 8 no. 4 (August 1971): 1–7. Palmer's famous encounter with the Russian explorer Bellingshausen is described in the latter's *Voyage … 1819–1821* 2:425–16. The log of the *Huron* for 1821–2 may be read on microfilm reel no. 2, G.W. Blunt White Library; that of the *Hero* is in the Palmer-Loper Papers, Library of Congress, box 4.

66 James W.S. Marr, "The South Orkney Islands," *Discovery Reports* 10 (November 1935): 282–382, gives much information on these islands; L. Harrison Matthews, *Penguins, Whalers, and Sealers: A Voyage of Discovery*, 87.

67 The discoveries of Bouvet, Marion, Crozet, and Kerguelen are discussed in John Dunmore, *French Explorers in the Pacific*, 2:166–249.

68 Auguste Toussaint's *History of the Indian Ocean*, for example, relegates these islands to an appendix on "the Southern Sea," 249–51, while Russell Owen's *The Antarctic Ocean*, is a sample of the other extreme.

69 [Great Britain, Challenger Office,] *Report on the Scientific Results of the Voyage of H.M.S. Challenger during the Years 1873–76*, ed. Sir C. Wyville Thomson, 1 (London 1885): 296.

70 Fanning, *Voyages and Discoveries*, chap. 17, discusses the voyage of the *Catherine*, which visited the Prince Edwards in 1803 to make contact with groups then ashore.

71 Ibid., 247.

72 Charles Medyett Goodridge, *Narrative of a Voyage to the South Seas, and the Shipwreck of the "Princess of Wales" Cutter, with an Account of Two Years Residence on An Uninhabited Island …*, 78.

73 P.R. Condy, "Distribution, Abundance, and Annual Cycle of Fur Seals (*Arctocephalus* spp.) on the Prince Edward Islands," *South African Journal of Wildlife Research* 8 (1978): 159–68.

74 [Great Britain, Challenger Office.] *Report*, 320–2, discusses the Crozets (p. 322 quoted, including Ross's visit).

75 On Desolation, see Briton C. Busch, ed., *Master of Desolation: The Reminiscences of Capt. Joseph J. Fuller*, especially xxv–xxix and references there cited; [Great Britain, Challenger Office] *Report*, vol. I, chap. 9. Americans at Desolation are discussed below, chap. 6.

76 Howay, "Metcalfe", Kirker, *Adventurers*, 60–1, Bertrand, *Americans*, 235–7; Stackpole, *Sea-Hunters*, chap. 14, all discuss the first sealers at Desolation. See also Rhys Richards, "The Journal of Erasmus Darwin Rogers, The First Man on Heard Island," *The American Neptune* 41 (1981): 280–305, who discusses the *Hillsborough*.

77 Toussant, *History*, 250; Kirker, *Adventurers*, 50–8.

78 Kirker, *Adventurers*, 55–6, for both the *Emily* and *Warren*.

79 J.S. Cumpston, *Kangaroo Island, 1800–1836*, 3. General sources for this section are John Beck, *A Maritime History of Australia*, and C. Hartley Grattan, *The Southwest Pacific to 1900*.

80 Kirker, *Adventurers*, 57, 105.

81 Ibid., 107; Ian S. Kerr, *Campbell Island: A History*, 7; A. Charles Begg and Neil C. Begg, *Dusky Bay*, 69, 74; Thomas Dunbabin, "Whalers, Sealers and Buccaneers," *Journal of the Royal Australian Historical Society* 11 (1925): 8–9.

82 J.S. Cumpston, *First Visitors to Bass Strait*, gives a thorough account of the *Nautilus*, 7–12. These early voyages may also be studied through the same author's *Shipping Arrivals & Departures, Sydney, 1788–1825* (Canberra 1977); Helen Mary Micco, *King Island and the Sealing Trade*,

1802, and J.C.H. Gill, "Notes on the Sealing Industry of Early Australia," *Journal of the Royal Historical Society of Queensland* 8 (1967): 213–45; A.G.E. Jones, "The British Southern Whale and Seal Fisheries," *The Great Circle (Journal of the Australian Association for Maritime History)* 3 (1981): 20–9, 90–102; and Barbara Little, "The Sealing and Whaling Industry in Australia before 1850," *The Economic Record* (Burwood, Victoria) 45 (1969): 109–27.

83 Quoted in Kerr, *Campbell Island*, 7.

84 Cumpston, *First Visitors*; see also R.M. Fowler, *The Furneaux Group, Bass Strait: A History.*

85 William John Dakin, *Whaleman Adventurers: The Story of Whaling in Australian Waters*, 30; Cumpston, *Shipping Arrivals*, 32.

86 Kirker, 101–9 (101 quoted), 113–22; Fanning, *Voyages and Discoveries.*

87 Cumpston, *First Visitors*, 15–16, and *Kangaroo Island*, 112; Kirker, *Adventurers*, 108; Stackpole, *Sea-Hunters*, 233–5; Delano, *Voyages*, 460 (quoted in Cumpston, *First Visitors*, 23).

88 Kirker, *Adventurers*, 110.

89 Nigel Wace and Bessie Lovett, *Yankee Maritime Activities and the Early History of Australia*, 3.

90 Harry O'May, *Hobart River Craft and Sealers of the Bass Strait*, 12.

91 Bach, *Maritime History*, 50–2, 70–76; Kerr, *Campbell Island*, 25; D.R. Hainsworth, *Builders and Adventurers: The Traders and the Emergence of the Colony, 1788–1821*, 84 ff. See also Hainsworth's "Iron Men in Wooden Ships: The Sydney Sealers, 1800–1820," *Labour History* (Sydney), no. 13 (November 1967): 19–25, and "Exploring the Pacific Frontier: The New South Wales Sealing Industry, 1800–1821." *Journal of Pacific History* 2 (1967): 59–75.

92 See R.A. Falla, "The Outlying Islands of New Zealand," *New Zealand Geographer* 4 (1948): 127–54, and Kerr, *Campbell Island.* Detailed information on sealing on these islands is available only for the Chatham group in Rhys Richards's useful study of primary sources, *Whaling and Sealing at the Chatham Islands.*

93 Thomas Dunbabin, "New Light on the Earliest American Voyages to Australia," *American Neptune* 10 (1950): 52–64; Werner Levi, "The Earliest Relations between the United States of America and Australia," *Pacific Historical Review* 12 (1943): 351–61.

94 R.A. Falla, "Exploitation of Seals, Whales, and Penguins in New Zealand," *Proceedings of the New Zealand Ecological Society* 9 (1962): 36; see also D.E. Gaskin, *Whales, Dolphins and Seals with Special Reference to the New Zealand Region*, 45–51, 139–62.

95 J.S. Cumpston, *Macquarie Island*, is the principal source on Macquarie; see also Kerr, *Campbell Island*, chap. 2, and A.M. Gwynn, "Notes

on the Fur Seals at Macquarie Island and Heard Island," *Australian National Antarctic Research Expedition, Interim Reports*, no. 4 (January 1953).

96 Bellingshausen, *Voyage*, 363–73; also quoted in Cumpston, *Macquarie*, 42–6.

97 Quoted in Cumpston, *Macquarie*, 35.

98 Ibid., 26.

99 Kerr, *Campbell Island*, 153–4.

100 Mary E. Gillham, *Sub-Antarctic Sanctuary: Summertime on Macquarie Island*, 154.

101 Thomas Philbrick, *James Fenimore Cooper and the Development of American Sea Fiction*.

102 Calvin Martin, *Keepers of the Game: Indian-Animal Relationships and the Fur Trade*.

103 Kirker, *Adventurers*, 86, notes, for example, a New England China trader distributing Spanish translations of the Declaration of Independence in Valparáiso in 1802.

104 We know the first sealing at Más Afuera was in 1792–3, and it seems there was none again until 1798–1801. Another wave followed in 1802–5. The highest number of vessels was in 1800 (14) and 1803 (15). If there were 10 ships every year from 1792 to 1805 (14 years), and each took 20,000 the 3,000,000 would not quite be reached – yet it seems fairly clear that this is too high a total number of ship visits and a very high average cargo, leading to my doubt about Delano's estimate.

105 Kirker, *Adventurers*, 167.

CHAPTER TWO

1 Horace Beck, *Folklore and the Sea*, 203–20, 301; David Thomson, *The People of the Sea*, and, for a fictionalized account, Ronald Lockley. *Seal Woman*. For a short example, see Kipling's story, "The Conversion of St. Wilfrid," in *Rewards and Fairies* (1910), in which a "great grey-muzzled old dog-seal" called "Padda" rescues a man from a reef.

2 Cassie Brown, *Death on the Ice: The Great Newfoundland Sealing Disaster of 1914*.

3 Farley Mowat and John de Vissier, *This Rock Within the Sea: A Heritage Lost*; (with etchings by David Blackwood), *Wake of the Great Sealers*. See also Mowat's *A Whale for the Killing*. The essential critical examination is Patrick O'Flaherty, *The Rock Observed: Studies in the Literature of Newfoundland*.

4 There are many accounts of the life history of the harp seal (see bibliography), ranging from the popular (e.g., Fred Bruemmer, *The Life of the Harp Seal*) to the scientific. A convenient recent summary is

K. Ronald and J.L. Dougan, "The Ice Lover: Biology of the Harp Seal (*Phoca groenlandica*)," *Science* 215, no. 4535 (19 February 1982): 928–33. For recent information on populations, past and present, and harp seal biology in general, I am particularly indebted to Dr. David E. Sergeant of the Arctic Biological Station, Ste-Anne de Bellevue, Quebec. On migrations, see his "Migrations of Harp Seals *Pagophilus groenlandicus* (Erxleben) in The Northwest Atlantic," *Journal of the Fishery Resources Board of Canada* 22 (1965): 433–64.

5 Ronald and Dougan, "Ice Lover," 928.

6 William Howe Greene, *The Wooden Walls Among the Ice Floes*, 11.

7 Laurence Irving, "Temperature Regulation in Marine Mammals," in Harald T. Anderson, ed., *The Biology of Marine Mammals*, 147–74.

8 Ronald and Dougan, "Ice Lovers," 931.

9 C.W. Andrews, "The Hazardous Industry of North Atlantic Sealing," *Animal Kingdom* 54 (1951): 75. A good account of hood sealing is Capt. William Barron, *Old Whaling Days*, 68–9.

10 The standard source of the cod fishery is Harold A. Innis, *The Cod Fisheries: The History of an International Economy*; see Ralph Greenlee Lounsbury, *The British Fishery at Newfoundland, 1634–1763*. The *Marygold* is discussed in Lewis Amadeus Anspach, *A History of the Island of Newfoundland ...*, 75–6 (quoted).

11 Innis, *Cod Fisheries*, 153n.

12 See Michael Staveley, "Migration and Mobility in Newfoundland and Labrador: A Study in Population Geography" (PH D diss., University of Alberta, 1973).

13 An excellent account, upon which this paragraph is principally based, is Chesley Sanger, "The Evolution of Sealing and the Spread of Permanent Settlement in Northeastern Newfoundland," in John J. Mannion, ed., *The Peopling of Newfoundland*, 136–51.

14 In addition to Sanger, ibid., and Innis, *Cod Fisheries*, the account of net sealing is based upon George Cartwright, *A Journal of Transactions and Events, During a Residence of Nearly Sixteen Years on the Coast of Labrador*; D.W. Prowse, *A History of Newfoundland from the English, Colonial, and Foreign Records*, 419–20, reprinting a letter from J. Bland, Bonavista, to Governor Gambier, 26 September 1802, on the early seal fishery; James Murphy, "The Old Sealing Days" (typescript of articles in St. John's *Evening Herald*, 1916) Centre for Newfoundland Studies, Memorial University of Newfoundland (CNS/MUN), St. John's; A.M. Lysaght, *Joseph Banks in Newfoundland and Labrador, 1766* (London, 1971); Capt. Temple [otherwise unidentified], "Seal Fisheries," Great International Fisheries Exhibition, Metropolitan Central Library, Toronto (July 1883), 155–74, CNS/MUN; W.A. Black, "The Labrador Floater Codfishery," *Annals of the Association of American Geographers* 50

(1960): 267–93. Innis adds that "Irish Catholics ate their way into the fishery more effectively than Protestants" (*Cod Fisheries* 387n), though offering no data in support of what appears a logical assumption.

15 Murphy, "Old Sealing Days," 3.

16 L.G. Chafe, *Chafe's Sealing Book: A History of the Newfoundland Sealfishery from the Earliest Available Records ... to 1923*, 15.

17 Quoted in Prowse, *History of Newfoundland*, 420.

18 Innis, *Cod Fisheries*, 303.

19 Cartwright, *Journal*, passim. Other British sealers concentrated upon Greenland; see the useful account of Thomas Southwell, *The Seals and Whales of the British Seas*, 83–7.

20 J.S. Colman, "The Present State of the Newfoundland Seal Fishery," *Journal of Animal Ecology* 6 (1937): 146, 149–50.

21 Shannon Ryan, "The Newfoundland Cod Fishery in the Nineteenth Century" (MA thesis, Memorial University of Newfoundland, 1971), gives tables on cod and seal from Colonial Office and Newfoundland Journal sources from 1803 to 1900 which should be used in conjunction with Chafe's annual tabulation.

22 Fred Bruemmer, "A Year in the Life of the Harp Seal," *Natural History* 84 no. 4 (April 1975): 49.

23 Capt. Abram Kean, *Old and Young Ahead: A Millionaire in Seals*, 131.

24 Such statistics as do exist, including those on losses, are calculated in *Chafe's Sealing Book*.

25 Ibid., 39–44, (42 quoted).

26 Ibid., 39.

27 On the confusion of "beaver hats" and "jackass brigs," see G.M. Story, W.J. Kirwin, and J.D.A. Widdowson, eds., *Dictionary of Newfoundland English* (Toronto 1982), 36, 272. Remarks on the development of the schooner fishery are based, save where otherwise noted, on Chesley W. Sanger's valuable thesis, "Technological and Spatial Adaptation in the Newfoundland Seal Fishery during the 19th Century" (MA thesis, Memorial University of Newfoundland, 1973).

28 *Chafe's Sealing Book*; Kean, *Old and Young Ahead*, 129.

29 Shannon Ryan, "The Newfoundland Salt Cod Trade in the Nineteenth Century," in James Hiller and Peter Neary, eds., *Newfoundland in the Nineteenth and Twentieth Centuries: Essays in Interpretation*, 45.

30 Eric W. Sager, "The Port of St. John's, Newfoundland, 1840–1889: A Preliminary Analysis," in Keith Matthews and Gerald Panting, eds., *Ships and Shipbuilding in the North Atlantic Region*, 35.

31 Eric W. Sager, "The Merchants of Water Street and Capital Investment in Newfoundland's Traditional Economy," in Lewis R. Fischer and Eric W. Sager, eds., *The Enterprising Canadians: Entrepreneurs and Economic Development in Eastern Canada, 1820–1924*, 79.

32 Sager, "Port of St. John's," 21.

33 Ibid., 24.

34 Discussion of the truck and credit systems is based upon Robert J. Brym and Barbara Neis, "Regional Factors in the Formation of the Fisherman's Protective Union of Newfoundland," in R.J. Brym and R. James Sacouman, eds., *Underdevelopment and Social Movements in Atlantic Canada*, 203–18; S.J.B. Noel, *Politics in Newfoundland*; in addition to Sanger, "Technological and Spatial Adaptation," and Black, "Labrador Floater Codfishery" ("lingering feudalism," 259). Eric W. Sager has made the apt point that it is very difficult to make any simple characterization of the people of Newfoundland, given the province's religious, ethnic, and social divisions, and despite the "muddled posturing of modern romantics," meaning writers such as Farley Mowat. "Newfoundland's Historical Revival and the Legacy of David Alexander," *Acadiensis* 9 (1981), 104–15 (110 quoted).

35 Innis, *Cod Fisheries*, 155–6.

36 John Feltham, "The Development of F.P.U. in Newfoundland (1908–1923)" (MA thesis, Memorial University of Newfoundland, 1959), 1–14.

37 Richard Gwyn, *Smallwood: The Unlikely Revolutionary*, 22.

38 J.G. Millais, *Newfoundland and Its Untrodden Ways*, 42, quoting Bob Saunders of Green's Pond, Bonavista Bay.

39 Peter Cashin, *My Life and Times, 1890–1919*.

40 Figures from Innis, *Cod Fisheries*, 304.

41 Discussion of an average voyage is based upon Sanger, "Technological and Spatial Adaptation"; Ryan, "Salt Cod Trade"; Philip Toque, *Newfoundland: As It Was, As It Is in 1877*, 3–4; Michael Carroll, *The Seal and Herring Fisheries of Newfoundland*; and Ottar Brox, *Newfoundland Fishermen in the Age of Industry: A Sociology of Economic Dualism*.

42 Interview by Margaret M. Ashley with George Batten, Spring 1974, typescript, Folklore Archive, Memorial University of Newfoundland, no. 74–146.

43 Berth money and strikes: Feltham, "Development of the F.P.U."; Murphy, "Old Sealing Days," 9–11; Philip Toque, *Wandering Thoughts, or Solitary Hours*, 188; R. Hattenhauer, "A Brief Labour History of Newfoundland," ms. report for Royal Commission on Labour Legislation in Newfoundland and Labrador (1970, copy in CNS/MUN), 95; Pol Chantraine, *The Living Ice*, 143–4; *Chafe's Sealing Book*, 38.

44 David King, quoted by Murphy, "Old Sealing Days," 9.

45 The phrase is Cassie Brown's, *Death on the Ice*, 20.

46 Murphy, "Old Sealing Days," 47.

47 J.K. Hiller, "A History of Newfoundland, 1874–1901" (PH D diss. Cambridge University, 1971), 8.

48 Robert Hunter, *Warriors of the Rainbow: A Chronicle of the Greenpeace Movement* 251; Newfoundland House of Assembly, 23rd sess., app. 516, 19 May 1913, Newfoundland Provincial Archives.

49 Joan Roper Scott, "The Function of Folklore in the Interrelationship of

the Newfoundland Seal Fishery and the Home Communities of the Sealers" (MA thesis, Memorial University of Newfoundland, 1975), 175; Loretta Marie Hayes, "Stories about the Labrador Fishery and the Seal Fishery by John Fitzgerald of South River, Conception Bay," Winter 1976, Folklore Archive, Memorial University of Newfoundland, no. 76–316.

50 Nicholas Smith, *Fifty-two Years at the Labrador Fishery*, is an example of what was probably typical; in the fifty-two years, he went only ten times to the ice.

51 Quoted by Ralph Barrett, Happy Valley, Labrador, *Decks Awash* (Memorial University of Newfoundland Extension Service), 7, no. 1 (February 1978): 51. The verse is common in many forms; another is quoted in Cynthia Lamson, *"Bloody Decks and a Bumper Crop": The Rhetoric of Sealing Counter-Protest*, 13:

Harbour Grace is a pretty place
And so is Peeley's Island
Daddy's going to buy me a brand new dress
When the boys come home from swilin'.

Lamson (p. 25) notes the control meaning of "harvest" and "bumper crop."

52 The account of men, gear, and methods is based upon Sanger, "Technology": Brown, *Death on the Ice*; Greene, *Wooden Walls Among the Ice-Floes*; George Allen England, *Vikings of the Ice: Being the Log of a Tenderfoot on the Great Newfoundland Seal Hunt*; Cashin, *My Life and Times*; Wilfred T. Grenfell, *Adrift on an Ice-Pan* and *Vikings of Today*.

53 Philip Sheppard, "Duncan Parsons of Harbour Grace South – Reminiscences of the Seal Fishery," Winter 1976; Parsons, interestingly, thought that at least half the crew on his first sealing trip in 1912 or 1913 had never been to the ice before. Another equally interesting account is Pauline G. Snow, "Albert Snow: Some Personal Experiences of the Seal Fishery in Newfoundland," November 1975. Both manuscripts are in the Folklore Archive, Memorial University of Newfoundland, no. 76–148.

54 Captain Robert A. Bartlett, "The Sealing Saga of Newfoundland," *National Geographic* 55 (1929): 95: "A man snogging seals has to be light-footed, running and jumping over the ice like a deer."

55 J.B. Jukes, *Excursions in and About Newfoundland, During the Years 1839 and 1840*, 273, 291, 176, all quoted in order.

56 Toque, *Wandering Thoughts*, 196.

57 Hayes, "Stories about the Labrador Fishery."

58 Wilfred T. Grenfell, "The Seal Hunters of Newfoundland," *Leisure Hour* (London) (1897–8) : 290 (copy at CNS/MUN).

59 Kaare Rodahl, " 'Spekk-finger' or Sealer's Finger," *Arctic* 5, no. 4 (December 1952): 235–40.

60 On the oil preparation, Anspach, *History of Newfoundland*, 424–5;
    Carroll, *Seal and Herring Fisheries*, 30; Daniel George Archibald, *Some
    Account of the Seal Fishery of Newfoundland, and the role of Preparing Seal
    Oil* (copy at CNS/MUN); Charles H. Stevenson, "Oil from Seals, Walrus,
    etc.," *Scientific American* 57 (1904) 23614–5; Newfoundland House of
    Assembly, 23rd sess., app., report of 19 May 1913, 515 (acid
    content); W. John Earle, "Seal Oil Operations in Twillingate, 1905–
    1955," Winter 1978, Folklore Archive, Memorial University of
    Newfoundland, no. 78–242. Sealskinners: James J. Fogarty. "The Seal-
    Skinners' Union," in J.R. Smallwood, ed., *The Book of Newfoundland*
    2: 100; Hattenhauer, "Brief Labour History," 93–108.
61 Archibald, "Some Account of the Seal Fishery," 6.
62 It was also used in Norway to adulterate "cod liver oil," unknown to
    the prospective consumer, according to Hugh M. Smith, "The
    Uruguayan Fur-Seal Islands," 289.
63 John R. Arnold, *Hides and Skins*, 518–21; Charles H. Stevenson,
    "Leather from Seal Skins," *Scientific American* 59 (1905): 24334–5, and
    "Utilization of the Skins of Aquatic Animals," *Report of the Commissioner
    of Fish and Fisheries for the Year ending 30 June 1902* (Washington,
    D.C., 1904), 28: 281–352.

CHAPTER THREE

1 *Chafe's Sealing Book* as usual supplies the data, along with Sanger,
   "Technology," on technical changes; an essential source on the Scottish
   vessels is Basil Lubbock, *The Arctic Whalers*. See also Chesley W.
   Sanger, "The 19th Century Newfoundland Seal Fishery and the
   Influence of Scottish Whalemen," *Polar Record* 20 (1980), 231–52. An
   appendix in Greene, *Wooden Walls*, lists all the steamers, wooden and
   steel, with data on tonnage, etc.
2 Aside from the sources listed in note 1, for the summary history of the
   individual firms I have used Paul O'Neill, *The Story of St. John's,
   Newfoundland*.
3 In addition to O'Neill, *St. John's*, see David Keir, *The Bowring Story*;
   Arthur C. Wardle, *Benjamin Bowring and His Descendants*; and Robert
   Brown Job, *John Job's Family*; all are uncritical.
4 Donald K. Regular, "The Commercial History of Munn & Co.,
   Harbour Grace" (research paper, Maritime History Group, Memorial
   University of Newfoundland, St. John's, 1971).
5 Sager, "Merchants of Water Street."
6 Cashin, *My Life and Times*, 44–5.
7 James K. Hiller, *The Newfoundland Railway 1881–1949*; Sanger,
   "Technology," chap. 7, and "Evolution of Sealing," 46–8.
8 D.W. Prowse, *A History of Newfoundland*, 452.
9 Ian McDonald, "W.F. Coaker and the Balance of Power Strategy: The

Fisherman's Protective Union in Newfoundland Politics," in James Hiller and Peter Neary, eds., *Newfoundland in the Nineteenth and Twentieth Centuries: Essays in Interpretation*, 149.

10 Feltham "Development of the F.P.U.," 14.

11 On vessels and conditions aboard, Sanger, "Technology"; Greene, *Wooden Walls*; Brown, *Death on the Ice*; England, *Vikings of the Ice*; statistics from *Chafe's Sealing Book*.

12 *Chafe's Sealing Book*, 36.

13 William Bixby, *Track of the "Bear"*; Polly Burroughs, *The Great Ice Ship "Bear"*; of lasting value, however, is M.A. Ransom, *Sea of the "Bear": Journal of a Voyage to Alaska to the Arctic, 1921*.

14 Kean, *Old and Young Ahead*, 134.

15 England, Grenfell, and Jukes all went as observers to the ice; see also Sir M.G. Winter, "Trip to the Seal Fishery, March 1907," *Newfoundland Quarterly* 75, no. 2 (February 1979): 21–7.

16 Clarence Deming, *By-ways of Nature and Life*, 99.

17 Wilfred T. Grenfell, "The Seal Hunters of Newfoundland," 286.

18 Once again, the best discussion of changing tactics is found in Sanger, "Technology," chap. 5. See also Scott, "Function of Folklore"; Scott, who has made a number of small-boat passages across the Atlantic, includes a thorough discussion of technique in his study.

19 Smith, *Fifth-two Years*, was one such.

20 David Moore Lindsay, *A Voyage to the Arctic in the Whaler "Aurora"*, 48.

21 Smith, *Fifty-two Years*, 48–54, discusses his voyage with Blandford (50 quoted).

22 Newfoundland Assembly Act, 36 Vic., c. 9, 1873.

23 Sanger, "Technology," 192.

24 *Chafe's Sealing Book*, 87–105, gives complete records for all sealing captains to 1923.

25 In addition to Chafe, the *Diana's* voyage is discussed in Lubbock, *Arctic Whalers*, 394 (value in 1868), and 424–7 (1892–3 seasons). The figure of $20,000 for preparation charges was stated by Grieve during the course of the negotiations between owners and sealers in the 1902 strike, St. John's *Evening Herald*, 10 March 1902.

26 David Alexander, "Newfoundland's Traditional Economy and Development to 1934," *Acadiensis* 5, no. 2 (Spring 1976): 56–78; see also Patricia Thornton, "Some Preliminary Comments on the Extent and Consequences of Out-Migration from the Atlantic Region, 1870–1920," in Lewis R. Fischer and Eric W. Sager, eds., *Merchant Shipping and Economic Development in Atlantic Canada*.

27 General remarks in the following four paragraphs are based principally upon Noel, *Politics in Newfoundland*; Peter Neary, "Democracy in Newfoundland: A Comment," *Journal of Canadian Studies* 4 (1969): 89–108; Alexander, "Newfoundland's Traditional Economy," and other

articles (see particularly James Hiller, "The Railway and Local Politics in Newfoundland, 1870–1901," in Hiller and Neary, eds., *Newfoundland in the Nineteenth and Twentieth Centuries*).

28 Boyle to Colonial Secretary Joseph Chamberlain, 12 March 1902, Governor-General's Correspondence, G/1/1/7 (1902), Provincial Archives of Newfoundland.

29 David Alexander, "Literacy and Economic Development in Nineteenth Century Newfoundland," *Acadiensis* 10 (1980): 3–34.

30 Neary, "Democracy," 92.

31 Gwyn, *Smallwood*, xii.

32 Ibid., 48–9, quoting Smallwood from the *Book of Newfoundland*.

33 See, for example, Keith Thomas, *Man and the Natural World: A History of the Modern Sensibility* (New York 1983).

34 Brym and Neis, "Regional Factors," 208.

35 England, *Vikings of the Ice*, 53–4.

36 On sealskinners' union, see sources noted above, p. 279n63.

37 Cashin, *My Life and Times*, 28; Arthur Fox, *The Newfoundland Constabulary*, 91–2.

38 *Trade Review*, (St. John's), 15 March 1902. See David Frank and Nolan Reilly, "The Emergence of the Socialist Movement in the Maritimes, 1899–1916," *Labour/Le Travailleur* 4 (1979), who note in an appendix (pp. 112–13) that the socialist movement in Newfoundland began only in 1906.

39 The following account of the events of March 1902, is based upon three St. John's newspapers, the *Daily News, Evening Herald,* and *Evening Telegram* for the week of 7–14 March, as well as Boyle to Chamberlain, 12 March 1902. Further information may be found in Frank W. Graham, *"We Love Thee, Newfoundland": Biography of Sir Cavendish Boyle, KCMG, Governor of Newfoundland 1901–1904,* and "J.R. McCowen, J.P., A.D.C., Inspector-General Constabulary," *Newfoundland Quarterly* 2 (1902): 15; McCowen had a considerable role to play in the strike.

40 Quoted in Shannon Ryan and Larry Small, *Haulin' Rope & Gaff: Songs and Poetry in the History of the Newfoundland Seal Fishery*, 65.

41 Boyle to Chamberlain, 12 March 1902, Governor-General's Correspondence.

42 Henry Youmans Mott, ed., *Newfoundland Men*, 21; *Who's Who In and From Newfoundland*, 119–20. Morine was elected for Bonavista in 1886, 1889, 1893, 1897, 1900, and 1904. His role is mentioned at several points in Noel, *Politics in Newfoundland*.

43 Rev. E. Hunt, "The Great Sealers' Strike of 1902," *Newscene* (supplement to St. John's *Daily News*), 61 March 1970.

44 The quotation is from the *Evening Herald*, 11 March. Cashin, *My Life and Times*, 44–5, gives a portrait of Jackman. *Chafe's Sealing Book* gives voyages and totals for all sealing masters.

45 England, *Vikings of the Ice*, offers this among a selection, 255; most have a clearly utilitarian background ("an empty craft always looms high"; "in a leaky punt with a broken oar, it's always best to hug the shore," and so on).

46 The *Daily News*, 11 March 1902, is the only paper to mention this point, without identifying the magistrate who acted or the time and place.

47 Boyle to Chamberlain, 12 March 1902, Governor-General's Correspondence.

48 Quoted in Graham, "*We Love Thee, Newfoundland*," 142–3.

49 On the FPU and Coaker, see Feltham, "Development of the FPU"; Ian McDonald, "W.F. Coaker and the Balance of Power Strategy: The Fishermen's Protective Union in Newfoundland Politics," in Hiller and Neary, *Newfoundland in the Nineteenth and Twentieth Centuries*, 148–80; and J.R. Smallwood's panegyric, *Coaker of Newfoundland: The Man who Led the Deep-Sea Fisherman to Political Power*. Coaker's *The History of the Fishermen's Protective Union of Newfoundland* is unfortunately only a journal of proceedings.

50 Noel, *Politics*, 81–6.

51 G.E. Panting, "The Fishermen's Protective Union of Newfoundland and the Farmer's Organizations in Western Canada," *Canadian Historical Association Report* (1963), 143, makes this point.

52 Hattenhauer, "Brief Labour History," 104–5.

53 Noel, *Politics*, 88.

54 The acts, respectively, are 38 Vic., c. 9 (1873), 46 Vic., c. 1 (1879); 50 Vic., c 23 (1887), and 51 Vic., c. 4 (1898).

55 3 Edw. 7, c. 17 (1899).

56 4 Geo. 5, c. 19 (1914).

57 Brown, *Death on the Ice*, is the best account of the 1914 disasters.

58 6 Geo. 5, c. 24 (1916).

59 Winter, "Trip to Seal Fishery," stresses the need for a doctor on each vessel, owing to the number of accidents.

60 St. John's *Evening Telegram*, 13 March 1903 (clipping file on sealing, Newfoundland Historical Society, St. John's).

61 Greene, *Wooden Walls*, app. lists the steel vessels; Sanger, "Technology," is the best study of their effect.

62 J.S. Colman, "The Present State of the Newfoundland Seal Fishery," *Journal of Animal Ecology* 6 (1937):157.

CHAPTER FOUR

1 The literature on the northern fur seal is vast, as the bibliography indicates. I have relied particularly upon the works of Victor B. Scheffer (a popular but accurate account, for example, is his *The Year of the*

Seal), Clifford Fiscus, and Karl Kenyon. To all three I owe a personal debt of obligation; the description which follows is, I hope, a fair composite of their views. A recent valuable overview is W. Nigel Bonner, *Seals and Man: A Study of Interactions*, especially 12–16, 46–53. Relative dimorphism depends very much upon the stage at which the measurements are made and what figure is used for adult males. A two-ton elephant seal male, for example, weighs nearly 7.5 times as much as a three-year old female after lactation (I am indebted to Burney Le Boeuf for this information).

2 George A. Bartholomew, "A Model for the Evolution of Pinniped Polygyny," *Evolution* 24 (1970): 546–59, discusses dimorphism and relates aggression to size; see also his "Behavioral Factors Affecting Social Structure in the Alaska Fur Seal," *Transactions of the 18th North American Wildlife Conference* (1953), 481–502. Bartholomew's experience with Pacific marine mammals is considerable; as one discussant jokingly but admiringly put it when the latter paper was delivered, Bartholomew "was one of the very few men alive who has ever attempted, and actually successfully taken the rectal temperature of an elephant seal" (p. 499). See also Victor B. Scheffer, *Adventures of a Zoologist*, 118.

3 The pregnancy rate has been much debated. A common assumption of near-100 per cent (or at least 80–90 per cent) was made over the years, but detailed scientific studies seem to indicate a rate closer to 60 per cent. Victor Scheffer and Ethel I. Todd, "History of Scientific Study of the Alaskan Fur Seal, 1786–1964," typescript, 2 vols. [1967], 3. (A copy is available in the Marine Mammal Division, Northwest and Alaska Fisheries Center, National Marine Fisheries Service, Seattle.)

4 Victor B. Scheffer, "Sealskins Alive," *Pacific Discovery* 13 (June 1960): 2–9.

5 Ian A. McLaren, "Seals and Group Selection," *Ecology* 48 (1960): 104–10.

6 Scheffer, *Adventures*, 88–9.

7 Karl W. Kenyon and Victor B. Scheffer, *A Population Study of the Alaska Fur-Seal Herd*.

8 Victor B. Scheffer and Karl W. Kenyon, "The Fur Seal Herd Comes of Age," *National Geographic* 101 (1952): 491–512.

9 Karl W. Kenyon and Ford Wilke, "Migration of the Northern Fur Seal, *Callorhinus ursinus*," *Journal of Mammalogy* 34 (1953): 86–98; Bartholomew, "A Model"; Karl W. Kenyon, "Diving Depths of the Steller Sea Lion and Alaska Fur Seal," *Journal of Mammalogy* 33 (1952): 245–6; Victor B. Scheffer, "Probing the Life Secrets of the Alaska Fur Seal," *Pacific Discovery* 3 (1950): 22–30.

10 Feeding habits and prey are discussed in D.J. Spaulding, "Comparative Feeding Habits of the Fur Seal, Sea Lion, and Harbour Seal on the

British Columbia Coast"; H. Kajimura, *The Opportunistic Feeding of Northern Fur Seals off California*; Ford Wilke and Karl W. Kenyon, "Migration and Food of the Northern Fur Seal," *Transactions of the 19th North American Wildlife Conference* (1954), 430–40; Ralph C. Baker, Ford Wilke, and C. Howard Baltzo, *The Northern Fur Seal*; Scheffer and Todd, "History of Scientific Study"; and, on the Japanese side, Ford Wilke, *Pelagic Fur Seal Research off Japan in 1950*, and F.H.C. Taylor, M. Fujinaga, and Ford Wilke, *Distribution and Food Habits of the Fur Seals of the North Pacific Ocean: Report of Cooperative Investigations by the Governments of Canada, Japan, and the United States of America.*

11 C.E. Gustafson, "Prehistoric Use of Fur Seals: Evidence from the Olympic Coast of Washington," *Science* 161 (1968): 49–51; Karl W. Kenyon, "Last of the Tlingit Sealers," *Natural History* 64 (1955): 294–8.

12 General sources on Russian America are listed in chap. 1, note 13 above; Russian sealing discussed in the next four paragraphs is based upon Scheffer and Todd, "History of Scientific Study," 4–19.

13 Peter A. Tikhmenev, *Historical Review of the Formation of the Russian-American Company and Its Activity up to the Present Time* (St. Petersburg, 1861; translation in the Bancroft Library, Berkeley), as used by Dorothy M. Jones, "A History of United States Administration in the Pribilof Islands, 1867–1946," 17. See also Martin, *Silver Fleece*, and, on Rezanov, Hector Chevigny, *Lost Empire: The Life and Adventures of Nikolai Petrovich Rezanov.*

14 Scheffer and Todd, "History," 14.

15 Gibson, *Imperial Russia in Frontier America*, 35.

16 William S. Laughlin, *Aleuts: Survivors of the Bering Land Bridge*, 21; 1816 and 1834 figures from Jones, "U.S. Administration," 24–5.

17 Jones, "U.S. Administration," 12–23; Laughlin, *Aleuts*, 12–14. The following paragraph is based upon the same sources.

18 Dorothy Knee Jones, *A Century of Servitude: Pribilof Aleuts under U.S. Rule*, 7.

19 The discussion of Russian-American relations in mid-century and prior to the purchase of Alaska is based upon Kushner, *Conflict on the Northwest Coast*; Gibson, *Imperial Russia*; and Chevigny, *Russian America.*

20 Kushner, *Conflict*, chap. 2.

21 Gibson, *Imperial Russia*, 24.

22 Kushner, *Conflict*, 121; ice: E.L. Keithahn, "Alaska Ice, Inc.," in Morgan B. Sherwood, ed., *Alaska and Its History*, 173–86.

23 Chevigny, *Russian America*, 223. Kushner, *Conflict*, provides a valuable study of the several aspects of the purchase and substantial notes for further reading.

24 Kushner, *Conflict*, 130–1; see also Harold Taggart, "Sealing on St. George Island, 1868," *Pacific Historical Review* 28 (1959): 352.

25 I have used in this narrative, except where otherwise noted, the following: Frank H. Sloss, "Who Owned the Alaska Commercial Company?" *Pacific Northwest Quarterly* 68 (1977): 120–30, and, with Richard A. Pierce, "The Hutchinson, Kohl Story: A Fresh Look," *Pacific Northwest Quarterly* 62 (1971): 1–6; Gerstle Mack, *Lewis and Hannah Gerstle*; Richard A. Pierce, "Prince D.P. Maksutov: Last Governor of Russian America," *Journal of the West* 6 (1967): 395–416; L.D. Kitchener, *Flag over the North: The Story of the Northern Commercial Company*, chap. 3; Frank H. Sloss, *Only on Monday: Papers Delivered before the Chit-Chat Club*, 73–86; Samuel F. Johnson, ed., *Alaska Commercial Company, 1868–1940*.

26 Taggart, "Sealing on St. George Island."

27 Gustave Niebaum, "Autobiographical Statement," 22 September 1885, and "Statement on Interests of Alaska, 8 October 1885," Bancroft Library, PK-38:1 and 40.

28 Niebaum, "Autobiographical Statement."

29 This point is mentioned only in Dunn and Bradstreet Reports, 14 (1870–5), "San Francisco," 171, which also reported, confidentially, that Wasserman was an extravagant speculator (Baker Library, Harvard Business School).

30 Mack, *Lewis and Hannah Gerstle*, 33.

31 Niebaum, "Autobiographical Statement."

32 On Williams, Haven, and the New London sealing and whaling industry, see Busch, *Master of Desolation*, Introduction, and sources there cited, particularly Robert Owen Decker, *Whaling Industry of New London*. On the *Peru*'s voyage, Alexander Starbuck, *History of the American Whale Fishery from Its Earliest Scenes in Many Lands* 2: 602–3.

33 Mystic, Conn., *Pioneer*, 9 May 1868, repeating item from San Francisco *Bulletin* (my thanks to Bill Peterson for this reference), reports Chappel's movements; on Morgan, Barnard L. Colby, *New London Whaling Captains*, chap. 1.

34 Taggart, "Sealing on St. George Island," 358. Martin, *Silver Fleece*, 137–8, describes events on St. George, where Parrott was the first to arrive.

35 William H. Dall to Dr. B.W. Evermann, 7 December 1910, RG 22, sec. 91, file 290 (Pribilofs), National Archives of the United States, Washington D.C.

36 J.W. White, "Statement on Alaska" (1885), Bancroft Library, P-K 42.

37 Emil Teichmann, *A Journey to Alaska in the Year 1868: Being a Diary ...*, 179 ff.

38 On Miller and his role, John F. Miller, "Statement concerning lease of Pribilof Islands," Bancroft Library, PK 41:1 (there are also a very few relevant papers in the Miller Collection, Bancroft Library, 81/106 CP);

Roland L. DeLorme, "The Alaska Commercial Company and the U.S. Customs Bureau in Alaska" (Paper delivered at conference on "The Sea in Alaska's Past," University of Alaska, Anchorage, 7–8 September 1979).

39 The best discussion of ownership is Sloss, "Who Owned the ACC?" See also Mack, *Lewis and Hannah Gerstle*, 33. Brief but interesting evaluations of the early ACC and some constituent parts are given in the Dunn and Bradstreet records, "San Francisco" volumes for the relevant years (Baker Library, Harvard Business School).

40 Sloss, "Who Owned the ACC?" 126–8.

41 Hiram Ketchum, Jr (Collector of Customs, Alaska District) to John Hogg, Jr, n.d. [1868], U.S. 55th Cong., 1st sess., H. doc. 92, pt. 1, 5 (Seal and Salmon Fisheries and General Resources of Alaska).

42 Scheffer and Todd, "History," 49.

43 The ACC's opponents claimed that Goldstone was the highest bidder, and the ACC the lowest, yet the lease was given to the ACC at Goldstone's price. Anti-Monopoly Association of the Pacific Coast, *A History of the Wrongs of Alaska*. The applicable congressional legislation was 16 Stat. 180 (1 July 1870).

44 Jones, "History," gives the text, 38–9.

45 Thomas Taylor (partner, Taylor and Bendel, San Francisco), to John S. McLean, Special Agent of the Treasury Department, 20 October 1869 (U.S. Doc. 92 1898), 8, gives these prices, adding that otters sold for $20 each at the islands, and a low $40–60 in San Francisco. The Aleuts were paid 40 cents each: "This seems to be a very low price, but the expense of fitting out vessels, the high price of wages and provision, and salary for agents," along with skins damaged by unskilled treatment, all made such prices necessary.

46 Scheffer and Todd, "History," 22, 34–6; George Rogers, "An Economic Analysis of the Pribilof Islands, 1870–1946," 24–5.

47 Scheffer and Todd, "History," 24, 27, 35.

48 Jones, "U.S. Administration," 42–4, and the same author's *A Century of Servitude*, 23.

49 Scheffer and Todd, "History," 31; Mack, *Lewis and Hannah Gerstle*, 37.

50 D.S. Jordan and G.A. Clark, "Truth About the Fur Seals of the Pribilof Islands," 2. On the drive and the killing, see, in addition to Scheffer and Todd, Craig A. Hansen, "Seals and Sealing," *Alaska Geographic* 9, no. 3 (1982): 41–73, a graphically illustrated article on the modern process (the entire issue is devoted to the Pribilofs).

51 Henry W. Elliott, *Our Arctic Province: Alaska and the Seal Islands*, 350.

52 Scheffer and Todd, "History," 15.

53 Rogers, "Economic Analysis," 19; Scheffer and Todd, "History," 46; H.W. Elliott to Secretary of the Treasury, 31 October 1873 (U.S.

Cong.' Doc. 92 pt. 1, 51), blamed particularly the tax, and estimated that without it some 40,000 gallons of fine seal oil could be produced on the islands (no basis for this calculation was provided).

54 Rogers, "Economic Analysis," is the principal source for this discussion; see note on sources for table 2.

55 Mack, *Hannah and Lewis Gerstle*, 36, 49, 56; Martin, *Silver Fleece*, 145–6, Rogers, "Economic Analysis," 31; [Anon.] "Sea Otter Hunting," *The Alaska Journal* 1 (1971, reprint from 6 May 1893): 46–8.

56 Rogers, "Economic Analysis," 29.

57 Based upon ibid. (1–2 quoted).

58 Eitchener, *Flag over the North*, 63, quoting Miller's letter of instruction to agent on St. Paul, 29 March 1872.

59 Ibid., 54, quoting special Treasury agent Charles Bryant.

60 Jones, "U.S. Administration," 42, and *Century*, 19–20, 41.

61 Jones, "U.S. Administration," 62–3, and *Century*, 54–8.

62 Special Agent Otis report for 1881, 30 July 1881, U.S. Cong. 1st sess., H. doc. 92, pt. 1, 147; bay rum and cologne: Emma Jane McIntyre, ms., July 1874, Pribilof Islands, Bancroft Library, PK 229.

63 Jones, *Century*, 25–7. Though a convenient recapitulation of highlights, Jones's samples do, in my opinion, accurately reflect the nature of these logs: U.S. Bureau of Commercial Fisheries, Pribilof Island log books, 1872–1945 (microfilm copies are available from the Alaska State Library, Juneau).

64 Jones, "U.S. Administration," 31, 69; see also Agent Bryant's report for 1875, U.S. 55th Cong., 1st sess., H. doc. 92, pt. 1, 69.

65 Jones, "U.S. Administration," 95; Pierce, "Maksutov," 410–11; Benjamin Franklin Gilbert, "Economic Developments in Alaska, 1867–1910," *Journal of the West* 4 (1965): 504–21; American-Russian Commercial Company, "The Alaska Fur Seal Bill," 18 March 1869 (letter quoted, with appendices); Anti-Monopoly Association, *Wrongs of Alaska*. (I have used copies of the last two tracts in the Provincial Archives of British Columbia, Victoria, respectively NWF 998.2/A518 and 972.2/A629.)

66 Jones, "U.S. Administration," 75; and *Century*, 36.

67 Mack, *Lewis and Hannah Gerstle*, 53–4 (quoted); Gilbert, "Economic Developments," 505–6. [Alaska Commercial Co.], *Reply of the Alaska Commercial Company to the Charges of Governor Alfred P. Swineford of Alaska* ... (n.p., c. 1888; copy in the Kellogg Library, Smithsonian).

68 Scheffer and Todd, "History," 43–5; James Thomas Gay, "Henry W. Elliott: Crusading Conservationist," *Alaska Journal* 3 (1973): 211–16.

69 Elliott, *Our Arctic Province*, 310–11.

70 Ibid.

71 G.C.L. Bertram, "Pribilof Fur Seals," *Arctic* 3 (1950): 83.

72 Henry W. Elliott, "The Loot and the Ruin of the Fur-Seal Herd of Alaska," *North American Review* (1907): 427.

73 Scheffer and Todd, "History," 34; Alton Y. Roppell and Stuart P. Davey, "Evolution of Fur Seal Management on the Pribilof Islands," *Journal of Wildlife Management* 29 (1965): 448–63.

74 Martin, *Silver Fleece*, 151.

75 Mack, *Lewis and Hannah Gerstle*, 54–5.

76 United States, *Fur Seal Arbitration,* vol. 3, Appendix to the Case of the United States before the Tribunal of Arbitration to Convene at Paris ... (1892), deposition of Leon Sloss, 7 May 1982, 91.

CHAPTER FIVE

1 In discussing the NACC, I have used Charles S. Campbell, Jr, "The Anglo-American Crisis in the Bering Sea, 1890–1891," in Morgan B. Sherwood, ed., *Alaska and Its History*, 315–40, and "The Bering Sea Settlements of 1912," *Pacific Historical Review* 32 (1963): 347–67.

2 Gilbert, "Economic Developments," 506.

3 Campbell, "Anglo-American Crisis."

4 Text of the lease in Jones, "U.S. Administration," 78–80; see also Mack, *Lewis and Hannah Gerstle*, 56, on date of the ACC lease expiration.

5 J.E. Ziebach to James Judge, 16 May 1897, James Judge Papers, Oregon Historical Society, Portland, Ore.

6 Rogers, "Economic Analysis," 79–82.

7 Jones, "U.S. Administration," 89–90.

8 Joseph Murray to James Judge, 4 May 1898, Judge Papers.

9 Rogers, "Economic Analysis," 83, 93–6, 105–6; Jones, *Century*, 40–4.

10 Quoted in Jones, *Century*, 41.

11 Discussion of the harvest is based upon Scheffer and Todd, "History," 53–60, 101; Ward T. Bower and E.C. Johnston, "Seals and Walruses," in Donald K. Tressler, ed., *Marine Products of Commerce*, 647–67 (Bower and Johnston both had considerable experience on the islands); and Rogers, "Economic Analysis," 61–4.

12 Rogers, "Economic Analysis," 74.

13 James Laver, *Taste and Fashion from the French Revolution to the Present Day*, 166–71.

14 Stevenson, "Utilization of the Skins of Aquatic Animals," 301.

15 The NACC's finances are broken down in Rogers, "Economic Analysis," 77–8.

16 F.W. Howay, *British Columbia from the Earliest Times to the Present* 2: 459.

17 Charles Spring (William Spring's son), "Origin of Pelagic Sealing," holograph ms., 12 April 1929, Provincial Archives of British Columbia, Victoria, I/BX/XP8. Howay, *British Columbia*, 2: 459–60, and

Spring's manuscript are the sources of the following three paragraphs
except where otherwise noted.

18 Howay, *British Columbia* 2: 461, makes the claim for the *City of San
Diego*; Victor Jacobsen, mate on the *Mary Ellen*, does the same for this
Canadian vessel in Ursula Jupp, ed., *Home Port Victoria*, 84.

19 J.M.S. Careless, "The Business Community in the Early Development
of Victoria, British Columbia," in J. Friesen and H.K. Ralston,
Historical Essays on British Columbia, 177–200; Margaret A. Ormsby,
British Columbia: A History, 74–7, 302.

20 Swan is quoted in Ivan Doig's memorable book, *Winter Brothers: A
Season at the Edge of America*, 166.

21 A. Alfred Mattsson, "Fur Seal Hunting in the South Atlantic," *American
Neptune* 2 (1942): 154–66 (not entirely on the South Atlantic); W.
Nigel Bonner, "International Legislation and the Protection of Seals,"
in National Swedish Environment Protection Board, *Proceedings from the
Symposium on the Seal in the Baltic*, 12–29.

22 Frederick Schwatka, *Nimrod of the North, or Hunting and Fishing
Adventures in the Arctic Regions*, 58.

23 Kenyon, "Last of the Tlingit Sealers;" see also "Last Indian Sealer Dies
at Neah Bay," *Marine Digest* (Seattle) 35 (27 July 1957): 13.

24 Mattsson, "Fur Seal Hunting," and "Sealing Boats," *American Neptune*
(1943): 327–32; R.R. Godden, "Frustrated Voyage, 1891," *Bulletin
of the Maritime Museum of British Columbia* (Fall 1978): 8–9.

25 Max Lohbrunner, typescript, n.d., British Columbia Provincial
Archives, Victoria, E/D/L83.

26 Carl Rydell, *Adventures ... Autobiography of a Seafaring Man*, 74–5.

27 Lohbrunner, typescript.

28 [Anon.], "Mutiny on the *Rand*: A Tale of Early Days Among the
Sealers," *Alaska Sportsman* (1958), clipping file, National Maritime
Museum, San Francisco.

29 Matt Mathieson, "When the Sandheads Lightship was Alive,"
typescript, 1945, Vancouver Maritime Museum.

30 *Monterey Trader*, 24 July 1935 ("gin"); Monterey *New Era*, 6 April
1898 (*Ainoko*), newspaper file, "Sealing," Monterey, California, Public
Library.

31 N. de Bertrand Lugrin, "Epic of the Seal Hunters," *Canadian
Geographical Journal* 4 (1932): 301.

32 Rydell, *Adventures*, 138.

33 On otter and seal hunting in the Kuriles, see Rydell, *Adventures*, chap.
11; Capt. H.J. Snow, *Notes on the Kurile Islands*, and *In Forbidden Seas:
Recollections of Sea-Otter Hunting in the Kuriles* (Kimberly, 39–52); the
latter remains the best source on the Kurile hunt but see also a lesser-
known work by a partner of Snow's, Alexander Allan, *Hunting the
Sea Otter*.

34 Oliver L. Austin, Jr, and Ford Wilke, *Japanese Fur Sealing*, is a comprehensive and valuable study, originally done by the Natural Resources Section, Central Headquarters, Supreme Command for Allied Powers, Tokyo. Austin and Wilke discuss Japanese attempts to control the trade.

35 Scheffer and Todd, "History," 26, 95–6; H.H. McIntyre, General Agent, ACC, to Gen. J.F. Miller, President, 15 August 1876, U.S. Congress, 1st sess., H. doc. 92, pt. 1, 89–90, gives details of the *Ocean Spray*'s raid.

36 The weapons are mentioned in U.S. Bureau of Commercial Fisheries, Pribilof Island log books, 23 August and 11 October 1907. Photographs in the Judge Papers, Oregon Historical Society, show artful poses (presumably of Mrs Judge) on the islands with both of these weapons (Judge was Treasury agent).

37 Snow, *Forbidden Seas*, 235.

38 C.E. Carrington, *The Life of Rudyard Kipling*, 156–7. See, however, J. Gordon Smith Papers (British Columbia Provincial Archives, Add MS 383, box 2, folder 57), noting that John Kernan, an Irish ex-sealer who ran the "Sailors' Home" in Yokohama, told Kipling – who visited the home – of the adventures of his schooner *Mystery*, which became Kipling's *Northern Light*.

39 Robert Dunn, "Alaska, the Seal-Warder, and the Japanese Raiders," *Harper's Weekly* 50 (15 September 1906): 1310–1.

40 The following account of the sealing fleet is based upon R.N. D'Armand and John Lyman, "The Sealing Fleet," *Marine Digest* 35–6 (1957–8), a series which appeared in forty parts; supplemented by E.D. Wright, ed., *Lewis & Dryden's Marine History of the Pacific Northwest …* (1897; reprint, Seattle, 1967), chap. 21; Snow, *In Forbidden Seas*; and "Annual Sealing Reports by Vessel, Victoria, 1889–1911," a handwritten compilation in the Provincial Archives of British Columbia, I/BS/SEI, all of which I have combined into my own list of vessels from which these conclusions have been drawn.

41 Mathieson, "Sandheads Lightship," 3.

42 Ibid.; Jacobsen's account is in Jupp, *Home Port Victoria*, 111.

43 *Fur Seal Arbitration*, 3: 498–9, gives prices, 1881–8.

44 Values of vessels for 1891 are given in "Annual Sealing Reports, Victoria."

45 "Evidence of Captain Victor Jacobson [sic]," typescript, n.d., Provincial Archives of British Columbia, I/BS/J15.

46 Public Archives of Canada, RG 33, GP 107, Royal Commission on Sealing, Testimony, vol. 1, contains a wealth of information on sealing; see also "Evidence of Captain Victor Jacobson," typescript.

47 Wright, *Lewis & Dryden's Marine History*, chap. 21, gives birthplaces

and brief biographical sketches of many masters. Individual careers are compounded from vessel lists (see n. 39 above), which commonly list the master's name.

48 Wright, *Lewis & Dryden's Marine History*, 427.

49 Laughrin, "Epic," 298.

50 On owners and their role in Victoria, Careless, "Business Community"; Harry Gregson, *A History of Victoria, 1942–1970*; Scholefield and Howay, *British Columbia*, vols, 3–4; J.B. Herr, *Biographical Dictionary of Well-Known British Columbians*; Derek Pethick, *Summer of Promise: Victoria, 1864–1914* (Victoria 1980), chaps. 4–6.

51 Scholefield and Howay, *British Columbia* 4: 1134–6.

52 Victoria Sealing Company Records, British Columbia Provincial Archives, Add. Ms. 16, vols. 1–2 (minutebook) show organization, shares, and various resolutions on wage scales and relations with Indians.

53 Victoria Sealing Co. Records, vol. 1, list vessels included in the company and their ownership. The R.P. Rithet Papers (Add. Ms. 504) in the same library, vol. 4, folders 10–18, contain bills of sale, indentures, and notes for a number of vessels. It should be added that vessels were normally divided into sixty-four shares, and could be owned by a number of individuals or firms. See also William Munsie's Letterbook 1895–1903 (Provincial Archives of British Columbia, MS 1/3s/M92), which adds further information; a Munsie letter of 19 December 1900 claimed that of the forty-five schooners on the coast at that time, forty-three were affiliated with the Victoria Sealing Company. Victor Jacobsen's remarks are in Public Archives of Canada, RC 33, GP 107, Royal Commission on Sealing, 1: 639–41.

54 Ormsby, *British Columbia*, 85, 215, 302; *Fur Seal Arbitration*, 2: 279, notes that while 1,100 men worked in sealing out of Victoria, 2,000–3,000 worked in the same fur industry in London.

55 *Fur Seal Arbitration*, 1892 depositions, 3: 318–26.

56 In the discussion of the Bering Sea crisis which follows, I have relied, as with the formation of the NACC, upon the work of Charles Campbell, "Anglo-American Crisis," and "Bering Sea Settlements," as well as Wright, *Lewis & Dryden's Marine History*, 428ff.; Robert Craig Brown, *Canada's National Policy, 1883–1900: A Study in Canadian-American Relations*; F.W. Howay, W.N. Sage, and H.F. Angus, *British Columbia and the United States*, chap. 13; Martin, *Silver Fleece*, 170–6; Albert T. Volwiler, "Harrison, Blaine, and American Foreign Policy, 1889–1893," *Proceedings of the American Philosophical Society* 79 (1938): 637–48; and James T. Gay, "Bering Sea Controversy: Harrison, Blaine, and Cronyism," *Alaska Journal* 3 (1973): 12–19. The actual seizures, according to a memorandum originally in the papers of Walter Q.

Gresham (secretary of state, 1893–4), may have originated in an 1881
order – itself the result of ACC pressure – from the acting secretary of
the treasury to San Francisco authorities to stop American pelagic
sealers. No action was taken in 1881, but five years later a new
secretary of the treasury forwarded a copy of the same letter to the
collector of customs in San Francisco, perhaps under the assumption
that it represented standing policy. The State Department, at least,
seems to have been thoroughly surprised when it learned of the seizures.
Memo by Prof. J.B. Moore, 24 January 1894, in University of Alaska,
Alaska History Research Project, "Alaska History Documents,"
typescript, Library of Congress, Manuscripts Division, F 901/A37, 6:
138–9.

57 Charles Hibbert Tupper, "Crocodile Tears and Fur Seals," *National
Review* 28 (September 1896): 89.

58 Campbell, "Anglo-American Crisis," 322.

59 Gay, "Bering Sea Controversy," 16, notes that Elkins dined with the
president in March. Elliott, for his part, never forgave Elkins, Blaine, or
Foster for their role in destroying the seal herd. "Alaska History
Documents," 6: 68, 82–3, 176 ff.

60 In addition to *Fur Seal Tribunal* (16 vols.), see the interesting account
of William Williams, "Reminiscences of the Bering Sea Arbitration,"
American Journal of International Law 37 (1943), 562–84.

61 Howay, *British Columbia and the U.S.*, 328–9; Brown, *Canada's National
Policy*, 123.

62 See the Alaska File of the Revenue Cutter Service, 1867–1914, National
Archives of the United States, microcopy no. 641, for details. The
Revenue Cutter service, forerunner of the Coast Guard, was established
in 1790; after 1871 it was known as the Revenue Marine Division,
and it was from this body that a special Fur Seal Patrol was organized.
Rogers, "Economic Analysis," 64–5; Dennis L. Noble, "Fog, Reindeer,
and the Bering Sea Patrol: 1867–1964" (Paper delivered at conference
on "The Sea in Alaska's Past," Anchorage, University of Alaska, 7–8
September 1979).

63 On the larger issues and negotiations, see Peter Neary, "Grey, Bryce,
and the Settlement of Canadian–American Differences, 1905–1911,"
Canadian Historical Review 49 (1968): 357–80; Thomas A. Bailey, "The
North Pacific Sealing Convention of 1911," *Pacific Historical Review*
4 (1935): 1–14; Gordon Ireland, "The North Pacific Fisheries,"
American Journal of International Law 36 (1942): 400–24.

64 Starting with a few vessels, by 1897 the Japanese were operating a
dozen, and by 1905–6 thirty, with 750 crewmen – thus possessing a
considerably larger fleet than Victoria. After the Russo-Japanese War
Japan assumed the role of warder of Robben Island, equipping it with a
guard and a radio station. C. Mcdonald, Tokyo, to Foreign Office

(Britain), 19 July 1906, enclosing report by Vice-Consul Hakodate, Public Archives of Canada, RG 7, GP 21, vol. 47, file 158.

65 Theodore Roosevelt, "The Fur Seal Fisheries," *Metropolitan Magazine* 25 (March 1907): 692. A bill was in fact introduced in the House to this end, but was rejected by the Senate, in 1896. Scheffer and Todd, "History," 59.

66 Neary, "Grey, Bryce," 379–80.

67 Testimony of Sir Charles H. Tupper on behalf of Victoria Sealing Co., Public Archives of Canada, RG 33, GP 107, Royal Commission on Sealing, 3: 9–10; Victoria Sealing Co. memo for Governor-General, September 1909, Public Archives of Canada, RG 7, GP 21, vol. 47, file 158. Catch figures from British Columbia list, Provincial Archives, Victoria; price from Gordon Newell, ed., *The H.W. McCurdy Marine History of the Pacific Northwest ... from 1895*, 178. The Victoria Sealing Co. issued 33,458 shares at $12.50 each (some $418,000 of $500,000 authorized) to fifty-five shareholders on the basis of the value of more than forty schooners and equipment. Only nine men held over 1,000 shares, including William Grant (a long-time sealing captain and the VSC's general manager), Rithet (largest shareholder with 4,278), Marvin (2,697), and Joseph Boscowitz (3,155).

68 C. Fox Smith, *Sailor Town Days*, 178.

69 Newell, *McCurdy Marine History*, 207, gives details on *Casco* and vessels in following paragraph; information on *Bayard* supplied by the director, Vancouver Maritime Museum.

70 Privy Council Report, 28 February 1916, Public Archives of Canada, RG 7, GP 21, vol. 47, file 158.

71 This total compares very closely to that of 8 million given in Ian MacAskie, *The Long Beaches: A Voyage in Search of the North Pacific Fur Seal*, 131.

72 Martin, *Silver Fleece*, 187.

73 In its reprint, *Matka and Kotik* became *The Story of Matka*. Full details on this and other Jordan writings are found in Alice N. Hays, *David Starr Jordan: A Bibliography of His Writings, 1871–1931*.

74 Jordan to Langley, 27 January 1897, D.S. Jordan Papers, Stanford University Archives, microfilm reel 17.

75 Jordan to F.A. Lucas, 26 January; Elliott to Jordan, 29 January; and Jordan to Elliott, 6 February 1897, ibid., reels 4 and 17. Elliott was fond of the "butchers" phrase, and used it again a decade later in "Loot and Ruin," 427.

76 Scheffer and Todd, "History," 77; Martin, *Silver Fleece*, 214–217.

77 Scheffer and Todd, "History," 101. Hornaday, regarded as something of a fanatic in the Bureau of Fisheries (unsigned memo, 25 May 1910, National Archives, RG 22, sec. 21, file 100, folder 2), did not speak for the zoo in this matter, since Charles S. Townsend, head of the New

York Aquarium, like the zoo a constituent part of the N.Y. Zoological Society, was a defender of sealing. President, N.Y. Zoological Society, to Charles Nagel, Secretary of Commerce and Industry, 1 June 1910, ibid.

78 John E. Lathrop, "The West and the National Capital," *Pacific Monthly* 26 (August 1911): 195–211.

79 U.S., 62nd Cong., 3rd sess., House Report 1425, preliminary report, 31 January 1913, 3. On Liebes, note the letter from H. Liebes & Co., San Francisco, to the Victoria Sealing Co., 2 May 1901 (Victoria Sealing Co. Archives, Correspondence, vol. 3, Add MS. 16, British Columbia Provincial Archives), which makes it clear that Liebes & Co. at least handled the Victoria Sealing Co. catch at San Francisco for some years before 1901, when the association collapsed because of a disagreement over prices. National Archives, RG 22, sec. 91, file 130, includes much correspondence on the Rothermel inquiry.

80 Jack London, *The Sea Wolf* (1903; reprint, New York: Heritage Press 1961), 280.

CHAPTER SIX

1 In discussing the life history of elephant seals, I have relied particularly on the works of R.M. Laws, and L.H. Matthews (southern species), and G.A. Bartholomew and Burney Le Boeuf (northern species) listed in the bibliography. My special thanks to Professor Le Boeuf for his comments on an early draft of this chapter.

2 Charles M. Scammon, *The Marine Mammals of the Northwestern Coast of North America*, 116.

3 L. Harrison Matthews, *Sea Elephant: The Life and Death of the Elephant Seal*; Robert Cushman Murphy, *Logbook for Grace: Whaling Brig 'Daisy' 1912–1913*.

4 For the northern species population, I am indebted to Professor Le Boeuf (December 1979), southern species estimates are from Scheffer, *Seals*, 3–5; B. Grzimek, *Grzimek's Animal Life Encyclopedia*, 12: 405–6.

5 A. Howard Clark, "The Antarctic Fur-Seal and Sea Elephant Industries," in George Brown Goode, ed., *The Fisheries and Fishery Industries of the United States*, 5, pt. 2: 400–60; Nathaniel W. Taylor, *Life on a Whaler or Antarctic Adventurers in the Isle of Desolation*, ed. Howard Palmer; Busch, *Master of Desolation*, Introduction, and "Elephants and Whales: New London and Desolation, 1840–1900," *American Neptune* 40 (1980): 117–26. Gavin Maxwell, *Seals of the World*, notes that the heaviest elephant seal may weigh "up to 4 tons."

6 In addition to the works referred to in note 1, see Robert Carrick and Susan E. Ingham, "Ecological Studies of the Southern Elephant Seal, *Mirounga leonina* at Macquarie Island and Heard Island," *Mammalia* 24

(1960): 325–42; Bonner, *Seals and Man*, chap. 5. I disagree with Bonner's assertion (p. 75) that rendering elephant blubber was somehow easier than whaling; the processes were less similar than he indicates.

7 B. Le Boeuf and C. Leo Ortiz, "Composition of Elephant Seal Milk," *Journal of Mammalogy* 58 (1977): 683–5; Le Boeuf, Ronald J. Whiting, and Richard F. Gantt, "Perinatal Behavior of Northern Elephant Seal Females and Their Young," *Behavior* 43 (1972): 121–56.

8 B. Le Boeuf and Richard S. Condit, "The Cost of Living in a Seal Harem," *Pacific Discovery* 36 (1983): 12–14; Robert L. DeLong, "Northern Elephant Seal," in Delphine Haley, ed., *Marine Mammals of Eastern North Pacific and Arctic Waters*, 211.

9 Richard Condit and Burney J. Le Boeuf, "Feeding Habits and Feeding Grounds of the Northern Elephant Seal," *Journal of Mammalogy*, in press; Ian A. McLaren, "Seals and Group Selection," *Ecology* 48 (1960): 104–110. Matthews, *Sea Elephant*, chap. 7, discusses leopard seals; they are known to take Weddell seals as big as themselves, and there would appear to be no reason for them to abstain from an opportune meal of elephant seal.

10 Carrick and Ingham, "Ecological Studies," 333; Le Boeuf, Whiting, and Gantt, "Perinatal Behaviour," 129.

11 Fanning, *Voyages and Discoveries*, 255.

12 For a good discussion of "snotters," see Matthews, *Sea Elephant*, 140–2.

13 See for example the account ledger of W.T. Russell, for the ship *Timoleon*, 1832 (Old Dartmouth Historical Society, New Bedford, ms. 62), the cargo of which sold at thirty-two cents a gallon for whale oil, forty cents for elephant oil, and seventy-five cents for sperm whale oil.

14 Stevenson, "Oil from Seals," 23614–5.

15 Fanning, *Voyages and Discoveries*, 215–6; see also 293–7 on the *Volunteer* and the *Sea Fox*, and Bellingshausen, *Voyage*, 88–93 (Bellingshausen saw elephant sealers here in late 1819).

16 James Weddell, *A Voyage Towards the South Pole Performed in the Years 1822–24*, 141.

17 Fanning, *Voyages and Discoveries*, 310.

18 Cawkell, Maling, and Cawkell, *The Falkland Islands*, 98.

19 Fanning, *Voyages and Discoveries*, 256.

20 Busch, *Master of Desolation*, 20–21; the lance was three feet of iron blade fixed onto a six-foot pole, not to be confused with the longer whale-lance.

21 See photograph section in Cumpston, *Macquarie*.

22 L. Harrison Matthews, "The Natural History of the Elephant Seal," *Discovery Reports* (Cambridge) 1 (1929): 246.

23 Fanning, *Voyages and Discoveries*, 257.

24 Taylor, *Life of a Whaler*, 147, mentions a scrap press; so does the log of

the ship *Isaac Hicks*, 30 October 1858, G.W. Blunt White Library, Mystic, Conn. On boiling time, Stevenson, "Oil from Seals," 23614–15.

25 Matthews, *Elephant Seal*, pp. 117–18. The *Pacific*'s log for 1829–31 is in the Phillips Library, Peabody Museum, Salem; her men lived ashore in a turf house, when not out on "boat patrol," their term for circumnavigating the island in search of seals for their double tryworks (each of 100 gallons).

26 R.M. Laws, "The Elephant Seal Industry at South Georgia," *Polar Record* 6 (1953): 746–54.

27 Laws, "Elephant Seal Industry," and "The Current Status of Seals in the Northern Hemisphere," *IUCN Papers*, no. 39, 144–57; J.E. Hamilton, "On the Present Status of the Elephant Seal in South Georgia," *Proceedings of the Zoological Society of London* 117 (1947): 272–5; George Sutton, *Glacier Island: The Official Account of the British South Georgia Expedition, 1954–55*, 66–8; Bonner, *Seals and Man*, 79–87.

28 Laws, "Elephant Seal Industry."

29 Stackpole, *Sea-Hunters*, chap. 14; Jorgenson is quoted in Richards, "Journal of Erasmus Darwin Rogers," 289.

30 [Phelps, William D.], *Fore and Aft; or, Leaves from the Life of an Old Sailor, by 'Webfoot'*, 49–50.

31 Clark, "Antarctic Fur-Seal," 418, 453. Starbuck, *History of the American Whale Fishery*, 197, gives the ship *Nancy* of New Bedford, sailing in 1789, as the first American whaler at Desolation. The New York brig *Athenian* fitted for elephants in 1838, but made for the Crozets; likewise the New London brig *Uxor* went to the Prince Edward Islands, 1838–40; the logs of both are at G.W. Blunt White Library, 0 Logs 4 and 31.

32 Voyages of E. Nash, bark *Bolton*, November–December 1843 (23 and 25 November quoted), G.W. Blunt White Library, Log 820; the *Bolton*'s own log is in the Stonington Historical Society.

33 Stevenson, "Oils from Seals," 23615.

34 Taylor, *Life on a Whaler*; Scammon, *Marine Mammals*; Busch, *Master of Desolation*; Clark, "Antarctic Fur-Seal." The following remarks on procedures at Desolation and Heard are based on the logs of the bark *Alert*, 1860–2; bark *Charles W. Morgan*, 1916–17; schooner *Charles Colgate*, 1863–77, 1883–7; schooner *Francis Allyn*, 1886–7; schooner *Golden West*, 1865–8, ship *Isaac Hicks*, 1868–71; schooner *Pacific*, 1858–61, 1862–63, and bark *Trinity*, 1869–70, all in the G.W. Blunt White Library, Mystic; schooner *Emeline*, 1842–4, Kendall Whaling Museum, Sharon, Mass.; and schooner *Exile*, 1852, Old Dartmouth Historical Society, New Bedford, Mass.

35 Richards, "Journal of Erasmus Darwin Rogers"; A.M. Gwynn, "Notes on the Fur Seals at Macquarie Island and Heard Island," *Australian National Antarctic Research Expedition, Interim Reports*, 1953. It is

possible, but unproven, that seals were taken on the rocks known as the McDonald Islands, which lie about twenty-five miles west of Heard; G.M. Budd, "Breeding of the Fur Seal at McDonald Islands, and Further Population Growth at Heard Island," *Mammalia* 36 (1972): 423–7.

36 The connection between Heard and other masters and Maury is discussed in Bertrand, *Americans in Antarctica*, chap. 12. On Maury, see Frances Leigh Williams, *Francis Fontaine Maury*.

37 Richards, "Journal of Erasmus Darwin Rogers"; Brian Roberts, "Historical Notes on Heard and McDonald Islands," *Polar Record* 5 (1940): 580–4; Bertrand, *Americans in Antarctica*, chap. 13.

38 Schooner *Charles Colgate*, logbook, 1875.

39 Richards, "Journal of Erasmus Darwin Rogers," 295.

40 Arthur Scholes, *Fourteen Men: The Story of the Antarctic Expedition to Heard Island*, 239.

41 Bertrand, *Americans in Antarctica*, 246.

42 Busch, *Master of Desolation*, 224–5; *Charles Colgate*, logbook, 2 October 1883; Mill, *Siege of the South Pole*, 420, blames the *Challenger*, but the *Challenger Report* (1: 321–2) shows clearly that the rabbits were there first. The dogs are mentioned in A. Grenfell Price, *The Winning of Australian Antarctica*, 1: 30.

43 William H. Macy, "Adventure with a Seal-Elephant," *Whaleman's Shipping List* 30, no. 8, 16 April 1873; a fictionalized account, it nevertheless has the ring of truth, and is featured in the "bible" of whalemen and sealers. This brief story is the only account I have found which attempts to make of the elephant something of a Moby Dick – a respectable and somewhat malevolent adversary in the half-light of dawn.

44 Raymond Rallier du Baty, *15,000 Miles in a Ketch* (London 1912), 325; Matthews, *Sea Elephant*, 63.

45 Allyn, *Old Sailor's Story*, 57–8.

46 H.N. Moseley, *Notes by a Naturalist on the "Challenger,"* 187.

47 Allyn, *Old Sailor's Story*, 59.

48 Edward Boykin, *Ghost Ship of the Confederacy*, 306–40.

49 On the New London trade, see Busch, "Elephants and Whales," and *Master of Desolation*, Introduction, and Decker, *Whaling Industry of New London*.

50 Bertrand, *Americans in Antarctica*, 248–9.

51 On this voyage, and the sources for these calculations, see Busch, *Master of Desolation*, Introduction.

52 Elmo Paul Hohman, *The American Whaleman*, 13.

53 See Briton C. Busch, "Cape Verdeans in the American Whaling and Sealing Industry, 1850–1900," *The American Neptune*, in press.

54 Letter of Instruction, unsigned, Lawrence & Co., to Capt. Simeon

Church, 31 July 1871, Lawrence Papers, col. 25, box 2/9, G.W. Blunt
White Library; the same collection, box 2/3, gives a crew list for the
same vessel paying green hands twenty-five cents a month.

55 *Charles Colgate*, logbook, 27 July 1871.

56 *Francis Allyn*, logbooks for 1891–2 and 1893–5, Old Dartmouth
Historical Society, New Bedford; logbook of bark *Charles W. Morgan*,
1916–17, G.W. Blunt White Library.

58 S.H. Wiley, American consul, Havre, to R.D. Murphy, American
consul, Paris, 8 February 1938, discusses the Boissière concession in
some detail. Wiley, who was approached for American aid, was told by
Boissière that his concession (for thirty years) had expired in 1937,
but was up for renewal. Boissière paid some 25,000 francs a year to the
colony of Madagascar (under which jurisdiction Kerguelen was included)
for rights to fishing and sealing on the Crozet and Kerguelen groups.
Smithsonian Archives, col. 7176, Fish & Wildlife Field Reports, box
132, folder 18. An earlier report of 1924 put the fee at 1,000 francs
per annum, and noted that Boissière had sublet to Irwin & Johnson
(South Africa), a Cape Town fishery operation, but then accused Irwin
& Johnson of unfair profits at his expense. Gordon Jackson, *The British
Whaling Trade*, 193, 210–11; D.C. Poole, consul general, Cape Town,
to State Department, 9 April 1924; Smithsonian Archives, col. 7074,
Leonhard Stejneger Papers, box 15, folder 4. See also Matthews,
Penguins, Whalers and Sealers, 75; J.R.L. Anderson *High Mountains and
Cold Seas: A Biography of H.W. Tilman*, 286 (reporting activities seen by
Tilman in the 1960s); A.G.E. Jones, "Island of Desolation," *Antarctica*
6 (1971): 22–6 (national park); G.M. Budd, "Population Increase in
the Kerguelen Fur Seal at Heard Island," *Mammalia* 34 (1970): 410–14.

59 On the Crozets and Prince Edward Islands, T.P.A. Ring, "The
Elephant Seals of Kerguelen Land," *Proceedings of the Zoological Society
of London* 103 (1923): 431–43; R.W. Rand, "Elephant Seals on Marion
Island," *African Wildlife* 16 (1962): 191–8, and "Notes on the Marion
Island Fur Seal," *Proceedings of the Zoological Society of London* 126
(1956): 65–82; Judith E. King, "The Northern and Southern
Populations of *Arcotocephalus gazella*," *Mammalia* 23 (1959): 19–40.

60 Laws, "Current Status," 155; Brewster, *Antarctica*, 35, and chap. 7.

61 Richards, "Journal of Erasmus Darwin Rogers," 302–3 (303n60 adds
that "Kerguelen Jim" Robinson, born midst the rain and snow of
Kerguelen, died of thirst fifty years later in a West Australian desert);
William L. Crowther, "Captain J.W. Robinson's Narrative of a Sealing
Voyage to Heard Island, 1858–60," *Polar Record* 15 (1970): 301–16,
and "A Surgeon as Whaleship Owner," *The Medical Journal of Australia*
(Sydney) 1 (1943): 549–54; A.J. Villiers, *Vanished Fleets*, 180 ff.; Hugh
B. Evans, "A Voyage to Kerguelen in the Sealer *Edward* in 1897–98,"
Polar Record 16 (1973): 789–91, records a much later visit.

62 Foster, *Furneaux Group*, 40, quoting the Supreme Court Papers, New South Wales Archives.

63 Macco, *King Island*, pt. 2, 45–6; Foster, *Furneaux Group*, 44, 55, 58. The calculations are based on South Georgia comparisons, in which six barrels of oil make one ton, or 189 gallons (6 × 31.5 gallons) equal one ton, hence 10.58 pounds a gallon. At a barrel an elephant (one ton = six elephants), 300 tons require about 1,800 elephants. Figure on barrels an elephant from Laws, "Elephant Seal Industry at South Georgia," *Polar Record* (1953): 746–54.

64 This and following paragraph are based on Cumpston, *Macquarie*, 40.

65 Bellingshausen, *Voyage*, 367; see also Cumpston, *Macquarie*, 44–5, which quotes the same account.

66 P. Cunningham, *Two Years in New South Wales* (1826): 2: 103, quoted in Cumpston, *Macquarie*, 67.

67 *Sydney Gazette*, 12 December 1822, quoted in Cumpston, *Macquarie*, 60.

68 Cumpston, *Macquarie*, 149.

69 Ibid., 220–1; the observer was Edgar R. Waite, curator of the Canterbury Museum, Christchurch.

70 Cumpston, *Macquarie*, 309–11, quoting Sydney *Morning Herald*, 17 August 1919.

71 Cumpston, *Macquarie*, 311–2, 324–5; John K. Ling, "A Review of Ecological Factors affecting the Annual Cycle in Island Populations of Seals," *Pacific Science* 23 (1969): 399–413.

72 Carrick and Ingham, "Ecological Studies," put the Macquarie population at 110,000 in 1960; I have added an estimated increase and additional stocks on other Australian/New Zealand islands for the rough figure of 150,000.

73 Decker, *Whaling Industry*, 112.

74 Schwatka, *Nimrod of the North*, 53.

75 Clark, "Antarctic Fur Seal," 402.

76 Stevenson, "Oil from Seals," 23614–15; Laws, "Elephant Seal Industry," 753; R.J. van Aarde, "Fluctuations in the Population of Southern Elephant Seals *Mirounga leonina* at Kerguelen Island," *South African Journal of Zoology* 15 (1980): 99–106.

77 Ling, "Review of Ecological Factors."

78 A.L. Kroeber, "Elements of Culture in Native California," in R.F. Heizer and M.A. Whipple, eds., *The California Indians: A Source Book*, 9–12; N.A. Mackintosh, *The Stocks of Whales*, 164. David A. Henderson, *Men and Whales at Scammon's Lagoon*, is the best account of greys; Henderson notes that they were not molested between 1795 and 1846, and even after they began to be exploited were known for producing a darker oil. On the hide and tallow trade, see Briton C. Busch, ed., *Alta California, 1840–1842*, Introduction.

79 Henry Delano Fitch to Abel Stearns, n.d. [1840], Stearns Papers, Huntington Library; I am indebted to Adele Ogden for this quotation.

80 Busch, *Alta California*, 90–1, 171–2, 321–2

81 Log of ship *Charles Phelps* (Stonington), 1844–7, 1 and 5 January 1846, Bancroft Library; log of ship *Stonington* (New London), 29 November 1846; and journal of Ellery Nash of ships *America* and *Magnolia*, 1846, both in G.W. Blunt White Library, logs 335 and 820. See Henderson, *Men and Whales*; Dave W. Rice and Allen A. Wolman, *The Life History and Ecology of the Grey Whale (Eschrichtius robustus)*, and Edwin C. Starks; *A History of California Shore Whaling* (Sacramento, 1922).

82 J. Ross Browne, *A Sketch of the Settlement and Exploration of Lower California*, 60; compare the unsuccessful sealing venture of the ship *Charles Phelps* of Stonington, December 1845, Journal of Gordon Hall, Bancroft Library.

83 Mulford, *Prentice Mulford's Own Story*.

84 Burney J. Le Boeuf, "Back from Extaction?" *Pacific Discovery* 30, no. 5 (September-October, 1977): 1–7 (2–3 quoted); Charles Haskins Townsend, "The Northern Elephant Seal and the Guadalupe Fur Seal," *Natural History* 24 (1924): 567–78 (570 quoted); Laurence M. Huey, "Past and Present Status of the Northern Elephant Seal with a Note on the Guadalupe Fur Seal," *Journal of Mammalogy* 11 (1930): 188–94; George A. Bartholomew, Jr., and Carl L. Hubbs, "Winter Population of Pinnipeds about Guadalupe, San Benito, and Cedros Islands, Baja California," *Journal of Mammalogy* 33 (1952): 160–71.

85 Charles Harris, "A Cruise after Sea Elephants," *Pacific Monthly* 21 (1909): 334, mentions seeing the convicts there in the 1880s; the sign is quoted in George Hugh Banning, *In Mexican Waters*, 16 (the English is a literal translation of the Spanish term, "elephantes marinos"). See also Laurence M. Huey, "A Trip to Guadalupe, the Isle of My Boyhood Dreams," *Natural History* 24 (1924): 578–88.

86 Le Boeuf, "Back from Extaction."

87 Daniel K. Odell, "Seasonal Occurrence of the Northern Elephant Seal, *Mirounga angustirostris*, on San Nicolas Island," *Journal of Mammalogy* 55 (1974): 81; Le Boeuf; "Back from Extaction," and with Michael L. Bonnell and Robert K. Selander, "Elephant Seals: Genetic Variation and Near Extinction," *Science* 184 (24 May 1974): 908–9; Keith W. Radford, Robert T. Orr, and Carl L. Hubbs, "Reestablishment of the Northern Elephant Seal (*Mirounga angustirostris*) off Central California," *Proceedings of the California Academy of Sciences*, 4th ser., 31 (1965): 601–12.

88 The problem of mainland breeding, and *Sunset*'s role, is discussed in Burney J. Le Boeuf and Kathy J. Panken, "Elephant Seals Breeding on the Mainland in California," *Proceedings of the California Academy of Sciences*, 4th ser., 41 (1977): 267–80.

89 Jacques-Yves Cousteau and P. Diole, *Diving Companions: Sea Lion, Elephant Seal, Walrus,* 143.

90 Daniel J. Miller and Ralph S. Collier, "Shark Attacks in California and Oregon, 1926–1979," *California Fish & Game* 67 (1981): 76–99.

CHAPTER SEVEN

1 Cumpston, *Kangaroo Island,* chap. 13, describes the voyage of the *General Gates;* Phelps, *Fore and Aft,* 67, for encounter.

2 Discussion of the monk seals is based upon Judith E. King and R.J. Harrison, "Some Notes on the Hawaiian Monk Seal," *Pacific Science* 15 (1961): 282–93, and "The Monk Seals (genus *Monachus*)," *Bulletin of the British Museum (Natural History: Zoology)* 2 (1956): 203–56; A.M. Bailey, "The Hawaiian Monk Seal," *Museum Pictorial (Denver Museum of Natural History)* (1952): 1–30, and "The Monk Seal of the Southern Pacific," *Natural History* (American Museum Journal) 18 (1918): 396–9; Peter M. Knudtson, "The Case of the Missing Monk Seal," *Natural History* 81 (1977): 78–83; Robert J. Shallenberger, "A Seal Slips Away," *Natural History* 91 (1982): 48–53; Karl W. Kenyon, "Man vs the Monk Seal," *Journal of Mammalogy* 53 (1972): 46–53, and with Dale W. Rice, "The Life History of the Hawaiian Monk Seal," *Pacific Science* 13 (1959): 212–52; U.S. Dept. of Commerce, NOAA, NMFS, "Proposed Designation of Critical Habitat for the Hawaiian Monk Seal in the Northwestern Hawaiian Islands, Draft Environmental Impact Statement," 1980. The Caribbean monk seal was reported seen in 1973, but the sighting remains unconfirmed. See *Oryx* 14 (1977): 50; my thanks to George Nichols, David Campbell, and David Sergeant for an unpublished report giving further details of a 1980 search.

3 G. Causey Whittow, "Tropical Seals," *Sea Frontiers* 17 (1971): 285–7.

4 William A. Bryan, *Natural History of Hawaii,* 303–4.

5 Kenyon and Rice, "Life History of the Hawaiian Monk Seal," 115; the *Polynesian* report was on 15 August 1859.

6 Bailey, "Hawaiian Monk Seal," 5.

7 Shallenberger, "Seal Slips Away," is the most recent account of the population and its protection. The little-known coaling-station episode is mentioned in newspaper accounts of the time; R.G. Ward, *American Activities in the Central Pacific,* 4: 537–71.

8 Robert T. Orr, "Galapagos Fur Seal (*Arctocephalos galapagoensis*)," in IUCN, *Seals: Proceedings of a Working Meeting of Seal Specialists ...,* 124–8, reviews the nomenclature problem. On the islands in general, see Ian Thornton, *Darwin's Islands: A Natural History of the Galápagos,* especially chap. 9.

9 New York *Times,* 27 March 1983, sec. 20.

10 Fowler, *Furneaux Group,* 95–6.

11 Fanning, *Voyages and Discoveries*, 287.

12 D.S. Jordan to John W. Foster, State Dept., 7 and 10 January 1898; Clark (Jordan's assistant) to Lucas, 10 January 1898; Jordan Papers, Stanford University Archives, reel 17, discuss these voyages and Jordan's conclusions. Charles H. Townsend, "The Fur Seal of the Galapagos Islands," *Zoologica* 18 (1934): 43–56, considered Noyes's voyage the last.

13 Scheffer, *Seals, Sea Lions, Walruses*, 77–8.

14 Bartholomew and Hubbs, "Winter Population of Pinnipeds"; Luis A. Fleischer, "Guadalupe Fur Seal," in Delphine Haley, ed., *Marine Mammals of Eastern North Pacific and Arctic Waters*, 160–5; Andrew Hamilton, "Is the Guadalupe Fur Seal Returning?" *Natural History* 60 (1951): 90–4, all discuss population figures.

15 George Little, *Life on the Ocean; or Twenty Years at Sea*, 106–7; "Shelrack" probably means Capt. George Shelvocke who sailed in these waters in the early eighteenth century (see his *Voyage Round the World*); Townsend, "Fur Seal of the Galapagos."

16 Carl L. Hubbs, "Back from Oblivion. Guadalupe Fur Seal: Still a Living Species," *Pacific Discovery* 9 (1956): 14–21.

17 Townsend, "Fur Seal of the Galapagos," 445. Townsend, who, it will be remembered, took seven of the eight elephant seals from the same island, is immortalized in the name of *Arctocephalus townsendi*.

18 Lawrence M. Huey, "Guadalupe Island: An Object Lession in Man-Caused Devastation," *Science* 61 (1925): 405–7.

19 Dr. W. Thoburn to D.S. Jordan, 3 July 1897, Jordan Papers, Stanford University Archives, reel 15.

20 Fleischer, "Guadalupe Fur Seal"; Hamilton, "Is the Guadalupe Fur Seal Returning?"; Hubbs, "Back from Oblivion."

21 R.M. Lockley, *The Saga of the Grey Seal*, 137–8.

22 Richard S. Peterson, Carl L. Hubbs, Roger L. Gentry, and Robert L. DeLong, "The Guadalupe Fur Seal: Habitat, Behavior, Population Size, and Field Identification," *Journal of Mammalogy* 49 (1968): 668.

23 Fleischer, "Guadalupe Fur Seal"; Karl W. Kenyon, "Guadalupe Fur Seal (*Arctocephalus townsendi*)," in IUCN, *Seals*, 82–7.

24 This and the following paragraph are based upon George Nidever, "Life and Adventures," typescript, 1878, Bancroft Library, C-D 133; Carl Dittmann, "Narrative of a Seafaring Life on the Coast of California ...," ms., 1878, Bancroft Library, C-D 67; Leo H. Harloe, "The Life of Isaac J. Sparks," typescript, n.d., Santa Barbara Historical Society, F867/s638 (Harloe is a descendant of Sparks); Arthur Woodward, "Sea Otter Hunting on the Pacific Coast," and "Isaac Sparks – Sea Otter Hunter," *Historical Society of Southern California Quarterly* 20 (1938): 43–59, 119–34; Little, *Life on the Ocean*, 107–10; Ogden, *California Sea Otter Trade*; C.M. Scammon, "The Sea Otters," *The American Naturalist* (1870):

62–75; and Margaret Holden Eaton, *Diary of a Sea Captain's Wife: Tales of Santa Cruz Island* (Frank Eaton was Nidever's grandson); Victor B. Scheffer, "The Sea Otter on the Washington Coast," *Pacific Northwest Quarterly* 31 (1940): 371–88 (shooting from towers).

25 Eugene F. Rogers, "Memoirs," *Noticias: Quarterly Bulletin of the Santa Barbara Historical Society* 26 (1980): 41–59.

26 Edna M. Fisher, "Prices of Sea Otter Pelts," *California Fish & Game* 27 (1941): 261–5.

27 Harold McCracken, *Hunters of the Stormy Sea*, 284–5; Kenyon, *The Sea Otter*, 136, gives population figures.

28 Earl W. Kenyon, "The Sea Otter in Alaska," *Alaska Sportsman* 34 (1961): 17–18; Curt Beckham, "The Historical Sea Otter Transplant," *Pacific Wilderness Journal* 2 (1974): 16–18; "Rejected Transplants," *Pacific Search* 13 (1978): 19; Bettie Hiscocks, "Sea Otters Return to Canada's West Coast," *Canadian Geographic Journal* 32 (1977): 20–7; I.B. MacAskie, "Sea Otters: A Third Transplant to British Columbia," *The Beaver* 304 (1975): 9–11.

29 Woolfenden, *The California Sea Otter*; Jane H. Bailey, *Sea Otter*; Edna M. Fisher, "Habits of the Southern Sea Otter," *Journal of Mammalogy* 20 (1939): 21–36; Hal Roth, "Sea Otter Population," *Smithsonian* 1, no. 12 (December 1970), 30–7; Thomas R. Luchlin, "Home Range and Territoriality of Sea Otters Near Monterey, California," *Journal of Wildlife Management* 44 (1980): 576–82; *The Otter Raft* (Carmel, Calif.) nos. 17–28 (summer 1977–winter 1982); Augustin S. Macdonald, 'The Sea Otter Returns to the California Coast," *California Historical Society Quarterly* 17 (1938) 243–4; *Monterey Peninsula Herald*, 28 February 1974.

30 Kenyon, *The Sea Otter*, 136; Betty S. Davis, "Sea Otter Irony: Overabundant Yet Endangered," *Oceans* 12 (1979): 60–2.

31 The discussion of otters, abalone, and urchins – and the relationship to Morro Bay – is based (in addition to sources listed in n. 29 above) upon Betty S. Davis, "The Southern Sea Otter Revisited," *Pacific Discovery* 30 (1977): 1–13; John F. Palmisano and James A. Estes, "Sea Otters: Pillars of the Nearshore Community," *Natural History* 85 (1976): 46–53; Karl W. Kenyon, "Sea Otter," in Haley, ed., *Marine Mammals*, 226–34; [Anon.] "Sea Otters Ambushed," *National Parks & Conservation Magazine* 45 (1971): 32–7; David O. Duggins, "Kelp Beds and Sea Otters: An Experimental Approach," *Ecology* 61 (1980): 447–53; L.F. Lowry and J.S. Pearse, "Abalones and Sea Urchins in an Area Inhabited by Sea Otters," *Marine Biology* 23 (1973): 312–16; Mabel M. Rockwell, *California's Sea Frontier*, 20–30 (on abalone industry); K.H. Mann and E.A. Breen, "The Relation Between Lobster Abundance, Sea Urchins, and Kelp Beds," *Journal of the Fishery Research Board of Canada* 29 (1972): 603–9; Chinese and abalone: *Monterey Peninsula Herald*, 19

September 1962; Steven J. Shimer, "The Underwater Foraging Habits of the Sea Otter, *Enhydra lutris*," *California Fish & Game* 63 (1977): 120–2.

32 Mark D. Stephenson, "Sea Otter Predation on Pismo Clams in Monterey Bay," *California Fish and Game* 63 (1977): 117–120; Daniel J. Miller, James E. Hardwicke, and Walter A. Dahlstrom, *Pismo Clams and Sea Otters.*

33 Scheffer, *Adventures*, 104.

34 *Otter Raft*, no. 28 (Winter 1982–3): 1, gives most recent data, likewise *Whale Center Bulletin* (Oakland, Calif.) 5, no. 4 (winter 1982). Chemicals and sharks: Jack P. Ames and G. Victor Morejohn, "Evidence of White Shark, *Carcharodon carcharias*, Attacks on Sea Otters, *Enhydra lutris*," *California Fish & Game* 66 (1980): 1916–209; G. Victor Morejohn, Jack A. Ames, and David B. Lewis, *Post Mortem Studies of Sea Otters, "Enhydra lutris L.," in California*, 68–74, mentions sharks as a cause of death only as an afterthought on p. 74. Woolfenden, *The California Sea Otter*, 7–11, provides further information on the otter controversy, particularly the role of the California Dept. of Fish and Game.

35 Gretchen M. Lyon, "Pinnipeds and a Sea Otter from the Point Mugu Shell Mound of California," *University of California Publications in Biological Science* 18 (1937): 133–68; Wayne Suttles, "Notes on Coast Salish Sea-Mammal Hunting," *Anthropology of British Columbia* 3 (1952): 10–20; Maria Rostworowski de Diez Canseco, *Recursos naturales renovables y pesca*, 112–13, discusses Quechua and Inca usages in Peru.

36 C.M. Scammon, "About Sea Lions," *Overland Monthly* 8 (1872): 270. Scammon estimated about five gallons of oil for each sea lion, or six sea lions a barrel. On sea lions in general, John Rowley, "Life History of the Sea-Lions on the California Coast," *Journal of Mammalogy* 10 (1929): 1–37; Bruce R. Mate, "California Sea Lion," and Roger L. Gentry and David E. Withrow, "Steller Sea Lion," in Haley, ed., *Marine Mammals*, 166–71 and 172–6; B.R. Mate, "History and Present Status of the Northern (Steller) Sea Lion, *Eumetopias jubatus*," in FAO Advisory Commitee on Marine Resources Research, *Mammals in the Seas* (Rome, 1982), 4: 311–17.

37 Scammon, "About Sea Lions." The Santa Barbara *Independent* was still using the terms black and barking, for example, in an article on sealing of 16 July 1908 (Santa Barbara Historical Society, clipping file).

38 Santa Barbara *Daily News and Independent*, 19 July 1917; Santa Barbara *News Press*, 6 August 1972 (SBHS file); Eaton, *Diary*, 181.

39 Richard S. Peterson and S.M. Cooper, "A History of the Fur Seals of California," University of California, Santa Cruz, *Año Nuevo Reports* 2 (1968): 47–54.

40 Paul Bonnot, "The Sea Lions, Seals and Sea Otter of the California Coast," *California Fish & Game* 37 (1951): 371–89; Santa Barbara *Weekly Press*, 7 July 1877 (SBHS); Scammon, "About Sea Lions," 269; Rowley, "Life History," 14–15.

41 Scammon, "About Sea Lions," 269.

42 Herbert C. Hensley, "Memoirs: the History of San Diego, City, County, and Region," typescript, 1952, San Diego Historical Society, Acc. 1750, 821; San Diego *Union*, 10 October 1892 (SDHS); Rowley, "Life History," 15–16.

43 Rowley, "Life History," 16.

44 Ibid.

45 Paul Bonnot, G.H. Clark, and S. Ross Hatton, "California Sea Lion Census for 1938," *California Fish & Game* 24 (1938): 415–19.

46 Mate, "California Sea Lion," 177.

47 Bonner, *Seals and Man*, 120–1 (sound and flippers); Victoria *Times*, 26 June 1929 (machine guns on patrol steamer *Givenchy*), clipping file, Provincial Archives of British Columbia, Victoria; Portland *Oregonian*, 14 April 1922 (dynamite on sandbars), clipping file, National Archives of the United States, RG 22, sec. 21. Harry Anderson, *Sea Lion – Creature of Controversy* (Fort Bragg, Calif. 1960; a pamphlet presented to the Senate Fact Finding Committee on natural resources), discusses the other alternatives such as depth charges and land mines.

48 Lou Barr, "Steller Sea Lion," *Oceans* 8, no. 4 (July–August 1975): 18–21; Gentry and Witherow, "Steller Sea Lion"; Mate, "Northern (Steller) Sea Lion" (species competition); D.G. Ainley, H.R. Huber, R.P. Henderson, and T.J. Lewis, "'Studies of Marine Mammals at the Farallon Islands, California, 1970–1975"; G.A. Bartholomew and Richard A. Bootlootian, "Numbers and Population Structure of the Pinnipeds on the California Channel Islands," *Journal of Mammalogy* 41 (1960): 366–75. For the figure of 156,000, I am indebted to Burney Le Boeuf (December 1983).

49 *Whale Center News Letter* 6, no. 1 (Spring 1983): 8.

50 Paul R. Ryan, "Oil and Gas Group Attacks Marine Sanctuary Program," *Oceanus* 26 (1983): 72–6.

51 Eaton, *Diary*, 15, 26, 54, 181–90; Henderson, *Men and Whales*, 93–4; Mulford, *Prentice Mulford's Story*; Jerry MacMullen, *They Came by Sea; A Pictorial History of San Diego Bay*, 50.

52 Logs of schooners *Chile*, Old Dartmouth Historical Society, log 869; *Thomas Hunt* (with which that of the *Penguin* is bound), *Hancox*, and *Rob Roy*, G.W. Blunt White Library, Mystic, logs 29, 731, and 842, and log of schooner *Talma*, 1832–3, Yale University Library, sealing logs, reel 3; Allyn, *Old Sailor's Story*, 33–41.

53 Schooner *Betsy, Whaleman's Shipping Paper* (printed form with additions, Stonington 1842), National Archives, Waltham, Mass., box 32556;

for whaling and trading, see log of the *Talma*, 1834–5; *Talma-Ann Howard* agreements of 1834 and 1835; log of *Ann Howard*, 17 September 1836 and 7 March 1838, and of *Betsey*, 5 April 1838 and 28 May 1841, all in G.W. Blunt White Library, respectively microfilm roll 1, and O log 6 and 7.

54 *Ann Howard* log, 8 January 1837, G.W. Blunt White Library, roll 1.

55 Log of brig *Athenian* (N.Y.) 1836–9 (title page quoted), G.W. Blunt White Library, O log 4.

56 Boston *Daily Evening Transcript*, 28 April 1833, quoted in Ward, ed., *American Activities*, 8: 61.

57 Southern sealing in the 1870s is based upon the following logs: in G.W. Blunt White Library: bark *Peru*, schooner *Franklin*, 1870–2 (col. 44); schooner *Golden West* 1871–6, 1878–9 (col. 25, vol. 119–20 and reel 1); schooner *Flying Fish*, 1872–8 (reels 32, 33, 36); schooner *Mary E. Higgins*, 1879–92 (col. 25, vol. 126); schooner *Sarah W. Hunt*, 1887–8 (log 43); schooner *Charles Colgate*, 1877–8 (roll 10). Old Dartmouth Historical Society: schooners *Flying Fish*, 1875–6 (log 876), *Thomas Hunt* 1878–80 (logs 842–3); and *Express*, 1885–6 (log 863, and in G.W. Blunt White Library, for 1886–7, log 185), and *Sarah W. Hunt*, 1889–92 (logs 838–9). In Nicolson Whaling Collection, Providence Public Library: journal of William Henry Appelman, schooner *Thomas Hunt*, 1873–5, 1875–6 (rolls 42 and 55); brig *Henry Trowbridge*, 1879–83 (roll 27), and schooner *Thomas Hunt*, 1882–4 (coll. 55, vol. 638). See also Edward Lee Dorsett, "Around the World for Seals: The Voyage of the Two-masted Schooner Sarah W. Hunt from New Bedford to Campbell Island, 1883–1884," *American Neptune* 11 (1951): 115–33.

58 Williams, Haven to Capt. Gilderdale, 16 June (quoted) and 17 August 1871, and to Capt. Holmes of the *Franklin*, 5 August 1871, G.W. Blunt White Library, col. 44, vol. 12 (filed with log of *Peru*).

59 Log of *Sarah W. Hunt*, Old Dartmouth Historical Society, log 839, 3 December 1891.

60 Log of *Peru*, 8 January 1872.

61 *Mystic Press*, 31 March 1876. (I am indebted to Bill Peterson of Mystic Seaport Museum for this and the following several newspaper references.)

62 Ibid., 13 June 1878.

63 Ibid., 8 February 1883.

64 Ibid., 17 April 1872.

65 Log of schooner *Charles Colgate*, 21 and 25 December 1877, and 18, 22, and 25 January, 30 March 1878, G.W. Blunt White Library, reel 10.

66 Dorsett, "Around the World for Seals."

67 Log of schooner *Flying Fish*, 9 and 28 September, and 6 October 1878, all quoted, G.W. Blunt White Library, reel 33.

68 Appleman journal, 24 January 1875, Nicolson Collection, roll 42.

69 Ibid., 16 November 1873.

70 Ibid., 21 January 1874.

71 Ibid., 1 November and 22 December 1974.

72 Log of *Thomas W. Hunt*, 18 January 1880, Old Dartmouth Historical Society, log 843.

73 Cawkell, Maling, and Cawkell, *The Falkland Islands*, 107–10; R.W. Coppinger, *Cruise of the "Alert," Four Years in Patagonian, Polynesian, and Mascarene Waters (1878–1892)* (London 1883): 112–16; R.N. Brown, R.C. Mossman, and J.H. Harvey Pirie, *The Voyage of the "Scotia": Being the Record of a Voyage of Exploration in Antarctic Seas ...* (Edinburgh 1906), 261; Temple, *Seal Fisheries*, 159; ms. on sealing in 1880s, Richard Rathbun Papers, Smithsonian Archives, col. 7078, box 2, folder 11 (reproducing an article from the Port Townsend, Wash., *Daily Argus*, 1 May 1889). The latter discusses an interesting conflict between American and local sealers during which some men burned coal oil on several of the rocks, including the Shag Rocks, in order to keep the seals away.

74 Wright, *Lewis & Dryden's Marine History*, 450.

75 Mattsson, "Fur Seal Hunting," 155.

76 Ibid.; Newell, *H.W. McCurdy Marine History*, 74, 87; the figure of 35,000 is from F.A. O'Gorman, "Fur Seals Breeding in the Falkland Islands Dependencies," *Nature* 192 (1961): 914–16.

77 Robert Cushman Murphy, *Logbook for Grace*; the *Daisy*'s log for 1908–9 is in the Nicolson Collection, vol. 192, and for 1912–13 (not kept by Cleveland), G.W. Blunt White Library, roll 65. See also Bertrand, *Americans in Antarctica*, chap. 14, which includes a full bibliography on the voyage.

78 Populations are discussed in W. Nigel Bonner, "Notes on the Southern Fur Seal in South Georgia," *Proceedings of the Zoological Society of London* 130 (1958); 241–52; "The Fur Seal of South Georgia," *British Antarctic Survey Scientific Reports*, no. 56 (1968), and *Seals and Man*, 64–9; Anelio Aguayo L., "The Present Status of the Antarctic Fur Seal *Arctocephalus gazella* at South Shetland Islands," *Polar Record* 19 (1978): 167–76, and "El lobo fino antartico, *Arctocephalus gazelle* (Peters) in el sector Antártico Chileno," *Ser. Cient. Inst. Antart. Chileno* 5 (1977): 5–16; F.R. Condy, "'Distribution, Abundance, and Annual Cycle of Fur Seals (*Arctocephalus* spp.) on the Prince Edward Islands," *South African Journal of Wildlife Research* 8 (1978): 159–68; Laws, "Current Status," and 'Antarctica: a Convergence of Life," *New Scientist* 40 (1983): 608–16; F.A. O'Gorman, "Fur Seals Breeding in the Falkland Islands Dependencies," and "The Return of the Antarctic Fur Seal," *New Scientist* 20 (1963): 374–6; M.R. Payne, "Growth of a Fur Seal Population," *Transactions of the Royal Society London* 279 (1977): 67–79; Barney Brewster, *Antarctica: Wilderness at Risk*, 83; D.W. Doidge, J.P.

Croxall, and J.R. Baker, "Density-dependent Pup Mortality in the Antarctic Fur Seal Arctocephalus gazella at South Georgia," *Journal of Zoology* (London), in press (my thanks to D.W. Doidge for this and other references on the population of *A. gazella*).

Other southern seal populations exploited by Americans have also revived. *A. pusillus pusillus* of South Africa and Namibia has been protected since the 1890s and harvested under government regulation (some 70,000 a year in the 1970s); the total population is about 850,000, regarded as the optimum sustainable level. *A. tropicalis* at Tristan da Cunha, Gough, Prince Edward islands, and Amsterdam and St. Paul numbered about 20,000 in the mid-1970s. *A. pusillus doriferus* of Australia has about the same population, mostly in the Bass Straits; in general it does not seem to be increasing. *A. forsteri* numbers some 40,000 in the New Zealand area; while a number of the outlying islands have been repopulated, this species has not returned to the Bass Strait where formerly it overlapped *A. pusillus doriferus*. On these seals, see P.D. Shaughnessy, "The Status of the Amsterdam Island Fur Seal" and "The Status of Seals in South Africa and Namibia," and R.M. Warneke, "The Distribution and Abundance of Seals in the Australasian Region, with Summaries of Biology and Current Research," all in FAO Advisory Committee on Marine Resources Research, *Mammals in the Seas* 4: 383–475.

79 On these species, in addition to the standard general works (King, Scheffer, Maxwell), I have used Richard Perry, *The Polar Worlds*, 144–56; G.C.L. Bertram, "The Biology of the Weddell and Crabeater Seals," *British Graham Land Expedition, 1934–37, Scientific Reports*, 1, no. 1 (1940); R.M. Laws, "The Seals of the Falkland Islands and Dependencies," *Oryx* 2 (1953): 87–97; G.L. Kooyman and H.T. Anderson, "Deep Diving," In Harald T. Anderson, ed., *The Biology of Marine Mammals*, 65–94; Alton Lindsey, "Notes on the Crabeater Seal," *Journal of Mammalogy* 19 (1938): 456–61.

80 Laws, "Current Status," 155. Bonner, *Seals and Man*, 101, gives an estimate of 14,858,000 for crabeater seals.

81 Condy, "Distribution, Abundance."

82 Mill, *Siege of the South Pole*, 373–5; W.G. Burn Murdoch, *From Edinburgh to the Antarctic ... during the Dundee Antarctic Expedition of 1892–93*.

83 *Brewster, Antarctica*, 83.

84 Randall R. Reeves, "Exploitation of Harp and Hooded Seals"; communication from David Sergeant, 47. The interests of scientists is a point made by L. Harrison Matthews in A.W. Mansfield, "Population Dynamics and Exploitation of Some Arctic Seals," in M.W. Holdgate, ed., *Antarctic Ecology*, 450.

85 Perry, *Polar Worlds*; Brewster, *Antarctica* ("Closing in for the Krill" is the title of chap. 6); Nigel Sitwell, "Sealing in the Antarctic," *Animals* 14 (1972): 156–7.

86 Sir Martin Conway, *No Man's Land: A History of Spitsbergen from Its Discovery in 1596 ...*, 30.

87 Ibid., 21. See also James Lamont, *Seasons with the Sea-Horses; or Sporting Adventures in the Northern Seas.*

88 Glover M. Allen, "The Walrus in New England," *Journal of Mammalogy* 11 (1930): 139–45; Harrison F. Lewis and J. Kenneth Doutt, "Records of the Atlantic Walrus and the Polar Bear in the Northern Part of the Gulf of St. Lawrence," *Journal of Mammalogy* 23 (1942): 365–75; M.J. Dunbar, "The Pinnipedia of the Arctic and Subarctic," *Bulletin of the Fisheries Research Board of Canada* 85 (1949): 1–22; Chantraine, *The Living Ice*, 93–5; Lubbock, *Arctic Whalers*, 426.

89 Randall R. Reeves, *Atlantic Walrus (Odobenus rosmarus rosmarus): A Literature Survey and Status Report.*

90 On the life history of the walrus, see Richard Perry, *The World of the Walrus*; John S. Burns, *The Walrus in Alaska: Its Ecology and Management*; Karl W. Kenyon, "Walrus," in Haley, ed., *Marine Mammals* 178–83; Francis H. Fay and Carleton Ray, "Influence of Climate on the Distribution of Walruses, *Odobenus rosmarus (L),*" *Zoologica* 53 (1968): 1–13, 19–32; George Reiger, "What Now for the Walrus?" *National Wildlife* 14 (1976): 50–7; G. Carleton Ray, "Learning the Ways of the Walrus," *National Geographic* 156 (1979): 464–580; Douglas Wartzok and G. Carleton Ray, *The Hauling-Out Behavior of the Pacific Walrus.*

91 Fay and Ray, "Influence of Climate," 19–32.

92 Reiger, "What Now," 53.

93 United States, Interagency Task Group, NMFS/NOAA and FWS, *Final Impact Statement: Consideration of a Waiver of the Moratorium and Return of Certain Marine Mammals to the State of Alaska* (Washington 1978), 1: 52–3; Kenyon, "Walrus." An example of a statement that the walrus digs up food with the tusks: Fred Bruemmer, "The North American Walrus," *Canadian Geographical Journal* 86 (1973): 90–5.

94 John J. Burns, "Remarks on the Distribution and Natural History of Pagophilic Pinnipeds in the Bering and Chukchi Seas," *Journal of Mammalogy* 51 (1970): 44–54.

95 Richard K. Nelson, *Hunters of the Northern Ice*, 364.

96 Log of brig *Victoria* (Honolulu), G.W. Blunt White Library, log 549.

97 I am indebted to the members of the bowhead census project at the Old Dartmouth Historical Society for this point.

98 On numbers, Perry, *World of the Walrus*, 124; Francis H. Fay, "History and Present Status of the Pacific Walrus Population," *Transactions of the North American Wildlife Conference* 22 (1957): 434–9.

99 Stevenson, "Oil from Seals"; Francis H. Fay, "Industrial Utilization of Marine Mammals," *Proceedings of the 29th Alaska Science Conference* (1979), 75–9.

100 Schwatka, *Nimrod of the North*, 35, 46; Capt. F.E. Kleinschmidt, "A Day of Blood," *Pacific Monthly* 23 (1910): 561–79 (prices); Lubbock, *Arctic Whalers*, 426 (bicycle seats); Reeves, *Atlantic Walrus*, 27 (billiard tips); Arnold, "Hides and Skins," 523.

101 Allen, "Walrus in New England," 12 (teeth); Perry, *World of the Walrus*, 122 (numbers), and 130 (Kenyon account).

102 U.S. Interagency Task Group, *Final Impact Statement* 1: 50.

103 Robert Frothingham, "Walrus Hunting with the Eskimos," in Frederick A. Blossom, ed., *Told at the Explorer's Club*, 123–33; Albert H. Tucker, "'Walrus Hunting in the Arctic," *Pacific Motor Boat* 17 (1925) (San Francisco Maritime Museum PSH 399/W3).

104 Log of steamer *Diana*, 1899 (and see *Esquimaux* and *Polynia*, 1880), Public Archives of Canada, MG 29, A 58, logs 3 and 4.

105 Perry, *World of the Walrus*, 124–30; A.W. Mansfield, "The Walrus in Canada's Arctic," *Canadian Geographic Journal* 73 (1966): 88–95; Fay, "History and Present Status," and Scammon, *Marine Mammals*, 181; A. Howard Clark, "The Pacific Walrus Fishery," in George Browne Goode, ed., *The Fisheries and Fishery Industries of the United States* 5, no. 2, particularly p. 318, noting that 1,260,000 walruses were taken in 1870–80 alone, which contrasts with Fay's 1,300,000 for 1800–50.

106 James A. Estes and J.R. Gilbert, "Evaluation of an Aerial Survey of Pacific Walruses (*Odobensus rosmarus divergens*)," *Journal of the Fishery Research Board of Canada* 35 (1978): 1130–40.

107 U.S. Interagency Task Force, *Final Impact Statement* 1: 50; Reiger, "What Now" (for 170,000); Perry, *World of the Walrus*, 130 (reproductive rate); Kenyon, "Walrus" (King Island). See also National Archives of the United States, RG 22, sec. 91, file 266 (Walrus protection), on Canadian efforts; Francis H. Fay, Howard M. Feder, and Samuel W. Stocker, *An Estimation of the Impact of the Pacific Walrus Population on its Food Resources in the Bering Sea*.

108 U.S. Interagency Task Force, *Final Impact Statement*; Bonner, *Seals and Man*, 33.

109 For an example of the unpredictable, see Francis H. Fay and Brendon P. Kelly, "Mass Natural Mortality of Walruses (*Odobenus rosmarus*) at St. Lawrence Island, Bering Sea, Autumn 1978," *Arctic* 33 (1980): 226–45.

CHAPTER EIGHT

1 Scheffer and Todd, "History," 105.

2 Ibid., 140–5; George Y. Harry, Jr., *The Effects of Management on the Pribilof Islands Fur Seal Herd*.

3 Scheffer and Todd, "History," 134–7.

4 Austin and Wilke, *Japanese Fur Sealing*, reviews Japanese policy; see also Rogers, "Economic Analysis," 115. My thanks to Walter Kirkness for information on Japanese sealing during World War II.

5 Scheffer and Todd, "History," 197–8.

6 Harry, *Effects of Management*, 2.

7 Scheffer and Todd, "History," 165.

8 Victor B. Scheffer, *A Voice for Wildlife*, 63; Bonner, *Seals and Man*, 50–2.

9 Scheffer and Todd, "History," 177 and 198.

10 Harry, *Effects of Management*, 8; L. Keith Miller, *Energetics of the Northern Fur Seal in Relation to Climate and Food Resources of the Bering Sea*.

11 Harry, *Effects of Management*, 10–14; Bonner, *Seals and Man*, 52 (quoted).

12 The following remarks on Greenpeace are based upon Greenpeace sealing correspondence in the Anchorage and Vancouver Offices (I am grateful to officers of the organization for access to these files), particularly John Frizell, Research Director, Greenpeace, to Larry Merculieff, President, Tanadqusix Corporation, 9 July 1979, and press release, 18 June 1979 (Anchorage); Will Anderson, Director, Greenpeace Anchorage, to Patrick Moore, Director, Greenpeace Vancouver, 5 and 30 May 1979.

13 Moore to Greenpeace Offices, 15 May 1979 (Greenpeace, Vancouver).

14 Moore memorandum to Pribilof Greenpeace expedition, 6 June 1979 (Greenpeace, Vancouver); each of these references has a special meaning for Greenpeace members and supporters, on which see Hunter, *Warriors of the Rainbow*.

15 Scheffer and Todd, "History," 112–22; Martin, *Silver Fleece*, 251; Philip B. Fouke to Charles Nagel, Secretary of Commerce and Labor, 22 September and 10 October 1910, National Archives of the United States, RG 22, sec. 91, file 110 (fur seals); Secretary of Commerce to Fouke, 11 June 1916; ibid., file 165 (Fouke Fur Co.); John R. Lyman, Bureau of Foreign and Domestic Commerce, survey report of Fouke Fur Co. (enclosing statement by Fouke), ibid., RG 22, sec. 285 (fur seal management). Lyman's report includes a reference to Wilson-Fouke conversation.

16 Lyman report, National Archives, RG 22, sec. 285. Without the transfer of the fur dressers and their incorporation, Fouke would have lost the sale contract, which stipulated that he set up a dressing and dying plant.

17 Rogers, "Economic Analysis," 100; Scheffer and Todd, "History," 127–8; "The Seal and the Treasury Department," *Fortune* 6 (1930): 60–72, 122, 124. Ralph C. Baker, Ford Wilke, and C. Howard Baltzo, *The Northern Fur Seal*.

18 "The Seal and the Treasury Dept."; William L. Finley, "Salmon, Seals and Skullduggery: Activities of the Commissioner of Fisheries Open to Question," *Nature Magazine* 28 (1936): 299–303.

19 "Fur Ring, Linked with U.S., Threatens All Seals: International
    Exploiters' Record of Deals Discloses Much Politics," *The Dearborn
    Independent: The Ford International Weekly*, 18 August 1923, 2–3.

20 [Anon.], "Furor over Alaskan Seals," *Business Week* 10 February 1962,
    60–4.

21 E. Wells Erbin to Assistant Secretary Orme Lewis (Fisheries), report
    (with enclosures) on Fouke Fur Co., 16 March 1954, National
    Archives, RG 22, sec. 285 (fur seal management); W.E. Corbin, chief,
    Division of Administration, F&WS, memo, 27 August 1953, and file
    on Fouke contracts, Office of Marine Mammals (OMM), Washington,
    D.C., Pribilof files.

22 Review of Fouke Co. Operations and Records, March 1962 (quoted),
    OMM, general Pribilof files, x-binder, doc. X-1. The weekly newsletter
    of Senator Frank E. Gruening of Alaska, 15 December 1961, discusses
    the pressures applied by the senator to persuade the government to
    cancel the lease when Fouke refused to shift some operations to Alaska
    (copy in Federal Archives, Seattle, Alaska Seal Program files, box
    49994).

23 Controller General's report to Secretary of Interior, 10 October 1963,
    OMM, Pribilof files, x-binder, X-12.

24 Notes on meeting of Fouke with representatives of Interior Dept., 29
    November 1961 (memo dated 7 December), and memo, March 1963,
    on Supara award; Fouke to Director of F & WS/BCI, 24 October 1961;
    all OMM, general Pribilof files; Fouke Contract, 1165, and Supura
    Contract, 1963, are x-binder, X-10, and X-11, OMM.

25 Reach, McClinton & Co., New York, report to Fouke Co., November
    1968, copy in OMM, general Pribilof files.

26 Rogers, "'Economic Analysis," 133.

27 See Alaska File of the Revenue Cutter Service, 1867–1914, National
    Archives, microcopy 641, and Federal Archives, Seattle, Dept. of
    Interior, Alaska Fur Seal Section, correspondence, box 12494, Bering
    Sea Patrol, particularly Asst. Secretary of Commerce to Secretary of
    Treasury, 21 January 1937. For some complaints, Frank T. Bell to H.J.
    Christoffers, Bureau of Fisheries, Seattle, 21 January 1938, ibid., box
    12498; box 51850, file 556, gives examples of Japanese illegal sealers.

28 Walter Kirkness, director, Pribilof Island Program to Will Anderson,
    executive director, Greenpeace Alaska, 27 July and 10 August 1979,
    Sealing file, Greenpeace Anchorage Alaska Office. My thanks to Mr
    Kirkness for discussing the fur seal administration with me.

29 Oran R. Young. *Natural Resources and the State: The Political Economy of
    Resource Management* (Berkeley 1981), 70–4. Young concludes that
    the St. George experiment serves little purpose save appeasement of
    conservation interests and should be abandoned; it is worth adding that
    the United States did not embark upon the experiment before acquiring
    the approval of the other treaty powers.

30 Rogers, "Economic Analysis," 161.

31 U.S. Bureau of Commercial Fisheries, Pribilof Island Log Books, St. Paul (Alaska State Library, Juneau), 29 December 1912.

32 Rogers, "Economic Analysis," 117, 157; Jones, Century, 63n10 and 69.

33 Jones, Century, 65 (from her 1975 field notes).

34 Pribilof Log Books, St. Paul, 31 January 1917.

35 Ibid., 28 March 1929.

36 E.T. Christoffers to Radioman-in-Charge, 22 August 1927, Federal Archives, Seattle, Pribilof files, box 12491, miscellaneous records.

37 Jones, Century, 74, and "U.S. Administration," 160.

38 Jones, Century, 84.

39 Jones, "U.S. Administration," 116, 147–8, and Century, 57–9.

40 Jones, "U.S. Administration," 125–9; Rogers, "Economic Analysis," 160.

41 McMillin to Johnston, 12 September 1942, box 12503. See also Johnston to Ward T. Bower, 10 October, same file, including petition and explaining what efforts had been made to improve living arrangements.

45 Bower to Block, 23 October 1943, ibid., box 12504 (Funter Bay Misc. file).

46 Johnston to Bower, 15 June 1942, ibid., box 12506, DI files "s". The water tank, at least, was burnt by accident. The adventures of the six-man garrison are described by the sergeant in charge in Lyman R. Ellsworth, Guys on Ice (New York 1952).

47 Jones, Century, 115. See also Rogers, "Economic Analysis," 165–7.

48 Martin, Silver Fleece, and in a revised edition, The Sea Bears: The Story of the Fur Seal, though essentially they are the same book. Jones, Century, chap. 7, discusses Martin's role; Jones has benefited from the use of Martin's recollections and papers (in Cuernavaca, Mexico).

49 Jones, Century, chap. 7.

50 Baker to C.L. Olson, general manager, 15 July 1959, Federal Archives, Seattle, Interior Files, box 49992; Jones, Century, chap. 8.

51 Jones, Century, 147–60; Young, Natural Resources and the State, chap. 3.

52 The title of a study sponsored by the Aleuts: Barbara Torrey, Slaves of the Harvest: The Story of the Pribilof Aleuts (St. Paul Island: Tanadgusix Corporation 1973).

53 As usual with Newfoundland data, the figures on catches and prices are from Chafe's Sealing Book, while Greene, Wooden Walls, app., lists vessels. See also Colman, "Present State of the Newfoundland Seal Fishery" (1937).

54 David Sergeant, "Harp Seals and the Sealing Industry," Canadian Audubon 25 (1963): 29–35.

55 E.L. Chicanot, "New Eyes for the Sealing Fleet," Scientific American 135 (1928): 409–11; Chafe's Sealing Book, 28–9. The first aerial

spotting of seals was actually in 1922, but from land, and the pilot and sealers could not reach a mutually satisfactory agreement in time on how much the spotting service was worth.

56 H.A. Ryan, *Seal Meat Handling and Processing* (St. John's 1977) (pamphlet, CNS/MUN).

57 A sample (from *Decks Awash* (St. John's) 7, no. 1 [1978]): Canned Seal Pie: put one can (large) of seal meat in large skillet (remove bones), add 1 onion, 1 beef Oxo cube, 3 or 4 each of potatoes and carrots, cut in cubes, seasoning; after cooking, add biscuit topping and bake.

58 E. Calvin Coish, *Season of the Seal: The International Storm over Canada's Seal Hunt*, 70. In addition to Ryan, *Seal Meat*, see for problems involved J.P. Botta, A.P. Downey, J.T. Laudes, and B.P. Noonan, *Utilization of Inshore Newfoundland-Caught Harp Seal ("Pagophilus groenlandicus"): Sensor Quality of Frozen, Stored, Salted, and Smoked Seal Meat*; and "Sealing: Preparing Seal Meat," Government of Canada, Fisheries and Oceans, publication 80/003E (1980).

59 C.W. Andrews, "The Hazardous Industry of North Atlantic Sealing," *Animal Kingdom* 54 (1951): 66–94; J.S. Colman, "The Newfoundland Seal Fishery and the Second World War," *Journal of Animal Ecology* 18 (1949), 40–6.

60 Colman, "Newfoundland Seal Fishery," 45.

61 Quoted in Ralph Matthews, "The Smallwood Legacy: The Development of Underdevelopment in Newfoundland, 1949–1972," *Journal of Canadian Studies* 13 (1978–9): 89–108.

62 Ibid.; see also Gwyn, *Smallwood*.

63 Sergeant, "Harp Seals and the Sealing Industry," and "History and Present Status of Populations of Harp and Hooded Seals," *Biological Conservation* 10 (1976): 95–118; *Chafe's Sealing Book*, supplement for 1937; Ole Friele Backer, "Seal Hunting off Jan Mayen," *National Geographic* 93 (1948): 57–72.

64 Sergeant, "History and Present Status," 106; Reeves, *Exploitation of Harp and Hooded Seals*, 27; Jon Barzdo, *International Trade in Harp and Hooded Seals* (London and Yarmouthport, Maine 1980), gives figures in detail for the 1970s. A much higher percentage than Canada's 2 per cent of hooded seals was taken by Norway at Jan Mayen.

65 St. John's *Daily News*, 12 and 13 March 1968, and 3 February 1972; St. John's *Telegraph Journal*, 14 March 1981 (Newfoundland Historical Society, clipping file, sealing). Barzdo, *International Trade*, gives details on where the skins go: the bulk are processed in Norway (G.C. Rieber Co. of Bergen, above all) and West Germany, but find their final market in East and West Germany and South Europe; perhaps 10 per cent find their way back to Canada – but 40 per cent go to make up novelty items.

66 Sergeant, "Harp Seals and the Sealing Industry."

67 Ibid.; for a history of recent conservation efforts, see Coish, *Season of the Seal*, especially 59–65.

68 Coish, *Season*, 81–2, 96.

69 Ibid., 81–2, 95; [Carter W. Andrews and Robert Parker], *Brief to the Special Advisory Committee to the Ministry of Fisheries and Forestry on Seals Presented to the Government of Newfoundland and Labrador, May 25, 1971* (CNS/MUN), 37–41; W. Nigel Bonner, *Humane Killing of Seals.*

70 See also Davies's *Seal Song*, with photos by Eliot Porter.

71 The role of various organizations may be traced in Coish, *Season*, and, from the Greenpeace viewpoint, in Paul Watson, *Sea Shepherd: My Fight for Whales and Seals*, chap. 10.

72 Tony Williamson, "Inuit, the Innocent Victims," *Decks Awash* 7, no. 1 (February 1978), 60–1; Coish, *Season*, 194, 258; ringed seal prices had risen from $2 in 1960–1 to $25 each in 1965.

73 Canada, Dept. of Fisheries and Oceans, information bulletin I-HQ-83-08E (1983).

74 J.R. Beddington and H.A. Williams, *The Status and Management of the Harp Seal in the North-West Atlantic*, 7–10; Ronald and Dougan, "The Ice Lover"; Lee M. Talbot, "Maximum Sustainable Yield: An Obsolete Management Concept," *Transactions of the North American Wildlife and Natural Resources Conference* 40 (1975): 91–6; D.M. Lavigne, "Management of Seals in the Northwest Atlantic Ocean," ibid. 44 (1979): 488–97. The 1983 Dept. of Fisheries and Oceans estimate of about two million harp seals is found in news release NR-HQ-083-016E and information bulletins I-HQ-06E, 07E, and 08E (all 1983).

75 D.E. Sergeant, "Estimating Numbers of Harp Seals," in K. Ronald and A.W. Mansfield, eds., *Biology of the Seal*, 274–80; W.D. Bowen and D.E. Sergeant, "Mark-Recapture Estimates of Harp Seal Pup (*Phoca groenlandica*) Production in the Northwest Atlantic," *Canadian Journal of Fisheries and Aquatic Sciences* 40 (1983): 274–80; David M. Lavigne, "Life or Death for the Harp Seal," *National Geographic* 149 (1976): 129–42; A.W. Mansfield, "Population Dynamics"; Beddington and Williams, *Status and Management*.

76 Reeves, *Exploitation of Harp and Hooded Seals*, 35.

77 For example, Sergeant, "History and Present Status," 113.

78 *Maclean's*, 22 March 1982, 123, and 14 March 1983, 15, both use $13,000,000; Canada, Dept. of Fisheries and Oceans, information bulletin I-HQ-83-08E (1983), gives $12,000,000 for 1981, but only $10,000,000 for 1982.

79 D.L. Dunn, *Canada's East Coast Sealing Industry, 1976: A Socio-Economic Review*, 25.

80 Ibid.

81 Hunter, *Warriors of the Rainbow*, 249–51.

82 On Codpeace and similar organizations, I have used clippings files,

Newfoundland Historical Society, St. John's. Particularly notable is the Moncton, New Brunswick, *Transcript*, 10 February 1979. On the Europarliament buttons, *Maclean's*, 22 March 1982, 23.

83 New Zealand *Herald*, 19 March 1979, Newfoundland Historical Society, clippings file.

84 See for example a pamphlet entitled *The Seal Hunt*, released by the Information Branch, Department of Fisheries and the Environment, Canada, Fisheries and Marine Service [c. 1976], with David Blackwood illustrations.

85 Watson, *Sea Shepherd*, chaps. 10–11.

86 David Letterman show, 21 July 1980 (NBC, Syracuse, N.Y.). Davies's point is well taken; landsmen in their small boats prefer to hunt bigger seals which have more meat and which do not require such delicate care to avoid pelt spoilage; B. Budgell, "Sealing," Great Northern Peninsula: St. Anthony," Folklore Archive, 76–368, Memorial University of Newfoundland.

87 *Maclean's*, 6 December 1982, 23.

88 Ibid., 14 March 1983, 15.

89 Canada, Department of Fisheries and Oceans, news releases NR-HQ-082-957E (25 November 1982), NR-HQ-082-059E (16 December 1982), NR-HQ-083-027E (9 May 1983).

90 *Le Monde*, 24 February 1983.

91 John Lister-Kaye, *Seal Cull: The Grey Seal Controversy*, 103–4; see also Bonner, *Seals and Man*, chap. 8.

92 Botta, Downey, Luades, Noonan, *Utilization*.

93 Greene to Miss Lefroy, 17 January 1937 (the year of Greene's death), Archives of the CNS/MUN, MF 46.

94 David Letterman show, 21 July 1980.

95 Lister-Kaye, *Seal Cull*, 13.

96 *Decks Awash* 7, no. 1 (February 1978): 11.

97 Scheffer, *Adventures*, 173–87. For an interesting example of the debate between conservationists and managers, see the points made in the exchange on harp seals between David Sergeant and Sidney J. Holt, *Nature* 17 February, 14 July, and 6 September 1983.

98 Scheffer, *Adventures*, 177.

99 Victor B. Scheffer, "Conservation of Marine Mammals," in Haley, ed., *Marine Mammals*, 244.

100 Scheffer, *Adventures*, 177.

101 Harry R. Lillie, *The Path Through Penguin City*, xvi.

102 Nor is this simply a question of the active conservationists mentioned above. The extent to which seals have a popular following can also be judged by the success of the many nonscientific books on seals which have appeared over the years, examples of which (excluding those written exclusively for children), are Rowena Farr, *Seal Morning*; R.H.

Pearson, *A Seal Flies By*; H.G. Hurrell, *Atlanta My Seal*; Nina Warner Hooke, *The Seal Summer*; Dean Jennings, *Valla: The Story of a Sea Lion*; Ken Jones, *Seal Doctor* (first published in 1970 as *Orphans of the Sea*); Lyn Hancock, *There's a Seal in My Sleeping Bag*; Harry Goodridge and Lew Dietz, *A Seal Called Andre*. Most of these, on inspection, prove to be about either grey seals, harbor seals, or sea lions, and not the species discussed in this work. They also focus upon seal-man interaction (often in a possessive way – "My Seal"). Nevertheless, consumers of such books and films are potential antisealing supporters. To this list should be added at least one motion picture: the Samuel Goldwyn Company's *The Golden Seal* (1983).

Bibliography

The bibliography lists books, substantial works in manuscript (such as theses) short monographs, and articles in books and periodicals. For other unpublished documents, logbooks, journals, and correspondence, consult chapter notes. Official government publications are listed in the bibliography only when the individual author or authors are clearly designated in the publication itself.

BOOKS, THESES, AND UNPUBLISHED MONOGRAPHS

Acton, Henry. *Catching of the Whale and Seal* ... Salem: Ives & Jewett 1838.

Adams, J.G.L. "Newfoundland Population Movements with Particular Reference to the Post-War Period." PHD diss., McGill University, 1970.

Alaska. University of Alaska History Research Project. Alaska History Documents. Vol. 6. "Material from the Library of Congress Relating to the Bering Sea Controversy, 1891–1912." n.d. Library of Congress. Manuscript Division. Typescript.

Allan, Alexander. *Hunting the Sea Otter*. London: Horace Cox 1910.

Allyn, Gurdon L. *The Old Sailor's Story* ... . Norwich, Conn.: Gordon Wilcox 1879.

Andersen, Harald T., ed. *The Biology of Marine Mammals*. New York: Academic Press 1969.

Anderson, J.R.L. *High Mountains and Cold Seas: A Biography of H.W. Tilman*. Seattle: The Mountaineers 1980.

Anspach, Rev. Lewis Amadeux. *A History of the Island of Newfoundland* ... .
   London: privately printed 1819.
Arnold, John R. *Hides and Skins*. Chicago: A.W. Shaw 1925.
Arsen'ev, V.A., and K.I. Panin, eds. *Pinnipeds of the North Pacific*.
   Washington: Israel Program for Scientific Translations, for U.S. Dept. of
   Interior and National Science Foundation 1971.
Atkinson, Robert. *Shillay and the Seals*. London: Collins & Harvill 1980.
Bach, John. *A Maritime History of Australia*. Sydney: Pan Books 1982.
Backhouse, K.M. *Seals*. London: Arthur Barker 1969.
Bailey, Jane H. *Sea Otter: Core of Conflict, Loved or Loathed*. Morro Bay,
   Calif. El Morro Publications 1979.
Balch, Edwin Swift. *Antarctica*. Philadelphia: Lane & Scott 1902.
Bancroft, Hubert Howe. *History of Alaska, 1730–1885*. San Francisco:
   A.L. Bancroft & Co. 1886.
– *History of British Columbia, 1792–1887*. San Francisco: History Co. 1887.
– *History of California*. 7 vols. Santa Barbara: R. Wallace Hebbard 1969.
Banning, George Hugh. *In Mexican Waters*. Boston: Charles E. Lauriat 1925.
Barabash-Nikiforov, Iliallich. *The Sea Otter (Kalan)*. Jerusalem: Israeli
   Program for Scientific Translation (for National Scientific Foundation)
   1962.
Barragy, Terrence Joseph. "American Maritime Otter Diplomacy."
   PHD diss., University of Wisconsin, 1974.
Barratt, Glynn. *Russia in Pacific Waters, 1715–1825*. Vancouver: University
   of British Columbia Press 1981.
Barron, Capt. William *Old Whaling Days*. 1895. Reprint. Cardiff: Conway
   Maritime Press 1970.
Bartlett, Capt. Robert A. *The Log of Bob Bartlett* ... . New York:
   G.P. Putnam's 1928.
– *Sails over Ice*. New York: Scribner's 1934.
Bateson, Charles. *Dire Strait: A History of Bass Strait*. Sydney: A.H. &
   A.W. Reed for Broken Hill Proprietary Co. 1973.
Beck, Horace. *Folklore and the Sea*. Middletown, Conn.: Wesleyan University
   Press 1973.
Beechey, F.W. *Voyage to the Pacific and Behring's Straits ... 1825–8*. Edited by
   Robert Huish. London: William Wright 1836.
Begg, A. Charles, and Neil C. Begg. *Dusky Bay*. N.Y.: Barnes & Noble 1969.
– *The World of John Boultbee, Including an Account of Sealing in Australia and
   New Zealand*. Christchurch, N.Z.: Whitcoulls 1979.
Belcher, Capt. Sir Edward. *Narrative of a Voyage round the World Performed in
   H.M.S. "Sulphur" during the years 1836–1842*. 2 vols, London: Henry
   Colburn 1843.
Belda, Daniel Lluca, Llovell Adams, and S.G. Losocki. *Dos mamiferos marinos
   de Baja California*. Mexico City: Instituto Mexicano de Recursos Naturales
   1969.

Bellingshausen, Capt. *The Voyage of ... to the Antarctic Seas, 1819–1821.* Edited by Frank Debenham. 2 vols. London: Hakluyt Society 1945. Reprint New York: Kraus 1976.

Bercuson, David, Jr, and Philip A. Buckner, eds. *Eastern and Western Perspectives: Papers from the Joint Atlantic Canada/Western Canadian Studies Conference.* Toronto: University of Toronto Press 1981.

Bertrand, Kenneth J. *Americans in Antarctica, 1775–1948.* New York: American Geographical Society 1971.

Bingham, Edwin R., and Glen A. Lowe. *Northwest Perspectives: Essays on the Culture of the Pacific Northwest.* Eugene, Ore.: University of Oregon Press 1979.

Bixby, William. *Track of the "Bear."* New York: David McKay 1965.

Boit, John. *Voyage of the "Columbia": Around the World with ... 1790–1793.* Edited by Dorothy O. Johansen. Portland, Ore.: Beaver Books 1960.

Bonner, W. Nigel. *Seals and Man: A Study of Interactions.* Seattle: University of Washington Press, for Washington Sea Grant Program 1982.

Boykin, Edward. *Ghost Ship of the Confederacy: The Story of the "Alabama" and her Captain, Raphael Semmes.* New York: Funk & Wagnalls 1957.

Bradley, Harold Whitman. *The American Frontier in Hawaii: The Pioneers, 1789–1843.* Stanford: Stanford University Press 1942.

Brewster, Barney, *Antarctica: Wilderness at Risk.* San Francisco: Friends of the Earth 1982.

Brown, Cassie. *Death on the Ice: The Great Newfoundland Sealing Disaster of 1914.* Toronto: Doubleday Canada 1972.

Brown, R.N., R.C. Mossman, and J.H. Harvey Pirie. *The Voyage of the 'Scotia': Being the Record of a Voyage of Exploration in Antarctic Seas by Three of the Staff.* Edinburgh: Blackwood 1906.

Brown, Robert Craig. *Canada's National Policy, 1883–1900.* Princeton: Princeton University Press 1964.

Brown, Vinson. *Sea Mammals and Reptiles of the Pacific Coast.* New York: Collier 1976.

Browne, J. Ross. *A Sketch of the Settlement and Exploration of Lower California.* New York: D. Appleton 1868.

Bruemmer, Fred. *The Life of the Harp Seal.* New York: Time Books 1977.

Bryan, William Alanson. *Natural History of Hawaii.* Honolulu: Hawaiian Gazette 1915.

Brym, Robert J., and R. James Sacouman. *Underdevelopment and Social Movements in Atlantic Canada.* Toronto: New Hogtown Press 1979.

Burns, John J. *The Walrus in Alaska: Its Ecology and Management.* Juneau: Alaska Dept. of Fish and Game 1965.

Burroughs, Polly. *The Great Ice Ship "Bear": Eighty-Nine Years in Polar Seas.* New York: Van Nostrand 1970.

Busch, Briton Cooper, ed. *Alta California, 1840–1842: The Journal and Observations of William Dane Phelps, Master of the Ship "Alert."* Western

Lands & Waters, vol. 13. Glendale, Calif.: Arthur H. Clark 1983.

– , ed. *Master of Desolation: The Reminiscences of Capt. Joseph J. Fuller.* American Maritime Library, vol. 9. Mystic: Mystic Seaport Museum 1980.

Campbell, Archibald. *A Voyage Round the World from 1806 to 1812.* Edinburgh: Archibald Constable 1816.

Carrara, Italo Santiago. *Zoorecursos naturales de la Antartica. Mamiferos marinos (pinnipedios).* La Plata, Argentina: Universidad Nacional de la Plata 1964.

– *Lobos marinos, pingüinos y guaneras de las costas del litoral maritimo e Isla adyacentes de la Republica Argentina.* La Plata, Argentina: Universidad Nacional de la Plata 1952.

Carrington, C.E. *The Life of Rudyard Kipling.* Garden City, N.Y.: Doubleday & Co. 1955.

Carroll, Michael. *The Seal and Herring Fisheries of Newfoundland.* Montreal: John Lovell 1873.

Cartwright, George. *A Journal of Transactions and Events, During a Residence of Nearly Sixteen Years on the Coast of Labrador.* 3 vols. Newark, England: Allen & Ridge 1792.

Cashin, Maj. Peter. *My Life and Times, 1890–1919.* Edited by R.E. Buehler. Portugal Cove, Nfld.: Breakwater Books 1976.

Cawkell, M.B.R., D.H. Maling, and E.M. Cawkell. *The Falkland Islands.* London: Macmillan 1960.

Chafe, L.G. *Chafe's Sealing Book: A History of Newfoundland Sealfishery from the Earliest Available Records down to and Including the Voyage of 1923.* St. John's Nfld.: Trade Printers and Publishers 1923 (with annual supplements to 1946 inclusive).

Chantraine, Pol. *The Living Ice: The Story of the Seals and the Men Who Hunt Them in the Gulf of St. Lawrence.* Translated by David Lobdell. Toronto: McClelland & Stewart 1980.

Chevigny, Hector. *Lord of Alaska: The Story of Baranov and the Russian Adventure.* Portland, Ore.: Binfords & Mort 1965.

– *Lost Empire: The Life and Adventures of Nikolai Petrovich Rezanov.* Portland, Ore.: Binfords & Mort, 1958.

– *Russian America: The Great Alaskan Venture, 1741–1867.* New York: Viking Press 1965.

Chilton, Charles, ed. *The Subantarctic Islands of New Zealand.* 2 vols. Wellington: Philosophical Institute of Canterbury 1909.

Christie, E.W. Hunter. *The Antarctic Problem: An Historical and Political Study.* London: Allen & Unwin 1951.

Coaker, W.F. *The History of the Fisherman's Protective Union of Newfoundland.* St. John's: Union Publishing Co. 1920.

Coffey, David J. *Dolphins, Whales and Porpoises: An Encyclopedia of Sea Mammals.* New York: Macmillan 1977.

Coish, E. Calvin. *Season of the Seal: The International Storm over Canada's Seal Hunt.* St. John's: Breakwater Books 1979.

Colby Varnard L. *New London Whaling Captains*. Mystic, Conn.: Marine
    Historical Association 1936.

Colnet, Capt. James. *Journal ... Aboard the "Argonaut", 1789–1791*. Edited
    by F.W. Howay, Toronto: Champlain Society 1940.

Condon, Michael E. *The Fisheries and Resources of Newfoundland: "The Mine
    of the Sea," National, International, and Co-Operative*. St. John's: n.p. 1925.

Connolly, Joe. *On the Front*. St. John's: Jesperson Printing 1978.

Conway, Sir Martin. *No Man's Land: A History of Spitsbergen from its Discovery
    in 1596...* . Cambridge: Cambridge University Press 1906.

Cook, Warren L. *Flood Tide of Empire: Spain and the Pacific Northwest, 1543–
    1819*. New Haven: Yale University Press 1973.

Cooper, James Fenimore. *The Sea Lions*. Lincoln: University of Nebraska
    Press 1965 (fiction).

Corbett, P.E. *The Settlement of Canadian-American Disputes: A Critical Study
    of Methods and Results*. New Haven: Yale University Press 1937.

Corney, Peter. *Voyages in the Northern Pacific ... 1813 to 1818*. Honolulu:
    Thomas G. Thrum 1896.

Coes, Elliott. *The Fur-Bearing Animals of North America*. 1877. Reprint. New
    York: Arno Press, 1970.

Cousteau, Jacques-Yves, and P. Diole. *Diving Companions: Sea Lion, Elephant
    Seal, Walrus*. New York: A. & W. Visual Library 1974.

Cumpston, J.S. *First Visitors to Bass Strait*. Canberra: Roebuck Society 1973.

– *Kangaroo Island, 1800–1936*. Canberra: Roebuck Society 1970.

– *Macquarie Island*. Melbourne: Antarctic Division, Dept. of External Affairs,
    Australia 1968.

Dakin, William John. *Whaleman Adventurers: The Story of Whaling in Australian
    Waters ...* . 2nd ed. Sydney: Angus & Robertson 1938.

Dale, Paul W., ed. *Seventy North to Fifty South: The Story of Captain Cook's
    Last Voyage*. Englewood Cliffs, N.J.: Prentice Hall 1969.

Davies Brian. *Savage Luxury: The Slaughter of the Baby Seals*. New York:
    Taplinger 1971.

– *Seal Song*. Photos by Eliot Porter. New York: Viking Press, 1978.

Decker, Robert Owen. *The Whaling City: A History of New London*. Chester,
    Conn.: Pequot Press 1976.

– *Whaling Industry of New London*. York, Pa.: Liberty Cap Books 1973.

Delano, Amasa. *A Narrative of Voyages and Travels in the Northern and Southern
    Hemispheres ...* . Boston: E.G. House 1817.

Deming, Clarence. *By-ways of Nature and Life*. New York: G.P. Putnam's
    Sons 1884.

Dodge Bertha S., ed. *Marooned: Being the Narrative of the Sufferings and
    Adventures of Capt. Charles H. Barnard ... 1812–1816*. Middletown,
    Conn.: Wesleyan University Press 1979.

Doig, Ivan. *Winter Brothers: A Season at the Edge of America*. N.Y.: Harcourt
    Brace Jovanovich 1980.

Doran, Adelaide LeMert. *Pieces of Eight Channel Islands: A Bibliographical Guide and Source Book.* Glendale: A.H. Clark 1980.

Dunmore, John. *French Explorers in the Pacific.* 2 vols. Oxford: Clarendon Press 1965.

Dunn, D.L. *Canada's East Coast Sealing Industry, 1976, a Socio-Economic Review.* Ottawa: Fishing Services Directorate, Fisheries and Marine Service, Dept. of Fisheries and Environment 1977.

Eaton, Margaret Holden. *Diary of a Sea Captain's Wife: Tales of Santa Cruz Island.* Santa Barbara: McNally & Loftin, West 1981.

Edwards, Deltus M. *The Toll of the Arctic Seas.* London: Chapman & Hall 1910.

England, George Allan. *Vikings of the Ice: Being the Log of a Tenderfoot on the Great Newfoundland Seal Hunt.* New York: Doubleday, Page 1924. Reprint, entitled *The Greatest Hunt in the World.* Montreal: Tundra Books 1969.

– *The White Wilderness. A Story of the Great Newfoundland Seal-Hunt ...* London: Cassell 1924 (fiction).

Fanning, Capt. Edmund. *Voyages and Discoveries in the South Seas, 1792–1832.* Salem: Marine Research Society 1924.

– *Voyages to the South Seas, Indian and Pacific Oceans, China Sea, Northwest Coast, Feejee Islands, South Shetlands, & &.* 2nd ed. New York: William H. Vermilye 1838. Reprint. Fairfield, Wash.: Ye Galleon Press 1970.

Farre, Rowena. *Seal Morning.* 1957. London: Arrow Books 1965 (fiction).

Fay, C.R. *Life and Labour in Newfoundland, Based on Lectures Delivered at the Memorial University of Newfoundland.* Toronto: University of Toronto Press 1956.

Feltham, John. "The Development of the F.P.U. [Fishermen's Protective Union] in Newfoundland (1908–1923)." MA thesis, Memorial University of Newfoundland, 1959.

Fisher, Raymond H. *The Russian Fur Trade, 1550-1700.* Berkeley: University of California Press 1943.

Fisher, Robin. *Contact and Conflict: Indian-European Relations in British Columbia, 1774–1890.* Vancouver: University of British Columbia Press 1977.

Flint, Sue. *Let the Seals Live.* Sandwick, Shetland, Scotland: Thule Press 1979.

Forbes, Alexander. *California: A History of Upper and Lower California ... .* London: Smith, Elder 1839. Reprint. New York: Kraus 1972.

Foust, Cliford M. *Muscovite and Mandarin: Russia's Trade with China and Its Setting 1727–1805.* Chapel Hill: University of North Carolina Press 1969.

Fox, Arthur. *The Newfoundland Constabulary.* St. John's: privately printed 1971.

Freuchen, Peter. *Arctic Adventure: My Life in the Frozen North.* New York: Farrar & Rinehart 1935.

– *Book of the Eskimos.* Cleveland: World Publishing Co. 1961.

– *Vagrant Viking: My Life and Adventures.* New York: Julian Messner, 1954.

– and Finn Salomonsen. *The Arctic Year*. New York: G.P. Putnam's Sons 1958.

Friesen, J., and H.K. Ralston, eds. *Historical Essays on British Columbia*. Toronto: McClelland & Stewart 1976.

Gasking, D.E. *Whales, Dolphins and Seals with Special Reference to the New Zealand Region*. London: Heinemann 1972.

Gay, James Thomas. "American Fur Seal Diplomacy." PHD diss., University of Georgia, 1971.

George, William. *A Sealer's Journal; or A Cruise of the Schooner "Umbrina."* Victoria: H.G. Waterson, 1895.

Gibson, James R. *Feeding the Russian Fur Trade: Provisionment of the Okhotsk Seaboard and the Kamchatka Peninsula, 1639–1895*. Madison: University of Wisconsin Press 1969.

– *Imperial Russia in Frontier America: The Changing Geography of Supply of Russian America, 1784–1867*. New York: Oxford University Press 1976.

Gillespie, T.H. *A Book of King Penguins*. London: Herbert Jenkins 1932.

Gillham, Mary E. *Sub-Antarctic Sanctuary: Summertime on Macquarie Island*. London: Gollancz 1967.

Gillsäter, Sven. *Wave after Wave*. London: Allen & Unwin 1964.

Gleason, Duncan. *Islands of California: Their History Resources and Physcial Characteristics*. Los Angeles: Sea Publications 1959.

Goebel, Julius. *The Struggle for the Falkland Islands: A Study in Legal and Diplomatic History*. 2nd ed. New Haven: Yale University Press 1982.

Golder, F.A., ed. *Bering's Voyages*. 2 vol. New York: American Geographical Society 1922.

– *Russian Expansion of the Pacific, 1641–1850*. Cleveland: Arthur H. Clark 1914.

Golovin, V.M. *Around the World in the "Kamchatka," 1817–1819*. Honolulu: Hawaiian Historical Society and the University Press of Hawaii 1979.

Goode George Brown, et al., *The Fisheries and Fishery Industries of the United States*. 5 vols. Washington, D.C.: USGPO 1884.

Goodridge, Charles Medyett. *Narrative of a Voyage to the South Seas, and the Shipwreck of the "Princess of Wales" Cutter ... .* Exeter: W.C. Featherstone 1843.

Goodridge, Harry, and Lew Dietz. *A Seal Called Andre*. New York: Warner Books 1976.

Gough, Barry M. *Distant Dominion: Britain and the Northwest Coast of North America, 1579–1809*. Vancouver: University of British Columbia Press 1980.

Graebner, Norman A. *Empire on the Pacific: A Study in American Continental Expansion*. New York: Ronald Press 1955.

Graham, Frank W. *"We Love Thee, Newfoundland": Biography of Sir Cavendish Boyle, KCMG, Governor of Newfoundland 1901–1904*. St. John's; privately printed 1979.

Grattan, C. Hartley. *The Southwest Pacific to 1900*. Ann Arbor: University of Michigan Press 1963.

Great Britain, Challenger Office. *Report on the Scientific Results of the Voyage of H.M.S. "Challenger" During the Years 1873–76 ...* . Edited by Sir C. Wyville Thomson and John Murray. Vol. 1, pt. 1. London: HMSO 1885.

Greenbie, Sydney and Marjorie. *Gold of Ophir*. London: Heinemann 1925.

Greene, Maj. William Howe. *The Wooden Walls Among the Ice Floes: Telling the Romance of the Newfoundland Seal Fishery*. London: Hutchinson 1933.

Greenhill, Basil. *The Merchant Schooners*. Vol. 1, 2nd ed. Newton Abbot, Devon: David & Charles 1968.

Gregson, Harry. *A History of Victoria 1842–1970*. Vancouver: J.J. Douglas 1970.

Grenfell, Wilfred Thomason. *A Labrador Doctor: Autobiography*. London: Hodder & Stoughton n.d.

– *Adrift on an Ice Pan*. Boston: Houghton Mifflin 1909.

– *Vikings of Today*. New York: Fleming H. Revell n.d.

Gwyn, Richard, *Smallwood: The Unlikely Revolutionary*. Toronto: McClelland & Stewart 1968.

Hainesworth, D.R., ed. *Builders and Adventurers: The Traders and the Emergence of the Colony, 1788–1821*. Melbourne: Cassell Australia 1968.

Haley, Delphine, ed. *Marine Mammals of Eastern North Pacific and Arctic Waters*. Seattle: Pacific Search Press 1978.

Hancock, Lyn. *There's a Seal in My Sleeping Bag*. Toronto: Collins 1972.

Harris, C.J. *Otters: A Study of the Recent Lutrinae*. London: Weidenfeld & Nicolson 1968.

Harvey, Rev. M. *Newfoundland at the Beginnings of the 20th Century ... A Treatise of History and Development*. New York: South Publishing Co. 1902.

Hattenhauer, R. "A Brief Labour History of Newfoundland." Report Prepared by the Royal Commission on Labour Legislation in Newfoundland and Labrador, 1970. Memorial University of Newfoundland, St. John's. Typescript.

Hayes, Williams. *Stonington Chronology, 1649–1976*. Chester, Conn.: Pequot Press, for Stonington Historical Society 1976.

Hays, Alice N. *David Starr Jordan: A Bibliography of His Writings 1871–1931*. Stanford: Stanford University Press 1952.

Heaps, Leo. *Log of the "Centurion": Based on the Original Papers of Capt. Philip Saumarez on board H.M.S. "Centurion" ... 1740–44*. New York: Macmillan 1975.

Henderson, David A. *Men and Whales at Scammon's Lagoon*. Los Angeles: Dawson's Book Shop 1972.

Henderson's City of Victoria and Suburban Directory. 12th ed. Victoria: Henderson Publishing Co. 1905.

Herr, J.B. *Biographical Dictionary of Well-Known British Columbians*. Vancouver: Kerr & Begg 1890.

Hewer, H.R. *British Seals.* New York: Taplinger 1975.

Hickling, Grace. *Grey Seals and the Farne Islands.* London: Routledge, Kegan Paul 1962.

Hiller, J.K. "A History of Newfoundland, 1874–1902." PH D diss., Cambridge University, 1971.

– and Peter Neary. *Newfoundland in the Nineteenth and Twentieth Centuries: Essays in Interpretation.* Toronto: University of Toronto Press 1980.

Hillinger, Charles. *The California Islands.* Los Angeles: Academy Publishers 1953.

Hochman, Elmo Paul. *The American Whaleman: A Study of Life and Labor in the Whaling Industry.* New York: Longmans, Green 1928.

Holdgate, Martin W., ed. *Antarctic Ecology.* 2 vols. New York and London: Scientific Committee on Antarctic Research 1970.

– *Mountains in the Sea: The Story of the Gough Island Expedition.* London: Macmillan 1958.

Hooke, Nina Warner. *The Seal Summer.* New York: Harcourt, Brace 1964.

Howay, F.W. *British Columbia from the Earliest Times to the Present.* Vol. 2. Vancouver: S.J. Clarke 1914.

– *A List of Trading Vessels in the Maritime Fur Trade, 1785–1825.* Edited by Richard A. Pierce. Kingston, Ontario: Limestone Press 1973.

– , W.N. Sage, and H.F. Angus. *British Columbia and the United States: The North Pacific Slope from Fur Trade to Aviation.* Toronto: The Ryerson Press 1942.

Hunt, William R. *Arctic Passage: The Turbulent History of the Land and People of the Bering Sea, 1697–1975.* New York: Scribner's 1975.

Hunter, Robert. *Warriors of the Rainbow: A Chronicle of the Greenpeace Movement.* New York: Holt, Rinehart, 1979.

Hurrel, H.G. *Atlanta My Seal.* London: William Kimber 1963.

Hutchins, John G.B. *The American Maritime Industries and Public Policy, 1789–1914.* Cambridge, Mass.: Harvard University Press 1941. Reprint. New York: Russell & Russell 1969.

Ingraham, Joseph. *Journal of the Brigantine "Hope": Voyage to the Northwest Coast of North America, 1790–2.* Edited by Mark D. Kaplanoff. Barre, Mass. Imprint Society 1971.

Innis, Harold A. *The Cod Fisheries: The History of an International Economy.* New Haven: Yale University Press 1940.

International Union for Conservation of Nature and Natural Resources. *Seals: Proceedings of a Working Meeting of Seal Specialists ... .* University of Guelph, Ontario, Canada [1972]. Morgues, Switzerland: IUCN 1973.

Jackson, Gordon. *The British Whaling Trade.* London: Adam & Charles Black 1978.

Jennings, Dean. *Valla: The Story of a Sea Lion.* New York: World Publishing Co. 1969 (fiction).

Job, Robert Brown, *John Job's Family: A Story of his Ancestors and Successors ...* *1730 to 1953.* 2nd ed. St. John's: privately printed 1954.

Johnson, Samuel P., ed. *Alaska Commercial Company, 1868–1940 ... .* [San Francisco]: E.E. Wachter [1940].

Johnson, Scott W. *An Annotated Catalog of Published and Unpublished Sources of Data on Populations, Life History, and Ecology of Coastal Marine Mammals of California.* La Jolla, Calif.: United States, NOAA/NFMS, Southwest Fisheries Center 1979.

Jones, Dorothy Knee. *A Century of Servitude: Pribilof Aleuts under U.S. Rule.* Washington, D.C.: University Press of America 1980.

– "A History of United States Administration in the Pribilof Islands, 1967–1946." Fairbanks: Institute of Social, Economic and Government Research, University of Alaska, for the [U.S.] Department of Justice, Indian Claims Commission Dockets 352 and 369, 1976.

Jones, Ken. *Seal Doctor.* 1970, entitled *Orphans of the Sea.* London: Collins, 1978.

Judd, Bernice. *Voyages to Hawaii before 1860 ... .* 1929. Reprint. Honolulu: University Press of Hawaii 1974.

Jukes, J.B. *Excursions in and about Newfoundland, During the Years 1839 and 1840.* London: John Murray 1842.

Jupp, Ursula, ed. *Home Port Victoria.* Victoria: privately printed 1967.

Kean, Capt. Abram, OBE. *Old and Young Ahead: A Millionaire in Seals, Being the Life History of ... .* London: Heath, Cranton 1935.

Keir, David. *The Bowring Story.* London: Bodley Head 1962.

Kenyon, Karl W. *The Sea Otter in the Eastern Pacific Ocean.* 1969. New York: Dover, 1975.

Kerr, Ian S. *Campbell Island: A History.* Wellington, N.Z.: A.H. and A.W. Reed 1976.

King, Judith E. *Seals of the World.* 2nd ed. London: British Museum 1983.

King, Richard. *Narrative of a Journey to the Shores of the Arctic Ocean in 1833, 1834, and 1835.* London: Richard Bentley 1836.

Kirker, James. *Adventurers to China: Americans in the Southern Oceans, 1792–1812.* New York: Oxford University Press 1970.

Kitchener, L.D. *Flag over the North: The Story of the Northern Commercial Company.* Seattle: Superior Publishing Co. 1954.

Kushner, Howard I. *Conflict on the Northwest Coast: American-Russian Rivalry in the Pacific Northwest, 1790–1867.* Westport, Conn.: Greenwood Press 1975.

Lamont, James. *Seasons with the Sea-Horses: or, Sporting Adventures in the Northern Seas.* London: Hurst & Blackett 1861.

Lamson, Cynthia. *"Bloody Decks and a Bumper Crop": The Rhetoric of Sealing Counter-Protest.* St. John's: Memorial University of Newfoundland, Institute of Social and Economic Research 1979.

Langscorff, G.H. von. *Voyages and Travels in Various Parts of the World, during the Years 1803, 1804, 1805, 1806, and 1807.* 2 vols. London: Henry Colburn 1813–14.

Laughin, William S. *Aleuts: Survivors of the Bering Land Bridge.* New York: Holt, Rinehart, Winston 1980.

Laver, James. *Taste and Fashion from the French Revolution to the Present Day.* 2nd ed. London: Harrap 1945.

Le Boeuf, Burney, and Stephanie Kaza, eds. *The Natural History of Año Nuevo.* Pacific Grove, Calif.: Boxwood Press 1981.

Ledyard, John. *John Ledyard's Journal of Captain Cook's Last Voyage.* Edited by James Kenneth Munford. Corvallis, Ore.: Oregon State University Press 1963.

Lillie, Harry R. *The Path Through Penguin City.* London: E. Benn 1955.

Lindsay, David Moore. *A Voyage to the Arctic in the Whaler "Aurora."* Boston: Dana Estes 1911.

Lister-Kaye, John. *Seal Cull: The Grey Seal Controversy.* Harmondsworth, Middlesex: Penguin Books 1979.

Little, George. *Life on the Ocean; or, Twenty Years at Sea ... .* 12th ed. Boston: Waite, Peirce 1946.

Lockley, Ronald M. *Grey Seal, Common Seal: An Account of the Life Histories of British Seals.* New York: October House 1966.

– *Seal Woman.* London: Methuen 1974 (fiction).

– *The Seals and the Curragh ... .* London: J.M. Dent 1954. American title: *The Saga of the Grey Seal.* New York: Devin-Adair 1954.

London, Jack. *The Sea Wolf.* 1903. Reprint. New York: Heritage Press 1961 (fiction).

Lounsbury, Ralph Greenlee. *The British Fishery at Newfoundland, 1643–1763.* New Haven: Yale University Press 1934. Reprinted. Hamden, Conn. Archon Books 1969.

Lubbock, Basil. *The Arctic Whalers.* Glasgow: Brown, Son & Ferguson 1937.

Lust, Peter. *The Last Seal Pup: The Story of Canada's Seal Hunt.* Montreal: Harvest House 1967.

Lysaght, A.M. *Joseph Banks in Newfoundland and Labrador, 1766: His Diary, Manuscripts and Collections.* London: Faber & Faber 1971.

MacAskie, Ian. *The Long Beaches: A Voyage in Search of the North Pacific Fur Seal.* Vancouver: Sono Nis Press 1979.

McClung, Robert M. *Hunted Mammals of the Sea.* New York: Morrow 1978.

McCracken, Harold. *Hunters of the Stormy Sea.* Garden City, N.Y.: Doubleday 1957.

McDonald, Ian. "Coaker the Reformer: A Brief Biographical Introduction." n.d. Memorial University of Newfoundland, St. John's. Typescript (another copy in Public Archives of Canada).

Mack, Gerstle. *Lewis and Hannah Gerstle.* [San Francisco]: privately printed 1953.

Mackay, R.A., ed. *Newfoundland: Economic, Diplomatic and Strategic Studies.*
    Toronto: Oxford Univesity Press 1946.
MacMullen, Jerry. *They Came by Sea: A Pictorial History of San Diego Bay.*
    San Diego: Ward Ritchie Press 1969.
McNab, Robert. *The Old Whaling Days: A History of Southern New Zealand
    from 1830–1840.* 1913. Reprint. Auckland: Golden Press 1975.
Mannion, John J., ed. *The Peopling of Newfoundland: Essays in Historical
    Geography.* St. John's: Memorial University of Newfoundland 1977.
Martin, Calvin. *Keepers of the Game: Indian-Animal Relationships and the Fur
    Trade.* Berkeley: University of California Press 1978.
Martin, Fredericka. *The Hunting of the Silver Fleece: Epic of the Fur Seal.* New
    York: Greenberg 1946.
– *The Sea Bears: The Story of the Fur Seal.* Philadelphia: Chilton 1960.
Martin, Richard Mark. *Mammals of the Oceans.* New York: Putnam's 1977.
Matthews, Keith. "A History of the West of England-Newfoundland
    Fishery." PHD thesis, Oxford University, 1968.
– "Lectures on the History of Newfoundland, 1500–1830." 1973. Maritime
    History Group, Memorial University of Newfoundland, St. John's.
    Photocopy.
– "A Who Was Who of Families Engaged in the Fishery and Settlement of
    Newfoundland 1660–1840." 1971. Memorial University of
    Newfoundland, St. John's. Photocopy.
Matthews, L. Harrison. *Penguins, Whalers and Sealers: A Voyage of Discovery.*
    New York: Universe Books 1978.
– *Sea Elephant: The Life and Death of the Elephant Seal.* London: MacGibbon
    & Kee 1952.
– *The Seals and the Scientists.* London: Peter Owen 1979.
– *South Georgia: The British Empire's Subantarctic Outpost.* Bristol: John Wright
    & Sons 1931.
– *Wandering Albatross: Adventure among the Albatrosses and Petrels in the Southern
    Ocean.* London: MacGibbon & Kee 1951.
Matthews, Ralph. *"There's No Better Place than Here": Social Change in Three
    Newfoundland Communities.* Toronto: Peter Martin Associates 1976.
Maxwell, Gavin. *Raven Seek Thy Brother.* New York: Dutton 1969.
– *Seals of the World.* Boston: Houghton Mifflin 1967.
Méanger, Francis M. *The Kingdom of the Seal,* Chicago: Loyola University
    Press 1962.
Micco, Helen Mary. *King Island and the Sealing Trade, 1802.* Canberra:
    Roebuck Society 1971.
Mill, Hugh R. *The Siege of the South Pole.* New York: Frederick A. Stokes
    1905.
Millais. J.G. *Newfoundland and Its Untrodden Ways.* 1907. Reprint. New York:
    Arno Press 1967.

Mitterling, Philip I. *America in the Antarctic to 1840*. Urbana: University of
Illinois Press 1959.

Morine, Sir Alfred B. *The Railway Contract, 1898 and Afterwards,
1883–1933*. St. John's: n.p. 1933.

Mortimer, Lt. George. *Observations and Remarks made During a Voyage to the
Islands of Teneriffe, Amsterdam, Sandwich Islands* ... . London: privately
printed 1791.

Moseley H.N. *Notes by a Naturalist on the "Challenger,"* ... *1872–1876*.
London: Macmillan 1979.

Mott, Henry Youmans, ed. *Newfoundland Men: A Collection of Biographical
Sketches*. Concord N.H.: T.W. & J.F. Cragg 1894.

Mowat, Farley. *Wake of the Great Sealers*. Boston: Little, Brown 1973.

– *A Whale for the Killing*. Boston: Little, Brown 1972.

– and Joan de Visser. *This Rock Within the Sea: A Heritage Lost*. Boston:
Little, Brown 1968.

Mulford Prentice. *Prentice Mulford's Story: Life by Land and Sea*. New York:
F.J. Needham 1889.

Munroe, Kirk. *The Fur-Seal's Tooth: A Story of Alaskan Adventure*. New
York: Harper & Bros. 1898 (fiction).

Murdoch, W.G. Burns *From Edinburgh to the Antarctic; An Artist's Notes and
Sketches during the Dundee Antarctic Expendition of 1892–93*. London:
Longmans, Green 1894.

– *Modern Whaling and Bear Hunting*. London: Seeley, Service & Co. 1917.

Murphy, James. "The Old Sealing Days." Articles from St. John's *Evening
Herald* 1916. Memorial University of Newfoundland, St. John's.
Typescript.

Murphy, Robert Cushman. *Bird Islands of Peru: The Record of a Sojourn on
The West Coast*. New York: G.P. Putnam's Sons 1925.

– *Logbook for Grace: Whaling Brig "Daisy," 1912–1913*. New York:
Macmillan 1947. Illustrated ed.: *A Dead Whale or a Stove Boat: Cruise of the
"Daisy" in the Atlantic Ocean, 1912–1913*. Boston: Houghton Mifflin
1967.

Nature Conservancy, The [Great Britain]. *Grey Seals and Fisheries*. Report of
the Consultative Committee on Grey Seals and Fisheries. London: HMSO
1963.

Neary, Peter, ed. *The Political Economy of Newfoundland, 1929–1972*.
Toronto: Copp Clark 1973.

Nelson, Richard K. *Hunters of the Northern Ice*. Chicago: University of
Chicago Press 1969.

Newell, Gordon, ed. *The H.W. McCurdy Marine History of the Pacific
Northwest ... from 1895* ... . Seattle: Superior Publishing Co. 1966.

Niedieck, Paul. *Cruises in the Bering Sea: Being Records of Further Sport and
Travel*. London: Rowland Ward 1909.

Noel, S.J.R. *Politics in Newfoundland*. Toronto: University of Toronto Press 1971.

Nunn, John. *Narrative of the Wreck of the "Favorite" on the Island of Desolation ... .* London: William Edward Painter 1850.

O'Flaherty, Patrick. *The Rock Observed: Studies in the Literature of Newfoundland*. Toronto: University of Toronto Press 1979.

Ogden, Adele. *The California Sea Otter Trade, 1784–1848*. 1941. Reprint. Berkeley: University of California Press 1975.

O'May, Harry. *Hobart River Craft and Sealers of Bass Strait*. Hobart: L.G. Shea n.d.

O'Neill, Paul *The Story of St. John's, Newfoundland*. 2 vols. Erin. Ont.: Press Porcépic 1976.

Ormsby, Margaret A. *British Columbia: A History*. [Toronto:] Macmillans in Canada 1958.

Owen, Russel. *The Antarctic Ocean*. New York: Whittlesey House 1941.

Pearson, R.H. *A Seal Flies By*. New York: Avon 1959.

Pernety, [Antoine]. *The History of a Voyage to the Malouine (or Falkland) Islands Made in 1763 and 1764 ... .* 2nd ed. London: William Goldsmith 1783.

Perry, Richard. *The Polar Worlds*. New York: Taplinger 1973.

– *The World of the Walrus*. New York: Taplinger 1968.

Peterson, Richard S., and George A. Bartholomew. *The Natural History and Behavior of the California Sea Lion*. Special Publication no. 1. New York: American Society of Mammalogists 1967.

Peterson, Roger Tory. *Penguins*. Boston: Houghton Mifflin 1979.

Pethick, Derek. *Summer of Promise: Victoria 1864–1914*. Victoria: Sono Nis Press 1980.

[Phelps, William D.] *Fore and Aft; or Leaves from the Life of an Old Sailor, by "Webfoot"*. Boston: Nichols & Hall 1871.

Philbrick, Thomas. *James Fenimore Cooper and the Development of American Sea Fiction*. Cambridge: Cambridge University Press 1961.

Pierce, Richard A. *Alaskan Shipping, 1867–1878*. Kingston, Ont. Limestone Press 1972.

Plante, Florent. *La chasse aux phoques*. Ottawa: Editions Lémeac 1978.

Price, A. Grenfell. *The Winning of Australian Antarctica: Mawson's* BANZARE *Voyages 1929–31* ... Vol. 1. Sydney: Angus & Robertson, for Mawson Institute for Antarctic Research, University of Adelaide 1962.

Prowse, D.W. *A History of Newfoundland from the English, Colonial, and Foreign Records*. 1895. Reprint. Belleville, Ont.: Mika Studio, 1972.

Pryde, Duncan. *Nunaga: Ten Years of Eskimo Life*. New York: Walker & Co. 1971.

Quam, Louis O., ed. *Research in the Antarctic*. Washington: American Association for the Advancement of Science 1971.

Rankin, Niall. *Antarctic Isle: Wild Life in South Georgia*. London: Collins 1951.

Ransom, M.A. *Sea of the "Bear": Journal of a Voyage to Alaska and the Arctic, 1921.* Annapolis: U.S. Naval Institute 1964.

Rawlyk, G.A., ed. *Historical Essays on the Atlantic Provinces.* Toronto: McClelland & Stewart 1967.

Reeves, Randall R. *Exploitation of Harp and Hooded Seals in the Western North Atlantic.* Washington: U.S. Marine Mammal Commission 1976.

Reynal, F.E. *Wrecked on a Reef; or, Twenty Months Among the Auckland Isles.* London: T. Nelson 1874.

Rice, Dave W., and Allen A. Wolman. *The Life History and Ecology of the Grey Whale ("Eschrichtius robustus").* Stillwater, Okla.: American Society of Mammalogists 1971.

Robinson, William Albert. *Voyage to Galapagos.* New York: Harcourt, Brace 1936.

Rockwell, Mabel M. *California's Sea Frontier.* Santa Barbara: McNally & Loftin 1962.

Rogers, George. "An Economic Analysis of the Pribilof Islands, 1870–1946." Typescript. Fairbanks: Institute of Social, Economic and Government Research, University of Alaska, for the [U.S.] Dept. of Justice, Indian Claims Commission Dockets 352 and 369, 1976.

Ronald, K., L.M. Hanley, P.J. Healey, and L.G. Selley. *An Annotated Bibliography of the Pinnipedia.* Charlottenlund, Denmark: International Council of the Exploration of the Sea 1976.

– and A.W. Mansfield, eds. *Biology of the Seal: Proceedings of a Symposium held in Guelph 14–17 August 1972.* Charlottenlund-Slot, Denmark: Conseil international pour l'exploration de la mer 1975.

Roquefeuil, Camille de. *A Voyage Round the World between the Years 1816–1819.* London: R. Phillips 1823.

Rostworowski de Diez Canseco, Maria. *Recursos naturales renovables y pesca.* Lima: Instituto de Estidios Perusanos 1981.

Rowe, Frederick W. *A History of Newfoundland and Labrador.* Toronto: McGraw-Hill Ryerson 1980.

Rowsell, Harry C. "Harp and Hood Seal Fisheries on the Front, March 17–30th 1975: A Report to the Committee on Seals and Sealing." [1975]. Memorial University of Newfoundland, St. John's. Typescript.

Ryan, Shannon. "The Newfoundland Cod Fishery in the Nineteenth Century." MA thesis, Memorial University of Newfoundland, 1971.

– and Larry Small. *Haulin' Rope & Gaff: Songs and Poetry in the History of the Newfoundland Seal Fishery.* St. John's: Breakwater Books 1978.

Rydell, Carl. *Adventures of ...: The Autobiography of a Seafaring Man.* Edited by Elmer Green. London: Edward Arnold 1924.

Rydell, Raymond A. *Cape Horn to the Pacific: the Rise and Decline of an Ocean Highway.* Berkeley: University of California Press 1952.

Sanger, Chesley W. "Technological and Spatial Adaptation in the Newfoundland Seal Fishery during the 19th Century." PH D. diss. Memorial University of Newfoundland, 1973.

Scammon, Charles M. *The Marine Mammals of the Northwest Coast of North America* .... 1874. Reprint. New York: Dover, 1968.

Scheffer, Victor B. *Adventures of a Zoologist.* New York: Scribner's, 1980.

– *Seals, Sea Lions and Walruses: A Review of the Pinnipedia.* Stanford: Stanford University Press 1958.

– *A Voice for Wildlife.* New York: Scribner's 1974.

– *The Year of the Seal.* New York: Scribner's 1970.

– and Ethel I. Todd. "History of Scientific Study of the Alaskan Fur Seal, 1786–1964." Edited by Clifford H. Fiscus. 1967. Marine Mammal Biology Laboratory, Seattle. Typescript.

Scholefield, E., and F.W. Howay. *British Columbia from the Earliest Times to the Present.* 4 vols. Vancouver: S.J. Clarke 1914.

Scholes, Arthur. *Fourteen Men: The Story of the Antarctic Expedition to Heard Island.* New York: E.P. Dutton 1952.

Scott, James W., ed. *Pacific Northwest Themes: Historical Essays in Honor of Keith A. Murray.* Bellingham Wash.: Center for Pacific Northwest Studies, Western Washington University 1978.

Scott, John Roper. "The Function of Folklore in the Interrelationship of the Newfoundland Seal Fishery and the Home Communities of the Sealers." MA thesis, Memorial University of Newfoundland 1974.

Schroeder, Joseph J., Jr., ed. *The Wonderful World of Ladies' Fashion, 1850–1920.* Chicago: Follett 1971.

Schwatka, Frederick. *Nimrod of the North, or Hunting and Fishing Adventures in the Arctic Regions.* New York: Cassell 1885.

Shelvocke, Capt. George. *A Voyage Round the World.* 1726. Reprint. London: Cassell 1928.

Sherwood, Morgan B. *Alaska and Its History.* Seattle: University of Washington Press 1967.

Simpson, George Gaylord. *Penguins Past and Present, Here and There.* New Haven: Yale University Press 1976.

Sinclair, Andrew. *Jack: A Biography of Jack London.* New York: Harper & Row 1977.

Sloss, Frank H. *Only on Monday: Papers Delivered before the Chit-Chat Club.* San Francisco: privately printed 1978.

Smallwood, J.R. *The Book of Newfoundland.* 2 vols. St. John's: Newfoundland Book Publishers 1937.

– *Coaker of Newfoundland: The Man who Led the Deep-Sea Fisherman to Political Power.* London: Labour Publishing Co. 1927.

Smeeton, Miles. *The Misty Islands.* New York: David McKay 1969.

Smith, C. Fox. *Sailor Town Days.* London: Methuen 1923.

Smith, Nicholas. *Fifty-two Years at the Labrador Fishery.* London: Arthur H. Stockwell 1936.

Smith, Thomas W. *A Narrative of the Life, Travels, and Sufferings of* ... . Boston: W.C. Hill 1844.

Snow, H.J. *In Forbidden Seas: Recollections of Sea-Otter Hunting in the Kurils.*
London: Edward Arnold 1910.
– *Notes on the Kuril Islands.* London: John Murray 1897.
Snow, N. Parker. *A Two Years' Cruise off Tierra del Fuego, the Falkland
Islands, Patagonia, and in the River Plate ... .* 2 vols. London: Longmans,
Brown, Green, Longmans 1857.
Southwell, Thomas. *The Seals and Whales of the British Seas.* London: Jarrold
& Sons 1881.
Spears, John R. *Captain Nathaniel Brown.Palmer: An Old-Time Sailor of the
Sea.* New York: Macmillan 1922.
Stackpole, Edouard A. *The Sea Hunters; The New England Whalemen during
Two Centuries, 1635–1835.* Philadelphia: Lippincott 1953.
– *The Voyage of the "Huron" and the "Huntress": The American Sealers and the
Discovery of the Continent of Antarctica.* Mystic, Conn.: Marine Historical
Association 1953.
– *Whales & Destiny: The Rivalry between America, France, and Britain for
Control of the Southern Whales Fishery, 1785–1824.* Amherst: University of
Massachusetts Press 1972.
Starbuck, Alexander. *History of the American Whale Fishery.* 2 vols. Reprint.
New York: Argosy-Antiquarian 1964.
Staunton, Sir George. *An Authentic Account of an Embassy from the King of
Great Britain to the Emperor of China ... .* 2 vols. London: privately printed
1797.
Staveley, Michael. "Migration and Mobility in Newfoundland and Labrador:
A Study in Population Geography." PH D diss., University of Alberta 1973.
Stefansson, Vilhjamur. *The Friendly Arctic: The Story of Five Years in Polar
Regions.* 2nd ed. New York: 1944.
Stejneger, Leonard. *The Russian Fur-Seal Islands.* Bulletin of the U.S. Fish
Commission (1896): 1–148. Washington: USGPO 1896.
Stonehouse, Bernard. *Animals of the Arctic: The Ecology of the Far North.* New
York Holt, Rinehart, & Winston 1971.
Stuart, Frank S. *A Seal's World: An Account of the First Three Years in the Life
of a Harp Seal.* New York: McGraw-Hill 1954.
Sturgis, William. *Journal.* Edited by S.W. Jackman. Victoria: Sono Nis Press
1978.
Sutton, George. *Glacier Island: The Official Account of the British South Georgia
Expedition, 1954–1955.* London: Chatto & Windus 1957.
Sweden National Swedish Environment Protection Board. *Proceedings from
the Symposium on the Seal in the Baltic, Lidingö, Sweden, June 4–6, 1974.*
Stockholm: NSEPB 1975.
Tansill, Charles Callan. *Canadian-American Relations, 1875–1911.* New
Haven: Yale University Press 1943.
Taylor, F.H.C., M. Fujinaga, and Ford Wilke. *Distribution and Food Habits of
the Fur Seals of the North Pacific Ocean: Report of Cooperative Investigations*

by the Governments of Canada, Japan and the United States of America, February-July 1952. Washington: U.S. Dept. of Interior, Fish & Wildlife Service 1955.

Taylor, Nathaniel W. *Life of a Whaler, or Antarctic Adventures in the Isle of Desolation*. Edited by Howard Palmer. 1929. Reprint. New London, Conn.: New London Country Historical Society 1977.

Teichmann, Emil. *A Journey to Alaska in the Year 1868: Being a Diary of the Late ... .* Edited by Oskar Teichmann. New York: Argosy-Antiquarian 1963.

Terhune, David. *The Harp Seal*. Toronto: Burns & MacEachern 1973.

Thornton, Ian. *Darwin's Islands: A Natural History of the Galápagos*. Garden City, N.Y.: The Natural History Press 1971.

Tocque, Philip. *Newfoundland: As It Was, and As It Is in 1877*. London: Sampson, Low, Marston, Seerle, Rivington 1878.

– *Wandering Thoughts, or Solitary Hours*. London: Thomas Richardson 1846.

Tomasevich, Jozo. *International Agreements on Conservation of Marine Resources, with Special Reference to the North Pacific*. Stanford: Food Research Institute 1943.

Tonnessen, J.N., and A.O. Johnsen. *The History of Modern Whaling*. Berkeley: University of California Press 1982.

Toussaint, Auguste. *History of the Indian Ocean*. Chicago: University of Chicago Press 1966.

Tutein, Peter. *The Sealers*. New York: G.P. Putnam 1938 (fiction).

Vessels Owned on the Pacific Coast. San Francisco: Commercial Publishing Co. 1888–1920.

Villiers, A.J. *Vanished Fleets: Ships and Men of Old Van Dieman's Land*. New York: Henry Holt 1931.

Wace, Nigel, and Bessie Lovett. *Yankee Maritime Activities and the Early History of Australia*. Canberra: Australian National University, Research School of Pacific Studies 1973.

Wade, Mason, ed. *Regionalism in the Canadian Community, 1867–1967*. Toronto: University of Toronto Press 1969.

Ward, R. Gerard, ed. *American Activities in the Central Pacific, 1790–1870*. 8 vols. Ridgewood, N.J.: The Gregg Press 1966–7.

Ward, W. Peter, and Robert A. J. McDonald, eds. *British Columbia: Historical Readings*. Vancouver: Douglas & McIntyre 1981.

Wardle, Arthur C. *Benjamin Bowring and his Descendants: A Record of Mercantile Achievement*. London: Hodder & Stoughton 1938.

Watson, Paul, with Warren Rogers. *Sea Shepherd: My Fight for Whale and Seals*. New York: Norton, 1982.

Watters, Reginald Eyre. *British Columbia: A Centennial Anthology*. Vancouver: McClelland & Stewart, 1958.

Weddell, James. *A Voyage Towards the South Pole, Performed in the Years 1822–24*. 1837. Reprint. Annapolis: U.S. Naval Institute 1970.

West, John. *The History of Tasmania*. Launceston, Tasmania: Henry Dowling, 1852.

Whiteley, George. *Northern Seas, Hardy Sailors*. New York: Norton, 1982.

Who's Who in and from Newfoundland. St. John's: R. Hobbs 1927 and 1937 editions.

Whympel, Frederick. *Travel and Adventure in the Territory of Alaska ...* . London: John Murray 1868.

Williams, Frances Leigh. *Matthew Fountaine Maury: Scientist of the Sea*. New Brunswick, N.J.: Rutgers University Press 1963.

Williamson, Thames. *North After Seals*. Boston: Houghton Mifflin 1934.

Woodward, Ralph Lee, Jr. *Robinson Crusoe's Island: A History of the Juan Fernandez Islands*. Chapel Hill, N.C.: University of North Carolina Press 1969.

Woolfenden, John. *The California Sea Otter: Saved or Doomed?* Pacific Grove, Calif.: Boxwood Press 1979.

Wright, E.D., ed. *Lewis & Dryden's Marine History of the Pacific Northwest ...* . 1897. Reprint. Seattle: Superior Publishing Co. 1967.

Zur Strassen, W.H. *The Fur Seal of Southern Africa*. Cape Town: Howard Timmins 1971.

SHORT MONOGRAPHS AND ARTICLES
IN BOOKS AND PERIODICALS

Aguayo L., Anello. "The Present Status of Antarctic Fur Seal 'Arctocephalus gazella' at South Shetland Islands." *Polar Record* 19 (1978): 167–76.

– "The Present Status of the Juan Fernandez Fur Seal." *Der Kongelige Norske Videnskabens Selskab* (Trondheim) 1 (1971): 1–4.

– "El lobo Fino de Juan Fernandez." *Revista de Biologica Marina* (Valparáiso) 14 (1971): 135–49.

– , René Maturana, and Daniel Torres. "El lobo fino anarctico, 'Arctocephalus gazelle' (Peters) en el sector Antártico Chileno (Pinnipedia-Oteriidae)." *Ser. Cient. Inst. Antártico Chileno* 5 (1977): 5–16.

Ainley, D.G., H.R. Huber, R.P. Henderson, and T.J. Lewis. *Studies of Marine Mammals at the Farallon Islands, California, 1970–1975*. Report MMC-74–04. Washington, D.C.: U.S. Marine Mammal Commission 1977.

Alexander, David. *The Decay of Trade: An Economic History of the Newfoundland Saltfish Trade, 1935–1965*. Social and Economic Studies, no. 19. St. John's: Memorial University of Newfoundland, Institute of Social and Economic Research 1977.

– "Development and Dependence in Newfoundland, 1880–1970." *Acadiensis* 4 (1974): 3–31.

– "Literacy and Economic Development in Nineteenth Century Newfoundland." *Acadiensis* 10 (1980): 3–34.

– "Newfoundland's Traditional Economy and Development to 1934." *Acadiensis* 5 (1976): 56–78.

– "The Political Economy of Fishing in Newfoundland." *Journal of Canadian Studies* 11 (1976): 32–40.

Allen, Glover M. "The Walrus in New England." *Journal of Mammalogy* 11 (1930): 139–45.

Allen, J.A. "The Hair Seals (Family Phocidae) of the North Pacific Ocean and Bering Sea." *Bulletin of the American Museum of Natural History* 16 (1902) 459–99.

Ames, Jack P., and G. Victor Morejohn. "Evidence of White Shark, 'Carcharodon carcharias,' Attacks on Sea Otters, 'Enhydra lutris'." *California Fish & Game* 66 (1980): 196–209.

Anderson, Harry. *Sea Lion–Creature of Controversy*. Fort Bragg, Calif.: Presented to the Senate Fact Finding Committee on Natural Resources 1960.

Andrews, C.W. "The Hazardous Industry of North Atlantic Sealing." *Animal Kingdom* 54 (1951): 66–94.

– *The Origin, Growth and Decline of the Newfoundland Sealfishery*. Harbour Grace. Nfld.: Conception Bay Museum, n.d.

[– and Robert Parker], *Brief to the Special Advisory Committee to the Minister of Fisheries and Forestry on Seals Presented to the Government of Newfoundland and Labrador, May 25, 1971*. St. John's: n.p. [1971].

Andrews, Clarence L. "Alaska Whaling." *Washington Historical Quarterly* 9 (1918): 3–10.

Anti-Monopoly Association of the Pacific Coast. *A History of the Wrongs of Alaska: An Appeal to the People and Press of America*. San Francisco: AMAPC 1875.

Antler, Steven. "The Capitalist Underdevelopment of Nineteenth-Century Newfoundland." In *Underdevelopment and Social Movements in Atlantic Canada*, edited by Robert J. Brym and R. James Sacouman, 179–202. Toronto: New Hogtown Press 1979.

Antonelis, George A., Jr., Stephen Leatherwood, and Daniel K. Odell. "Population Growth and Census of the Northern Elephant Seal, 'Mirounga angustirostris' on the California Channel Islands, 1958–78," *Fishery Bulletin* (U.S. Dept. of Commerce) 79 (181): 652–7.

Archibald, Daniel George. *Some Account of the Seal Fishery of Newfoundland, and the Mode of Preparing Seal Oil … *. Edinburgh: Murray & Gibb 1852.

Aretas, Raymond. "L'Eléphant de mer, 'Mirounga leonina' (L.)." *Mammalia* 15 (1951): 105–17.

Austin, Oliver L., Jr, and Ford Wilke. *Japanese Fur Sealing*. Special Scientific Report, Wildlife no. 6 Washington, D.C.: U.S. Dept. of Interior, Fish and Wildlife Service 1950.

Backer, Ole Friele. "Seal Hunting off Jan Mayen." *National Geographic* 93 (1948): 57–72.

Bailey, Alfred M. "The Monk Seal of the Southern Pacific." *Natural History* 18 (1918): 396–9.

– "The Hawaiian Monk Seal." *Museum Pictorial* (Denver Museum of Natural History), no. 7 (1952): 1–30.

Bailey, Thomas A. "The North Pacific Sealing Convention of 1911." *Pacific Historical Review* 4 (1935), 1–14.

Baker, Melvin. "The Politics of Municipal Reform in St. John's, Newfoundland, 1888–1892." *Urban History Review* 2 (1976): 12–29.

Baker, Ralph C. *Fur Seals of the Pribilof Islands.* Washington, D.C.: USGPO (U.S. Dept. of Interior, Fish and Wildlife Service) 1957.

– , Ford Wilke, and C. Howard Baltzo. *The Northern Fur Seal.* Bureau of Commercial Fisheries, circular 169; 2nd ed., circular 336. Washington, D.C.: U.S. Dept. of Interior, Fish and Wildlife Service 1963 and 1970.

Balch, Edwin Swift. "Antarctica Addenda." *Journal of the Franklin Institute* 157 (1904): 81–8.

– "Stonington Antarctic Explorers." *Bulletin of the American Geographic Society* 41 (1909): 473–92.

Balkwill, F.H. "On the Geographical Distribution of Seals." *The Zoologist,* 3rd ser., 12 (1888): 401–11.

Baltzo, C. Howard. *Living and Working Conditions on the Pribilof Islands, Alaska.* Bureau of Commercial Fisheries, Fishery leaflet no. 548. Washington: U.S. Dept. of Interior, Fish and Wildlife Service 1963.

Barr, Lou. "Steller Sea Lion." *Oceans* 8 (1975): 18–21.

Barragy, Terrence J. "The Trading Age, 1792–1844." *Oregon Historical Quarterly* 76 (1975): 196–224.

Bartholomew, George A., Jr. "A Model for the Evolution of Pinniped Polygyny." *Evolution* 24 (1970): 546–59.

– "Behavioral Factors Affecting Social Structure in the Alaska Fur Seal." *Transactions of the 18th North American Wildlife Conference* (1953): 481–502.

– "Reproductive and Social Behavior of the Northern Elephant Seal." *University of California Publications in Zoology* 47 (1952): 369–472.

– and Richard A. Bootlootian. "Numbers and Population Structure of the Pinnipeds on the California Channel Islands." *Journal of Mammalogy* 41 (1960): 366–75.

– and Carl A. Hubbs. "Population Growth and Season Movements of the Northern Elephant Seal, 'Mirounga angustirostris.'" *Mammalia* 24 (1960): 313–32.

– , – " Winter Population of Pinnipeds about Guadalupe, San Benito, and Cedros Islands, Baja California." *Journal of Mammalogy* 33 (1952): 160–71.

Bartlett, Capt. Robert A. "The Sealing Saga of Newfoundland." *National Geographic* 56 (1929): 91–130.

Beckham, Curt. "The Historic Sea Otter Transplant." *Pacific Wilderness Journal* 2 (1974): 16–18.

Beddington, J.R., and H.A. Williams. *The Status and Management of the Harp Seal in the North-West Atlantic: A Review and Evaluation.* Washington: U.S. Marine Mammal Commission 1980.

Belaze, George H., and G. Clusey Whitlow. *Bibliography of the Hawaiian Monk Seal 'Monachus schauinslaudi'* … . Technical report 35. Honolulu: Hawaii Institute of Marine Biology 1978.

Berdeque, Julio. *La foca fina, el elefante marino y la ballena gris en baja california* … . Mexico: Instituto Mexicano de Recursos Naturales Renouvables 1956.

Bertram, G.C.L. "The Biology of the Weddell and Crabeater Seals." *British Graham Land Expedition, 1934–37: Scientific Reports* (British Museum, Natural History) 1 (1940).

– "Pribilof Fur Seals." *Arctic* 3 (1950): 75–81, 83–5.

Best, Peter B. "Seals and Sealing in South and South West Africa." *South Africa Shipping News & Fishing Industry Review* 28 (1973).

– and P.D. Shaughnessy. "An Independent Account of Capt. Benjamin Morrell's Sealing Voyage to the South-West Coast of Africa in the 'Antarctic' 1828–29." *Fishery Bulletin of South Africa* 12 (1979): 1–19.

Black, W.A. "The Labrador Floater Codfishery." *Annals of the Association of American Geographers* 50 (1960): 267–93.

Bonnell, Michael L., and Robert K. Selander. "Elephant Seals: Genetic Variation and Near Extinction." *Science* 184 (1974): 908–9.

Bonner, W.N. "Exploitation and Conservation of Seals in South Georgia." *Oryx* 4 (1958): 373–80.

– "The Fur Seal of South Georgia." *British Antarctic Survey Scientific Reports*, no. 56, 1968.

– *Humane Killing of Seals.* London: Seal Research Unit, Natural Environment Research Council 1970.

– "International Legislation and the Protection of Seals." In National Swedish Environment Protection Board, *Proceedings from the Symposium on the Seal in the Baltic,* 12–29. Stockholm 1975.

– "Notes on the Southern Fur Seal in South Georgia." *Proceedings of the Zoological Society of London* 130 (1958): 241–52.

– *The Stocks of Grey Seals ('Halichoerus grypus') and Common Seals ('Phoca uitulina') in Great Britain: A Review of Present Knowledge.* London: Natural Environment Research Council 1976.

Bonnot, Paul. *Report on the Seals and Sea Lions of California, 1928.* Fish Bulletin no. 14. Sacramento: State of California, Division of Fish and Game 1929.

– "The Sea Lions, Seals, and Sea Otter of the California Coast." *California Fish & Game* 37 (1951): 371–89.

– G.H. Clark, and S. Ross Hatton. "California Sea Lion Census for 1938."
    California Fish & Game 24 (1938): 415–19.
Botta, J.P., A.P. Downey, J.T. Laudes, and B.P. Noonan. *Utilization of
    Inshore Newfoundland-Caught Harp Seal ('Pagophilus groenlandicus'): Sensor
    Quality of Frozen Stored, Salted, and Smoked Seal Meat*. Technical report no.
    916. St. John's: Seafood Technology Section, Inspection and Technology
    Branch, Dept. of Fisheries and Ocean, Fisheries and Marine Service 1980.
Boulva, J., and I.A. McLaren. "Biology of the Harbor Seal, 'Phoca vitulina,'
    in Eastern Canada." *Bulletin of the Fisheries Research Board of Canada*, no.
    200, 1979.
Bower, Ward T., and E.C. Johnston. "Seals and Walruses." In *Marine
    Products of Commerce*, edited by Donald K. Tressler, 647–67. New York
    1923.
Bradley, Harold Whitman. "Hawaii and the American Penetration of the
    Northeastern Pacific, 1800–1845." *Pacific Historical Review* 12 (1943):
    227–86.
Brandenberg, Fred G. "Notes on the Patagonian Sealion." *Journal of
    Mammalogy* 19 (1938): 44–7.
Brooks, James W. "The Management and Status of Marine Mammals in
    Alaska." *Transactions of the 28th North American Wildlife Conference* (1963):
    314–26.
– "The Pacific Walrus and Its Importance to the Eskimo Economy."
    Transactions of the 18th North American Wildlife Conference (1953): 503–10.
Brown, K.G. "The Leopard Seal at Heard Island.." *1941–54 Australian
    Antarctic Research Expedition (ANARE)*. Interim Reports no. 16. 1957.
Brown, R.N.R. "The Early Sealers of West Australia." *Nature* 143 (1939):
    731.
Brownell, Robert L., Jr., Christine Schonewald, and Randall R. Reeves.
    Preliminary Report on World Catches of Marine Mammals, 1966–1975.
    Washington, D.C.: U.S. Marine Mammal Commission 1979.
Brox, Ottar. *Newfoundland Fishermen in the Age of Industry: A Sociology of
    Economic Dualism*. Institute of Social and Economic Research, Social and
    Economic Studies, no. 9. St. John's, Nfld.: Memorial University of
    Newfoundland 1972.
Bruce, William S. "Cruise of the 'Balaena' and 'Active', Antarctic Seas,
    1892–93." *Geographical Journal* 7 (1896): 502–21, 625–43.
– "Measurements and Weights of Antarctic Seals taken by the Scottish
    National Antarctic Expedition." In *Report of the Scientific Results of the
    Voyage of s.s. 'Scotia'*, pt. 2, 159–74. Edinburgh 1915.
Bruemmer, Fred. "The North American Walrus." *Canadian Geographical
    Journal* 86 (1973): 90–95.
– "Seals at la Tabatière." *Canadian Geographical Journal* 62 (1966): 130–3.
– "A Year in the Life of the Harp Seal." *Natural History* 84 (1975): 42–9.

Brym, Robert J., and Barbara Neis. "Regional Factors in the Formation of the Fisherman's Protective Union of Newfoundland." In *Underdevelopment and Social Movements in Atlantic Canada*, edited by Robert J. Brym and R. James Sacouman, 203–18. Toronto 1979.

Budd, G.M. "Breeding of the Fur Seal at McDonald Islands, and Further Population Growth at Heard Island." *Mammalia* 36 (1972): 423–7.

– "Population Increase in the Kerguelen Fur Seal, 'Arctocephalus tropicalis gazella,' at Heard Island." *Mammalia* 34 (1970): 410–14.

Burns, John J. "Remarks on the Distribution and Natural History of Pagophilic Pinnipeds in the Bering and Chukchi Seas." *Journal of Mammalogy* 51 (1970): 445–54.

– *The Walrus in Alaska: Its Ecology and Management.* Juneau: Alaska Department of Fish & Game 1965.

Busch, Briton Cooper. "Cape Verdeans in the American Whaling and Sealing Industry, 1850–1900." *The American Neptune*, in press.

– "Elephants and Whales: New London and Desolation, 1840–1900." *American Neptune* 40 (1980): 117–26.

– "The Newfoundland Sealers' Strike of 1902." *Labour/Le Travailleur*, no. 14 (Fall 1984).

Camp, Charles Lewis. "The Chronicles of George C. Yount: California Pioneer of 1826." *California Historical Society Quarterly* 2 (1923): 3–66.

Campbell, Charles S. "The Anglo-American Crisis in the Bering Sea, 1890–1891." In *Alaska and Its History*, edited by Morgan B. Sherwood, 315–40. Seattle 1967.

– "The Bering Sea Settlements of 1912." *Pacific Historical Review* 32 (1963): 347–67.

Campbell, George L. "Nathaniel Brown Palmer and the Discovery of Antarctica." *Historical Footnotes: Bulletin of the Stonington Historical Society* 8 (1971): 1–7.

Careless, J.M.S. "The Business Community in the Early Development of Victoria, British Columbia." In *Historical Essays on British Columbia*, edited by J. Friesen and H.K. Ralston, 177–200. Toronto 1976.

Carrick, Robert, and Susan E. Ingham, "Ecological Studies of the Southern Elephant Seal, 'Mirounga leonina' (L.) at Macquarie Island and Heard Island." *Mammalia* 24 (1900): 352–42.

Carroll, Gerald F. "Born Again Seal." *Natural History* 91 (1982): 40–7.

Caves, R.E. and R.H. Holton, "An Outline of the Economic History of British Columbia, 1881–1951." In *Historical Essays on British Columbia*, edited by J. Friesen and H.K. Ralston, 152–66. Toronto 1976.

Chase, A.W. "The Sea-Lion at Home." *Overland Monthly* 3 (1869): 350–4.

Chicanot, E.L. "New Eyes for the Sealing Fleet." *Scientific American* 135 (1928): 409–11.

Clark, A. Howard. "The Antarctic Fur-Seal and Sea-Elephant Industries." In *The Fisheries and Fishery Industries of the United States*, edited by George Brown Goode, 5, pt. 2: 400–67. Washington 1882.

Clark, J.G.D. "Seal Hunting in the Stone Age of North Western Europe: A Study in Economic Prehistory." *Proceedings of the Prehistoric Society* (1946): 12–48.

Cleland, Robert G. "Asiatic Trade and American Occupation of the Pacific Coast." *American Historical Association Annual Report* 1 (1914): 281–9.

– "The Early Sentiment for the Annexation of California: An Account of the Growth of the American Interest in California, 1835–1846." *Southwestern Historical Quarterly* 1 (1914): 1–40; 2 (1914): 121–61; 3 (1915): 321–60.

Colmar, J.S. "The Newfoundland Seal Fishery and the Second World War." *Journal of Animal Ecology* 18 (1949): 40–6.

– "The Present State of the Newfoundland Seal Fishery." *Journal of Animal Ecology* 6 (1937): 145–59.

Condy, P.R. "Annual Cycle of the Southern Elephant Seal 'Mirounga leonina' (Linn.) at Marion Island." *South African Journal of Zoology* 14 (1979): 95–102.

– "Distribution, Abundance and Annual Cycle of Fur Seals (Arctocephalus spp.) on the Prince Edward Islands." *South African Journal of Wildlife Research* 8 (1978): 159–68.

Costa, Daniel. "The Sea Otter: Its Interaction with Man." *Oceanus* 21 (1978): 24–30.

Coughlin, Magdalen. "Boston Smugglers on the Coast (1797–1821): An Insight into the American Acquisition of California." *California Historical Society Quarterly* 46 (1967): 99–120.

– "Commercial Foundations of Political Interest in the Opening Pacific, 1789–1829." *California Historical Society Quarterly* 50 (1971): 15–33.

– "The Entrance of the Massachusetts Merchant into the Pacific." *Southern California Quarterly* 48 (1966): 327–52.

Cowdin, Elliot C. "The Northwest Fur Trade." *Hunt's Merchants' Magazine & Commercial Review* 14 (1846): 532–9.

Crowther, W.E.L.H., ed. "Captain J.N. Robinson's Narrative of a Sealing Voyage to Heard Island, 1858–60." *Polar Record* 15 (1970): 301–16.

– "A Surgeon as Whaleship Owner." *The Medical Journal of Australia* 1 (1943): 549–54.

D'Armand, R.N., and John Lyman. "The Sealing Fleet." *Marine Digest* 35–6 (1957–8): 40 parts.

Davis, Betty S. "Sea Otter Irony: Overabundance Yet Endangered." *Oceans* 12 (1979): 60–2.

– "The Southern Sea Otter Revisited." *Pacific Discovery* 30 (1977): 1–13.

Davis, Raymond, "The Fur Seal Killers." *Defenders* (Defenders of Wildlife) 50 (1975): 374–80.

Delong, Robert L. "Northern Elephant Seal." In *Mammals of Eastern North Pacific and Arctic Waters*, edited by Delphine Haley, 206–11. Seattle 1978.

Deutsch, Herman J. "Economic Imperialism in the Early Pacific Northwest," *Pacific Historical Review* 9 (1940): 377–88.

Dorsett, Edward Lee. "Around the World for Seals: The Voyage of the

Two-Masted Schooner 'Sarah W. Hunt' from New Bedford to Campbell Island, 1883–1884." *American Neptune* 11 (1951): 115–33.

Doughty, Robin W. "The Farallones and the Boston Men." *California Historical Society Quarterly* 53 (1974): 309–16.

Du Four, Clarence John, "The Russian Withdrawal from California." In *Alaska and Its History*, edited by Morgan B. Sherwood, 133–46. Seattle 1967.

Duggins, David O. "Kelp Beds and Sea Otters: An Experimental Approach." *Ecology* 61 (1980): 447–53.

Dunbabin, Thomas. "New Light on the Earliest American Voyages to Australia." *American Neptune* 10 (1950): 52–64.

– "Whalers, Sealers and Buccaneers." *Journal of the Royal Australian Historical Society* 9 (1925): 1–32.

Dunbar, M.J. "The Pinnipedia of the Arctic and Subarctic." *Bulletin of the Fisheries Research Board of Canada* 85 (1949): 1–22.

– "The Status of the Atlantic Walrus, 'Odobenuse rosmarus' (L.), in Canada." *The Arctic Circular* 8 (1954): 11–14.

Duncan, Bingham, "A Letter on the Fur-Seal in Canadian-American Diplomacy." *Canadian Historical Review* 43 (1962): 42–7.

Dunn, Robert. "Alaska, the Seal-Warder, and the Japanese Raiders." *Harper's Weekly* 50 (1906): 1310–1.

East, Ben. "Seal Management Has Paid Dividends." *Animal Kingdom* 50 (1947): 188–94.

– "Uncle Sam's Prize Fur Factory Closes Down." *Natural History* 51 (1943): 188–95.

Ebert, Earl E. "A Food Habitat Study of the Southern Sea Otter, 'Enhydra lutris nereis.' " *California Fish & Game* 54 (1960): 33–42.

Elliott, Henry W. "The Loot and the Ruin of the Fur-Seal Herd of Alaska." *North American Review* (1907): 426–36.

Ellsberg, Helen. "Furs that Launched a Thousand Ships." *American West* 2 (1974): 14–19.

– *Los Coronados Islands*. Glendale, Calif.: La Siesta Press, 1970.

Ely, Charles A., and Roger B. Clapp. "The Natural History of Laysan Island, Northwestern Hawaiian Islands." *Atoll Research Bulletin* 171 (1973).

English, Arthur S. *The Hair Seal: Some Investigations in the Life History of the Nfld. Seal*. St. John's: privately printed 1927.

Essig, E.Q. "The Russian Settlement at Ross." *California Historical Society Quarterly* 12 (1933): 191–209.

Estes, James A., and James R. Gilbert. "Evaluation of an Aerial Survey of Pacific Walruses ('Odobenus rosmarus divergens')." *Journal of the Fisheries Research Board of Canada* 35 (1978): 1130–40.

Evans, Hugh B. "A Voyage to Kerguelen in the Sealer 'Edward' in 1897–98." *Polar Record* 16 (1973): 189–91.

Evermann, Barton Warren. "The Northern Fur-Seal Problem as a Type of

Man Problems of Marine Zoology." *Bulletin of the Scripps Institute for Biological Research* 9 (1919): 13–26.

Falla, F.A. "Exploitation of Seals, Whales and Penguins in New Zealand." *New Zealand Ecological Society Proceedings* 9 (1962): 34–8.

– "The Outlying Islands of New Zealand." *New Zealand Geographer* 4 (1948): 127–34.

"Farallone Islands, The." *Hutchins California Magazine* 1 (1856): 49–57.

Fay, Francis H. "History and Present Status of the Pacific Walrus Population." *Transactions of the 22nd North American Wildlife Conference* (1957): 431–44.

– "Industrial Utilization of Marine Mammals." *Proceedings of the 29th Alaska Science Conference*, Alaska Division, American Association for the Advancement of Science, Sea Grant Report 79-6 (1979): 75–9.

– Howard M. Feder, and Samual W. Stoker. *An Estimation of the Impact of the Pacific Walrus Population on Its Food Resources in the Bering Sea.* Washington: U.S. Marine Mammal Commission 1977.

– and Brendon P. Kelly. "Mass Natural Mortality of Walruses ('Odobenus rosmarus') at St. Lawrence Island, Bering Sea, Autumn 1978." *Arctic* 33 (1980): 226–45.

– and Carleton Ray. "Influence of Climate on the Distribution of Walruses 'Odobenus rosmarus (linnaeas)'." *Zoologica* 53 (1968); 19–32.

Finley, William L. "Salmon, Seals and Skullduggery: Activities of Commissioner of Fisheries Open to Question." *Nature Magazine* 28 (1930): 229–303.

Fiscus, Clifford H. "Interactions of Marine Mammals and Pacific Hake." *Marine Fisheries Review* (1979): 1–9.

– *Marine Mammal-Salmonid Interactions: A Review.* Seattle: Marine Mammal Division Northwest and Alaska Fisheries Center 1978.

– *Northern Fur Seal–Steller's Sea Bear.* Seattle: Northwest and Alaska Fisheries Center 1977.

– and Gary A. Baines, "Food and Feeding Behavior of Steller and California Sea Lions." *Journal of Mammalogy* 47 (1966): 195–200.

Fisher, Edna M. "Habits of the Southern Sea Otter." *Journal of Mammalogy* 20 (1939): 231–36.

– "Prices of Sea Otter Pelts." *California Fish & Game* 27 (1941): 261–5.

– "The Sea Otter, Past and Present." *Proceedings of the Sixth Pacific Science Congress* 3 (1940): 221–36.

Fisher, H.D. *Harp Seals of the Northwest Atlantic.* Ottawa: Atlantic Biological Station, General series, circular no. 20. Ottawa: Fisheries Research Board of Canada 1952.

– "The Status of the Harbour Seal in British Columbia ... ." *Bulletin of the Canadian Fisheries Research Board* 93 (1952): 1–58.

Fisher, Robin. "Arms and Men on the Northwest Coast, 1774–1825." *British Columbia Studies*, no. 29 (1976): 3–18.

Fleischer, Luis A. "Guadalupe Fur Seal." In *Marine Mammals of Eastern North Pacific and Arctic Waters*, edited by Delphine Haley, 160–5. Seattle 1978.

Fleming, C.A. "Sea Lions as Geological Agents." *Journal of Sedimentary Petrology* 21 (1951): 22–5.

Fogarty, James J. "The Seal-Skinners Union." In *Book of Newfoundland*, edited by J.R. Smallwood, 2: 100. St. John's 1937.

Foote, Don Charles. "Remarks on Eskimo Sealing and the Harp Seal Controversy." *Arctic* 20 (1967): 267–8.

– , Victor Fischer, and George W. Rogers, *St. Paul Community Study*. Fairbanks: University of Alaska, Institute of Social, Economic and Governmental Research 1968.

Foster, Michael S., et al. *Towards an Understanding of the Effects of Sea Otter Foraging on Kelp Forest Communities in Central California*. Washington: U.S. Marine Mammal Commission 1979.

Fowler, Charles W. et al. *Comparative Population Dynamics of Large Mammals: A Search for Management Criteria*. Washington: U.S. Marine Mammal Commission 1980.

Frothingham, Robert. "Walrus Hunting with the Eskimos." In *Told at the Explorer's Club: True Tales of Modern Explorers,* edited by Frederick A. Blossom 123–33. London 1932.

"Furor over Alaskan Seals." *Business Week* 10 February 1962: 60–4.

Gaines, Sanford E., and Dale Schmidt. *Laws and Treaties of the United States Relevant to Marine Mammal Protection Policy*. Washington: U.S. Marine Mammal Commission 1978 (MMC75/09).

Galbraith, John S. "A Note on the British Fur Trade in California, 1821–1846." *Pacific Historical Review* 24 (1955): 253–60.

Gay, James T. "Bering Sea Controversy: Harrison, Blaine, and Cronyism." *Alaska Journal* 3 (1973): 12–19.

– "Henry W. Elliott: Crusading Conservationist." *Alaska Journal* 3 (1973): 211–16.

Gentry, Roger L., and David E. Withrow, "Steller Sea Lion." In *Marine Mammals of Eastern North Pacific and Arctic Waters*, edited by Delphine Haley, 166–71. Seattle 1978.

Gibbney, L.F. "The Seasonal Reproductive Cycle of the Female Elephant Seal, 'Mirounga leonina,' Linn., at Heard Island." *Australian National Antarctic Research Expeditions* 32 (1957).

Gilbert, Benjamin Franklin. "Economic Developments in Alaska, 1867–1910." *Journal of the West* 4 (1965): 504–21.

Gilbert, Bil. "A Marvel, a Madness." *Sports Illustrated* 43 (1975): 96–104.

Gilbert, James R., et al. *Grey Seals in New England: Present Status and Management Alternatives*. Washington, D.C.: U.S. Marine Mammal Commission 1979.

Gill, J.C.H. "Notes on the Sealing Industry of Early Australia." *Journal of the Royal Historical Society of Queensland* 8 (1967): 218–45.

Godden, R.R. "Aftermath." *The Bulletin: Journal of the Maritime Museum of British Columbia* (Summer 1979): 9–10.
– "Frustrated voyage, 1891." *The Bulletin: Journal of the Maritime Museum of British Columbia* (Fall 1978): 8–9.
– "Sealing Voyage, 1897." *The Bulletin: Journal of the Maritime Museum of British Columbia* (Summer 1979): 21–3.
Gough, Barry M. "James Cook and the Origins of the Maritime Fur Trade." *American Neptune* 38 (1978): 217–24.
Graham, Gerald S. "The Maritime Foundations of Imperial History." *Canadian Historical Review* 31 (1950): 113–24.
Grenfell, Wilfred T. "The Seal Hunters of Newfoundland." *Leisure Hour* (1897–8): 284–92.
Guillemand, F.H.H. "The Fur Seal and the Award." *Blackwell's* (November 1893): 745–54.
Gustafson, C.E. "Prehistoric Use of Fur Seals: Evidence from the Olympic Coast of Washington." *Science* 161 (1968): 49–51.
Gwynn, A.M. "Notes on the Fur Seals at Macquarie Island and Heard Island." *Australian National Antarctic Research Expedition, Interim Reports 4* (1954).
Hainsworth, D.R. "Exploring the Pacific Frontier: The New South Wales Sealing Industry, 1800–1821." *Journal of Pacific History* 2 (1967): 59–75.
– "Iron Men in Wooden Ships: The Sydney Sealers, 1800–1820." *Labour History* (Sydney) 13 (1967): 19–25.
Hall, K.R.L., and G.B. Schaller. "Tool-Using Behavior of the California Sea Otter." *Journal of Mammalogy* 45 (1964): 287–98.
Hall, Raymond E. "Chase Littlejohn, 1854–1943: Observations by Littlejohn on Hunting Sea Otters." *Journal of Mammalogy* 26 (1945): 89–91.
Hamilton, Andrew. "Is the Guadalupe Fur Seal Returning?" *Natural History* 60 (1951): 90–4.
Hamilton, J.E. "On the Present Status of the Elephant Seal in South Georgia." *Proceedings of the Zoological Society of London* 117 (1947): 272–5.
– "The Leopard Seal, 'Hydrurga leptonys' (de Blainville)." *Discovery Reports* 18 (1939): 239–64.
– "The Southern Sea Lion, 'Otario byronia.'" *Discovery Reports* 8 (1934): 269–318.
Hamman, Mary. "Conservation Ruffles Fur in Fashion World." *Smithsonian* 1 (1970): 54.
Hansen, Craig A. "Seals and Sealing." *Alaska Geographic* 9 (1982): 41–72.
Harris, Charles. "A Cruise after Sea Elephants." *Pacific Monthly* 21 (1909): 331–9.
Harry, George Y., Jr. *The Effects of Management on the Pribilof Islands Fur Seal Herd.* Seattle: U.S. Dept. of Commerce, NOAA, NMFS, Marine Mammal Laboratory 1979.

Hauser, Hillary. "Seals and Sea Lions: The Watchdogs of San Miguel." *Santa Barbara Magazine* 3 (1977): 16–24.

Hiller, James K. *The Newfoundland Railway 1881–1949.* Pamphlet no. 6. St. John's: Newfoundland Historical Society 1981.

Hinckley, Theodore C. "Rustlers of the North Pacific." *Journal of the West* 2 (1963): 22–30.

Hiscocks, Bettie. "Sea Otters Return to Canada's West Coast." *Canadian Geographical Journal* 32 (1977): 20–7.

Holdgate, M.W. "Terrestrial Ecosystems in the Antarctic." *Transactions of the Royal Society, London* 279 (1977): 5–25.

Hope, M.S.E. "Newfoundland Seal Fishery: Past-Present-Future." *Newfoundland Journal of Commerce* 33 (1966): 11, 18–21.

Howarth, Peter C. "The Seals of San Miguel Island." *Oceans* 9 (1976): 38–43.

Howay, F.W. "Capt. Simon Metcalfe and the Brig 'Eleanora'." *Washington Historical Quarterly* 16 (1925): 114–21.

– "Captains Grey and Kendrick: The Barrell Letters." *Washington Historical Quarterly* 12 (1921): 243–71.

– "Early Days of the Maritime Fur-Trade on the Northwest Coast." *Canadian Historical Review* 4 (1923): 26–44.

– "The Fur Trade in Northwestern Development." In *The Pacific Ocean in History: Papers and Addresses Presented at the Panama Pacific Historical Congress ... 1915* edited by H. Morse Stephens and Herbert E. Bolton, 276–86. New York 1917.

– "International Aspects of the Maritime Fur Trade." *Proceedings of the Royal Society of Canada* 36 (1942): 59–78.

– "An Outline Sketch of the Maritime Fur Trade." *Annual Report of the Canadian Historical Association, 1932* (Presidential Address), 5–14.

– "The Settlement and Progress of British Columbia, 1871–1814." In *The Cambridge History of the British Empire*, edited by J.H. Rose, A.P. Newton, and E.A. Benians, 6: 548–65. New York 1930.

– "The Voyage of the 'Hope', 1790–1792." *Washington Historical Quarterly* 11 (1920): 3–28.

– "William Sturgis: The Northwest Fur Trade." *British Columbia Historical Quarterly* 8 (1944): 11–25.

– "A Yankee Trader on the Northwest Coast, 1791–1795." *Washington Historical Quarterly* 21 (1930): 83–94.

Hubbs, Carl L. "Back from Oblivion. Guadalupe Fur Seal: Still a Living Species." *Pacific Discovery* 9 (1956): 14–21.

Huber, Harriet R., David G. Ainley, Stephen H. Morrell, Robert J. Beokelheide, and R. Philip Henders. *Studies of Marine Mammals at the Farallon Islands, California, 1978–1979.* Washington, D.C.: U.S. Marine Mammal Commission 1980.

Huey, Laurence M. "Guadalupe Island: an Object Lesson in Man-Caused Devastation." *Science* 61 (1925): 405–7.

– "Past and Present Status of the Northern Elephant Seal with a Note on the Guadalupe Fur Seal." *Journal of Mammalogy* 11 (1930): 188–94.

– "Pribilof Fur Seal Taken in San Diego County, California." *Journal of Mammalogy* 23 (1942): 95–6.

– "A Trip to Guadalupe, the Isle of My Boyhood Dreams." *Natural History* 24 (1924): 578–88.

Hunt, Rev. E. "The Great Sealers' Strike of 1902." *Newscene* (supp. to St. John's *Daily News*), 6 March 1970.

Imler, Ralph H., and Hosea R. Sarber. *Harbor Seals and Sea Lions in Alaska.* Special Scientific Report no. 28. Washington: U.S. Dept. of Interior 1947.

Ingham, Susan E. "Elephant Seals on the Antarctic Continent." *Nature* 180 (1957): 1215–16.

Ireland, Gordon. "The North Pacific Fisheries." *American Journal of International Law* 36 (1942): 400–24.

Irving, Laurence. "Temperature Regulations in Marine Mammals." In *The Biology of Marine Mammals*, edited by Harald T. Anderson, 147–74. New York 1969.

"Islands of the Seals: The Pribilofs." *Alaska Geographic* 9, no. 3 (1982) (entire issue)

"J.R. McCowen, J.P., and A.D.C., Inspector-General Constabulary." *Newfoundland Quarterly* 2 (1902): 15.

Johnson, Susan Hackley. "New Choices for the People of the Pribilofs." *Alaska Magazine* (May 1979): 6–9, 89–90.

– *The Pribilof Islands: A Guide to St. Paul, Alaska.* St. Paul: Tanadgusix Corporation 1978.

Jones, A.G.E. "The British Southern Whale and Seal Fisheries." *The Great Circle* (Australian Association for Maritime History) 3, pt. 1 (1981): 20–9.

– "Island of Desolation." *Antarctica* 6 (1971): 22–6.

Jones, Richard M. "Sealing and Stonington: A Short-Lived Bonanza." *The Log of Mystic Seaport* 28 (1977): 119–26.

Jordan, David S. *Observations on the Fur Seals of the Pribilof Islands.* Washington: GPO 1896.

–, and G.A. Clark. *Truth About the Fur Seals on the Pribilof Islands.* Bureau of Fisheries, Economic Circular no. 4. Washington, D.C.: Dept. of Commerce and Labor 1912.

Kajimura, H. *The Oportunistic Feeding of Northern Fur Seals off California.* Seattle: NOAA, NMFS, Marine Laboratory 1981.

Kean, Capt. Abram. "Commentary on the Seal Hunt." In *The Book of Newfoundland*, edited by J.R. Smallwood, 1:73–6. St. John's 1937.

Keithahn, E.L. "Alaska Ice, Inc." In *Alaska and Its History*, edited by Morgan B. Sherwood, 173–86. Seattle 1967.

Kemble, John Haskell. "The Cruise of the Schooner 'Tamana,' 1806–1807: An Episode in the American Penetration of the Pacific Ocean." *Proceedings of the American Antiquarian Society* 78 (1967): 283–98.

Kenyon, Karl W. "Diving Depths of the Steller Sea Lion and Alaska Fur Seal." *Journal of Mammalogy* 33(1952): 245–6.

– "Guadalupe Fur Seal ('Arctocephalus townsendi')." In International Union for the Conservation of Nature, *Seals: Proceedings of a Working Meeting* 82–7. Morgues, Switzerland 1973.

– "Hawaiian Monk Seal ('Monachus schauinslandi')" In International Union for the Conservation of Nature, *Seals: Proceedings of a Working Meeting*, 88–97. Morgues, Switzerland 1973.

– "History of the Steller Sea Lion at the Pribilof Islands, Alaska." *Journal of Mammalogy* 43 (1962): 68–75.

– "Last of the Tlingit Sealers." *Natural History* (June 1955): 294–8.

– "Man vs. the Monk Seal." *Journal of Mammalogy* 53 (1972): 687–96.

– "Sea Otter." In *Marine Mammals of Eastern North Pacific and Arctic Waters*, edited by Delphine Haley, 226–34. Seattle 1978.

– "The Sea Otter in Alaska." *Alaska Sportsman* 34 (1967): 17–18.

– "The Steller Sea Lion." *Pacific Discovery* 5 (1952): 4–13.

– "Walrus." in Delphine Haley, ed., *Marine Mammals of Eastern North Pacific and Arctic Waters*, Seattle 1978. 178–83.

– , and Dale W. Rice. "Abundance and Distribution of the Steller Sea Lion." *Journal of Mammalogy* 42 (1961): 223–34.

– , – "The Life History of the Hawaiian Monk Seal." *Pacific Science* 13 (1959): 215–52.

– , and Victor B. Scheffer. *A Population Study of the Alaska Fur-Seal Herd*. Special Scientific Report, Wildlife no. 12. Washington, D.C.: U.S. Dept. of Interior, Fish and Wildlife Service 1954.

– , – *The Seals, Sea-Lions, and Sea Otter of the Pacific Coast*. Fish and Wildlife Service, Circular no. 32. Washington, D.C.: U.S. Dept. of Interior 1955.

– , and Ford Wilke. "Migration of the Northern Fur Seal, 'Callorhinus ursinus'." *Journal of Mammalogy* 34 (1953): 86–98.

Kihn, Phyllis. "The Sea Journal of Captain Ebenezer Hooker Mix, 1817–1818." *Connecticut Historical Society Proceedings* 40 (1975): 8–18.

Kim, Ke Chung, Richard C. Chu, and George P. Barron. "Mercury in Tissues and Lice of Northern Fur Seals." *Bulletin of Environmental Contamination and Toxology* 11 (1974): 218–24.

King, Judith E. "The Monk Seals (genus 'Monachus')." *Bulletin of the British Museum (Natural History: Zoology)* 3 (1956): 203–56.

– "The Northern and Southern Populations of 'Arctocephalus gazella'." *Mammalia* 23 (1959): 19–40.

– "The Otariid Seals of the Pacific Coast of America." *Bulletin of the British Museum (Natural History: Zoology)* 2 (1954): 309–37.

– "Sea-Lions of the Genera Neophoca and Phocarctos." *Mammalia* 24 (1960): 444–56.

– , and R.J. Harrison, "Some Notes on the Hawaiian Monk Seal." *Pacific Science* 15 (1961): 282–93.

Kleinschmidt, Capt. F.E. "A Day of Blood." *Pacific Monthly* 23 (1910): 561–79.

Knox, G.A. "The Subantarctic Islands: Past, Present, and Future." *Proceedings of the New Zealand Ecological Society* 12 (1965): 69–72.

Knudtson, Peter M. "The Case of the Missing Monk Seal." *Natural History* 86 (1977): 78–83.

Kooyman, G.L., and H.T. Anderson, "Deep Diving." In *The Biology of Marine Mammals*, edited by Harald T. Anderson, 65–94. New York 1969.

Kroeber, A.L. "Elements of Culture in Native California." In *The California Indians: A Source Book*, edited by R.F. Heizer and M.A. Whipple, 3–65. 2nd ed. Berkeley 1971.

Kuykerdall, Ralph S. "A Northwest Trader at the Hawaiian Islands." *Oregon Historical Quarterly* 24 (1923): 111–31.

"Last Indian Sealer Dies at Neah Bay." *Marine Digest* 35, no. 47 (27 July 1957): 13.

Lathrop, John E. "The West and the National Capital." *Pacific Monthly* 26 (1911): 195–201.

Latourette, Kenneth Scott. "Voyages of American Sealing Ships to China, 1784–1844." *Transactions of the Connecticut Academy of Arts & Sciences* 18 (1927): 237–71.

Laughlin, William S. "Eskimos and Aleuts: Their Origins and Evolution." *Science* 142 (1963): 633–45.

Lavigne, David M. "Life or Death for the Harp Seal." *National Geographic* 149 (1976): 129–42.

– "Management of Seals in the Northwest Atlantic Ocean." *Transactions of the 44th North American Wildlife and Natural Resources Conference* (1979): 488–97.

Law, P.G., and T. Burstall. "Heard Island." *Australian National Antarctic Research Expeditions, Reports*, ser. B, vol. 2, no. 7 (1953).

Laws, R.M. "The Current Status of Seals in the Southern Hemisphere." In International Union for the Conservation of Nature, *Conference on Seals*, Paper no. 39, 144–57. Guelph 1973.

– "The Elephant Seal ('Mirounga leonina,' Linn.)." *Falkland Islands Dependency Survey, Scientific Reports* 1 (1953), no. 8; 2 (1956), no. 13; 3 (1956), no. 15.

– "The Elephant Seal Industry at South Georgia." *Polar Record* 6 (1953): 746–54.

– "Population Increase of Fur Seals at South Georgia." *Polar Record* 16 (1973): 856–8.

– "The Southern Elephant Seal ('Mirounga leonina,' Linn.) at South Georgia." *Norsk Hvalfangst-Tidende* 49 (1960): 466–76, 520–42.

Le Boeuf, Burney J. "Back from Extaction?" *Pacific Discovery* 30 (1977): 1–7.

– "Male-male Competition and Reproductive Success in Elephant Seals." *American Zoologist* 14 (1974): 163–70.

– , David Aurioles, Richard Condit, Claudio Fox, Robert Gisiner, Rigoberto Romero, and Francisco Sinsel. "Size and Distribution of the California Sea Lion Population in Mexico." *Proceedings of the California Academy of Sciences* 43 (1983): 77–85.

– and Kenneth T. Briggs. "The Cost of Living in a Seal Harem." *Mammalia* 41 (1977): 167–95.

– and Richard S. Condit. "The High Cost of Living on the Beach." *Pacific Discovery* 36 (1983): 12–14.

– , Donald A. Countryman, and Carl L. Hubbs. "Records of Elephant Seals, 'Mirounga angustirostris,' on Los Coronados Islands, Baja California, Mexico, with Recent Analyses of the Breeding Population." *San Diego Society of Natural History, Transactions* 18 (1975): 1–7.

– , Raymond S. Keyes, and Marianne Riedman. "White Shark Predation of Pinnipeds in California Coastal Water." *Fishery Bulletin* 80 (1982): 891–5.

– , and Kathy J. Panken. "Elephant Seals Breeding on the Mainland in California." *Proceedings of the California Academy of Sciences,* 4th ser., 41 (1977): 267–80.

– , Ronald J. Whiting, and Richard F. Gantt, "Perinatal Behavior of Northern Elephant Seal Females and Their Young." *Behavior* 43 (1972): 1–4.

Levi, Werner. "The Earliest Relations between the United States of America and Australia." *Pacific Historical Review* 12 (1943): 351–61.

Lewis, F. "Notes on Australian Seals." *Victoria Naturalist* 59 (1942): 24–6.

Lewis, Harrison F., and J. Kenneth Doutt. "Records of the Atlantic Walrus and the Polar Bear in the Northern Part of the Gulf of St. Lawrence." *Journal of Mammalogy* 23 (1942), 365–375.

Lindsey, Alton A. "Notes on the Crabeater Seal." *Journal of Mammalogy* 19 (1938): 456–61.

– "The Weddell Seal in the Bay of Whales, Antarctica." *Journal of Mammalogy* 18 (1937): 127–44.

Ling, John K. "A Review of Ecological Factors Affecting the Annual Cycle in Island Populations of Seals." *Pacific Science* 23 (1969): 399–413.

Little, Barbara. "The Sealing and Whaling Industry in Australia Before 1850." *The Economic Record* (Victoria, NSW) 45 (1969): 109–27.

Loughlin, Thomas R. "Home Range and Territoriality of Sea Otters Near Monterey, California." *Journal of Wildlife Management* 44 (1980): 576–82.

Lowry, L.F., and J.S. Pearse. "Abalones and Sea Urchins in an Area Inhabited by Sea Otters." *Marine Biology* 23 (1973): 312–19.

Lugrin, N. de Bertrand. "Epic of the Seal Hunters." *Canadian Geographical Journal* 4 (1932): 295–304.

Lyon, Gretchen M. "Pinnipeds and a Sea Otter from the Point Mugu Shell Mound of California." *University of California at Los Angeles Publications in Biological Sciences* 1 (1937): 133–68.

MacAskie, I.B. "Sea Otters: A Third Transplant to British Columbia." *The Beaver* 305 (1975): 9–11.

McCloskey, William. "Bitter Fight Still Rages over the Seal Killing in Canada." *Smithsonian* 10 (1979): 54–63.

McDonald, Ian. "W.F. Coaker and the Balance of Power Strategy: The Fishermen's Protective Union in Newfoundland Politics." In *Newfoundland in the Nineteenth and Twentieth Centuries: Essays in Interpretation*, edited by James Hiller and Peter Neary, 148–80. Toronto 1980.

McEvoy, Arthur F. "In Places Men Reject: Chinese Fishermen at San Diego, 1870–1893." *Journal of San Diego History* 23 (1977): 12–24.

MacKay, Corday. "Pacific Coast Fur Trade." *The Beaver* 286 (1955): 38–42.

McLaren, Ian A. "Are the Pinnipedia Biphylectic?" *Systematic Zoology* 9 (1960): 18–28.

– "Seals and Group Selection." *Ecology* 48 (1960): 104–10.

Macy, William H. "Adventure with a Sea-Elephant." *Whaleman's Shipping List and Merchant's Transcript* 30 (16 April 1872): 1.

Mann, K.H., and P.A. Breen. "The Relation Between Lobster Abundance, Sea Urchins, and Kelp Beds." *Journal of the Fisheries Research Board of Canada* 29 (1972): 603–9.

Mansfield, A.W. "Population Dynamics and Exploitation of Some Arctic Seals." In *Antarctic Ecology*, edited by M.W. Holdgate, 1: 428–50. London 1970.

– *Seals of Arctic and Eastern Canada*. Ottawa: Fisheries Research Board of Canada 1967.

– "The Walrus in Canada's Arctic." *Canadian Geographic Journal* 73 (1966): 88–95.

Marr, James W.S. "The South Orkney Islands." *Discovery Reports* 10 (1935): 282–382.

Martin, Col. Lawrence. "Antarctica Discovered by a Connecticut Yankee, Capt. Nathaniel Brown Palmer." *Geographical Review* 30 (1940): 529–52.

Mate, Bruce R. "California Sea Lion." In *Marine Mammals of Eastern North Pacific and Arctic Waters*, edited by Delphine Haley, 172–76. Seattle 1978.

Mathieson, Capt. Matt, "When the Sandheads Lightship was Alive." *Canadian Merchant Service Guild Annual* (1945): 25–6.

Matthews, L. Harrison. "The Natural History of the Elephant Seal with Notes on Other Seals Found in South Georgia." *Discovery Reports* 1 (1929): 233–356.

Matthews, Ralph. "The Smallwood Legacy: The Development of Underdevelopment in Newfoundland, 1949–1972." *Journal of Canadian Studies* 13 (1978–9): 89–108.

Mattsson, A. Alfred. "Fur Seal Hunting in the South Atlantic." *American Neptune* 2 (1942): 154–66.

– "Sealing Boats." *American Neptune* 3 (1943): 327–32.

Mawson, Sir D. "Macquarie Island, Its Geography and Geology." *Australian National Antarctic Research Expedition Reports* (ANARE), ser. A, 5 (1943): 1–194.

Melville, Herman. "The 'Gees." *Harper's New Monthly Magazine* (1856): 507–9.

Merculieff, Larry. "Traditional Living in a Modern Society." *Alaska Geographic* 9, no. 3 (1982): 97–106.

Miller, Daniel J. *The Sea Otter: Enhydra lutris.* Sacramento: State of California Resources Agency, Dept. of Fish and Game 1974.

– and Ralph S. Collier, "Shark Attacks in California and Oregon, 1926–1979." *California Fish & Game* 67 (1981): 76–99.

– , James E. Hardwicke, and Walter A. Dahlstrom. *Pismo Clams and Sea Otters.* Marine Resources Technical Report no. 31. California Dept. of Fish and Game 1975.

Miller, L. Keith. *Energetics of the Northern Fur Seal in Relation to Climate and Food Resources of the Bering Sea.* Washington: U.S. Marine Mammal Commission 1978.

Morejohn, G. Victor, Jack A. Ames, and David B. Lewis. *Post Mortem Studies of Sea Otters, 'Enhydra lutris' L. in California.* Marine Resources Technical Report, no. 30. California Dept. of Fish and Game 1975.

Morris, Isaac C. "Our Local Strikes." *Newfoundland Quarterly* 2 (1902): 20.

Muller-Schwarze, D., E.C. Waltz, W. Trivelpiece, and N.J. Volkman. "Breeding Status of Southern Elephant Seals at King George Island." *Antarctic Journal* (1978): 157–8.

Murphy, M.F. "Sea Otter–Past and Present." *Nature Magazine* 32 (1939): 425–8.

Murphy, Roberts Cushman. "A Desolate Island of the Antarctic." *Scientific American* supp. 1986 (24 January 1914): 60–4.

– "South Georgia." *National Geographic* 41 (1922): 408–44.

"Mutiny on the 'Rand': A Tale of Early Days Among the Sealers." *Alaska Sportsman* 25 (1958).

Neary, Peter. "Democracy in Newfoundland: A Comment." *Journal of Canadian Studies* 4 (1969): 89–108.

– "Grey, Bryce, and the Settlement of Canadian-American Differences, 1905–1911." *Candian Historical Review* 49 (1968): 357–80.

Nichols, Robert H. "Seal-Hunters of the North Pacific." In *British Columbia: A Centennial Anthology*, edited by Reginald Eyre Watters, 123–8. Vancouver 1958.

Odell, Daniel K. "Seasonal Occurrence of the Northern Elephant Seal, 'Mirounga angustirostris,' on San Nicolas Island." *Journal of Mammalogy* 55 (1974): 81–95.

Ogden, Adele. "The Californias in Spain's Pacific Otter Trade, 1775–1795." *Pacific Historical Review* 1 (1932): 444–69.

– "Russian Sea-Otter and Seal-Hunting on the California Coast, 1803–1841." *California Historical Society Quarterly* 12 (1933): 217–39.

O'Gorman, Fergus A. "Fur Seals Breeding in the Falkland Islands Dependencies." *Nature* 192 (1961): 914–16.

– "The Return of the Antarctic Fur Seal." *New Scientist* 20 (1963): 374–6.

Olds, J.M. "Notes on the Hood Seal ('Cystophora cristata')." *Journal of Mammalogy* 31 (1950): 450–2.

O'Neil, Marion. "The Maritime Activities of the North West Company 1813 to 1821." *Washington Historical Quarterly* 21 (1930): 243–71.

Orr, Robert T. "Galapagos Fur Seal ('Arctocephalos galapagoensis')." In International Union for the Conservation of Nature, *Seals: Proceedings of a Working Conference*, 124–8. Morgues, Switzerland 1973.

– "The Galapagos Sea Lion." *Journal of Mammalogy* 48 (1967): 62–9.

Osgood, Wilfred H., Edward A. Preble, and George H. Parker. "The Fur Seals and Other Life of the Pribilof Islands, Alaska, in 1914." *Bulletin of the U.S. Bureau of Fisheries* 34 (1915): 1–172.

Palmisano, John F., and James A. Estes. "Otters: Pillars of the Nearshore Community." *Natural History* 85 (1976): 46–53.

Panting, G.E. "The Fishermen's Protective Union of Newfoundland and the Farmer's Organization in Western Canada." *Canadian Historical Association Annual Report* (1963): 141–51.

Paulian, Patrice. "Note sur les phoques des Isles Amsterdam et Saint-Paul." *Mammalia* 21 (1957): 210–25.

Payne, M.R. "Growth of a Fur Seal Population." *Philosophical Transactions of the Royal Society of London,* ser. B, 279 (1977): 67–79.

Peterson, Richard S., and S.M. Cooper, "A History of the Fur Seals of California." *University of California, Santa Cruz, Año Nuevo Reports* 2 (1968): 47–54.

– and Burney J. Le Boeuf, "The Fur Seals are Coming Back to California." *Science Digest* (April 1971): 74–9.

– , Carl L. Hubbs, Robert L. Gentry, and Robert L. DeLong, "The Guadalupe Fur Seal: Habitat, Behavior, Population Size, and Field Identification." *Journal of Mammalogy* 49 (1968): 665–75.

Phelps, E.J. "The Behring Sea Controversy." *Harper's Monthly* 82 (1891): 766–74.

Pierce, Richard A. "Prince D.P. Maksutov: Last Governor of Russian America." *Journal of the West* 6 (1967): 395–416.

"Protest, Priorities, and the Alaska Fur Seal." *Audubon* 72 (1970): 114–15.

Radford, Keith W., Robert T. Orr, and Carl L. Hubbs. "Re-establishment of the Northern Elephant Seal ('Mirounga angustirostris') off Central

California." *Proceedings of the California Academy of Sciences* 4th ser., 31 (1965): 601–12.

Rand, R.W. "Elephant Seals on Marion Island." *African Wildlife* 16 (1962): 191–98.

— "Notes on the Marion Island Fur Seal." *Proceedings of the Zoological Society of London* 126 (1956): 65–82.

— *Studies on the Cape Fur seal*. Government of the Guano Islands Administration, Progress Report. Cape Town: Union of South Africa, Dept. of Agriculture 1949.

Ray, G. Carleton. "Learning the Ways of the Walrus." *National Geographic* 156 (1979): 564–80.

Reeves, Randall, R. *Atlantic Walrus ('Odobenus rosmarus rosmarus'): A Literature Survey and Status Report*. Wildlife Research Report no. 10. Washington, D.C.: U.S. Dept. of Interior, Fish and Wildlife Service 1978.

Reiger, George. "Song of the Seal." *Audubon* 77 (1975): 6–27.

— "What Now for the Walrus?" *National Wildlife* 14 (1976): 50–7.

Repenning, Charles A., Clayton E. Ray, and Dan Grigorescu. "Pinniped Biogeography." In *Historical Biography, Plate Tectonics, and the Changing Environment*, edited by Jane Gray and Arthur J. Boucet, 357–69. Eugene, Ore. 1979.

Richards, Rhys. "The Journal of Erasmus Darwin Rogers, The First Man on Heard Island." *American Neptune* 41 (1981): 180–305.

Rickard, T.A. "The Sea-Otter in History." *British Columbia Historical Quarterly* 11 (1947): 15–32.

Riedman, Marianne L. "Practice Makes Perfect: The Difficult Art of Mothering in an Elephant Seal Rookery." *Pacific Discovery* 36 (1983): 4–11.

Riley, Francis. *Fur Seal Industry of the Pribilof Islands, 1786–1965*. Washington, D.C.: Bureau of Commercial Fisheries, circular no. 275. U.S. Dept. of Interior, Fish and Wildlife Service 1967.

Ring, T.P.A. "The Elephant Seals of Kerguelen Land." *Proceedings of the Zoological Society of London* 93 (1923): 431–43.

Roberts, Brian. "Historical Notes on Heard and McDonald Islands." *Polar Record* 5 (1950): 580–4.

Rodahl, Kaare. "Spekk-finger or Sealer's Finger." *Arctic* 5 (1952): 235–40.

Rogers, Eugene F. "Memoirs." Edited by Joan A. Canby with Stella Rouse. *Noticias: Quarterly Bulletin of the Santa Barbara Historical Society* 26 (1980): 41–59.

Rogers, T.B. "The Last Voyage of the 'Southern Cross'." *Newfoundland Quarterly* 76 (1980): 21–30.

Ronald, K., and J.L. Dougan, "The Ice Lover: Biology of the Harp Seal ('Phoca groenlandica')." *Science* 215 (1892): 928–33.

Roosevelt, Theodore. "The Fur Seal Fisheries." *Metropolitan Magazine* 25 (1907): 687–98.

Root, Joel. "Narrative of a Sealing and Trading Voyage in the Ship
'Huron' ... 1802–1806." *Papers of the New Haven Colony Historical Society*
5 (1894): 144–71.

Roppel, Alton Y., and Stuart P. Davey. "Evolution of Fur Seal Management
on the Pribilof Islands." *Journal of Wildlife Management* 29 (1965): 448–63.

Ross, Frank E. "American Adventurers in the Early Marine Fur Trade
with China." *Chinese Social and Political Review* (Peking) 21 (1937): 221–
67.

Roth, Hal. "Sea Otter Population Now Booming Modestly in Western
Waters." *Smithsonian* 1 (1970): 30–7.

Rüe, E. Aubert de la. "Notes sur les Isles Crozet." *Bulletin du Musée National
d'Histoire Naturelle*, 2ème sér., 22 (1950): 197–203.

Rowley, John. "Life History of the Sea-Lions on the California Coast."
Journal of Mammalogy 10 (1929): 1–37.

Ryan, Paul R. "Oil and Gas Group Attacks Marine Sanctuary Program."
Oceanus 26 (1983): 72–6.

Ryan, Shannon. "The Newfoundland Salt Cod Trade in the Nineteenth
Century." In *Newfoundland in the Nineteenth and Twentieth Centuries: Essays
in Interpretation*, edited by James Hiller and Peter Neary, 39–66. Toronto
1980.

Sager, Eric W. "The Merchants of Water Street and Capital Investment in
Newfoundland's Traditional Economy." In *Enterprising Canadians:
Entrepreneurs and Economic Development in Eastern Canada, 1820–1914*,
edited by Lewis R. Fischer and Eric W. Sager, 75–96. Proceedings of the
Second Conference of the Atlantic Canadian Shipping Project. St. John's
1978.

– "Newfoundland's Historical Revival and the Legacy of David Alexander."
Acadiensis 9 (1981): 104–15.

– "The Port of St. John's, Newfoundland, 1840–1889: A Preliminary
Analysis." In *Ships and Shipbuilding in the North Atlantic Region*, edited by
Keith Matthews and Gerald Painting, 21–39. St. John's 1977.

Sanger, Chesley. "The Evolution of Sealing and the Spread of Permanent
Settlement in Northeastern Newfoundland." In *The Peopling of
Newfoundland: Essays in Historical Geography*, edited by John J. Mannion,
136–51. St. John's: Memorial University of Newfoundland 1977.

– "The 19th Century Newfoundland Seal Fishery and the Influence of
Scottish Whalemen." *Polar Record* 20 (1980): 231–52.

Scammon, C.M. "About Sea Lions." *Overland Monthly* 8 (1872): 266–72.

– "Sea-Elephant Hunting." *Overland Monthly* 6 (1870): 112–7.

– "The Sea Otters." *The American Naturalist* 4 (1870): 65–74.

– "Seal Islands of Alaska." *The Overland Monthly* 5 (1870): 297–301.

Scheffer, Victor B. "Conservation of Marine Mammals." In *Marine Mammals
of Eastern North Pacific and Arctic Waters*, edited by Delphine Haley, 242–
4. Seattle 1978.

– "Conserving the Alaska Fur Seals." *Proceedings of the Pacific Science Congress* 4 (1953): 615–19.
– *The Food of the Alaska Fur Seal.* Wildlife leaflet no. 329. Washington, D.C.: U.S. Dept. of Interior, Fish and Wildlife Service 1950.
– "New Life Burgeons at an Alaskan Fur-Seal Rookery." *Smithsonian* 1 (1970): 50–7.
– "Precarious Status of the Seal and Sea Lion on our Northwest Coast." *Journal of Mammalogy* 9 (1928): 10–16.
– "Probing the Life Secrets of the Alaska Fur Seal." *Pacific Discovery* 3 (1950): 22–30.
– "Sealskins Alive." *Pacific Discovery* 13 (1960): 2–9.
– "The Sea Otter on the Washington Coast." *Pacific Northwest Quarterly* 31 (1940): 371–88.
– "Use of Fur-Seal Carcasses by Natives of the Pribilof Islands, Alaska." *Pacific Northwest Quarterly* 39 (1948): 131–3.
– , and Karl W. Kenyon. "The Fur Seal Herd Comes of Age." *National Geographic* 101 (1952): 491–512.
Schuhmacher, W. Wilfried. "South African Light on American Fur Trade Vessels." *American Neptune* 40 (1980): 46–9.
– "Merchant Captain of the Pacific." *American Neptune* 41 (1981): 224–30.
Schwatka, Frederick. "The Fur-Seal Fishery Dispute." *North American Review* 146 (1888): 390–9.
Scott, Dafila. "Man's Influence on the Auckland Islands." *Animals* 13 (1971): 820–5.
Scott, Robert F., Karl W. Kenyon, J.L. Buckley, and Sigurd T. Olson. "Status and Management of the Polar Bear and Pacific Walrus." *Transactions of the 24th North American Wildlife Conference* (1959): 366–74.
"Sea Otter Hunting." *The Alaska Journal* 1, (1971): 46–8.
"Sea Otters Ambushed." *National Parks & Conservation Magazine* 45 (1971): 32–7.
"Seal and the Treasury Department, The." *Fortune* 66 (1930): 69–72, 122, 124.
"Seal Hunt, The." *Decks Awash* (Extension Service, Memorial University of Newfoundland) 7, no. 1 (February 1978), special issue.
Sergeant, D.E. "Estimating Numbers of Harp Seals," In *Biology of the Seal ... Guelph Conference, 1972,* edited by K. Ronald and A.W. Mansfield, 274–80. Charlottenlund slot, Denmark 1975.
– "Exploitation and Conservation of Harp and Hood Seals." *Polar Record* 12 (1965): 541–51.
– "Harp Seals and the Sealing Industry." *Canadian Audubon* 25 (1963): 29–35.
– "History and Present Status of Populations of Harp and Hooded Seals." *Biological Conservation* (London) 10 (1976): 95–118.
Severinghaus, Nancy C., and Mary K. Nerini. *An Annotated Bibliography on*

Marine Mammals of Alaska. Seattle: NOAA/NMFS, Marine Mammal
Laboratory 1977.

Shallenberger, Robert J. "A Seal Slips Away." *Natural History* 91 (1982):
48–53.

Sherwood, Morgan B. "George Davidson and the Acquisition of Alaska." In
Alaska and Its History, edited by Morgan B. Sherwood, 253–70. Seattle
1967.

Shimer, Steven J. "The Underwater Foraging Habits of the Sea Otter,
'Enhydra lutris.'" *California Fish & Game* 63 (1977): 120–2.

Simenstad, Charles A., James A. Estes, and Karl W. Kenyon. "Aleuts, Sea
Otters, and Alternate Stable-State Communities." *Science* 200 (1978):
403–11.

Sitwell, Nigel. "Sealing in the Antarctic." *Animals* 14 (1972): 156–7.

Sloss, Frank H. "Who Owned the Alaska Commercial Company?" *Pacific
Northwest Quarterly* 68 (1977): 120–30.

– , and Richard A. Pierce. "The Hutchinson Kohl Story: A Fresh Look."
Pacific Northwest Quarterly 62 (1971): 1–6.

Smith, Hugh M. "The Uruguayan Fur-Seal Islands." *Zoologica* 9 (1927):
271–84.

Smith, Thomas G. "Seal Hunters." *Oceans* 14 (1981): 16–17.

Sorensen, J.H. "Elephant Seals of Campbell Island." *Cape Expedition Series*
(New Zealand Sub-Antarctic Expedition, 1941–45), Bulletin no. 6, 1950.

Spaulding, D.J. "Comparative Feeding Habits of the Fur Seal, Sea Lion,
and Harbour Seal on the British Columbia Coast." *Bulletin of the Fishery
Research Board of Canada*, no. 146, 1964.

Stackpole, Edouard A. "The First Recognition of Antarctica." *Boston Public
Library Quarterly* 4 (1952): 3–19.

Starks, Edwin C. "A History of California Shore Whaling." *California Fish
& Game Commission, Fish Bulletin* no. 6, 1922.

– "Records of the Capture of Fur Seals on Land in California." *California
Fish & Game* 8 (1922): 155–60.

– "The Sea Lions of California." *American Museum Journal* 18 (1918):
226–37.

Stejneger, Leonhard. "Fur-seal Industry of the Commander Islands, 1897–
1922." *Bulletin of the U.S. Bureau of Fisheries* 41 (1925): 289–332.

Stephenson, Mark D. "Sea Otter Predation on Pismo Clams in Monterey
Bay." *California Fish & Game* 63 (1977): 117–20.

Stevens, Thomas A. "The Discovery of Antarctica." *The Log of Mystic
Seaport* 28 (1977): 106–14.

Stevenson, Charles H. "Aquatic Products in Arts and Industries: Fish Oils,
Fats, and Waxes." In *Report of Commissioner of Fish & Fisheries for Year
Ending 30 June 1902* 28 (1904): 177–279.

– "Leather from Seal Skins." *Scientific American* 59 (1905): 24334–5.

– "Oil from Seals, Walrus, etc.": *Scientific American* 57 (1904): 23614–5.

– "Utilization of the Skins of Aquatic Animals." *Report of Commissioner of Fish & Fisheries for Year Ending June 30, 1902* 28 (1904): 283–352.

Sturgis, William. "The Northwest Fur Trade." *Hunt's Merchants' Magazine and Commercial Review* 14 (1846): 532–9.

Suttles, Wayne. "Notes on Coast Salish Sea-Mammal Hunting." *Anthropology of British Columbia* 3 (1952): 10–20.

Taggart, Harold F. "Sealing on St. George Island, 1868." *Pacific Historical Review* 28 (1959): 351–60.

Talbot, Lee M. "Maximum Sustainable Yield: An Obsolete Management Concept." *Transactions of the 40th North American Wildlife and Natural Resources Conference* (1975): 91–6.

Tansill, C.C. "The Fur Seal Fisheries and the Doctrine of the Freedom of the Seas," Canadian Historical Association *Annual Report* (1942), 71–81.

Taylor, F.H.C., M. Fujinaga, and Ford Wilke. *Distribution and Food Habits of the Fur Seals of the North Pacific Ocean.* Washington, D.C.: U.S. Dept. of Interior, Fish and Wildlife Service 1955.

Thompson, Seton H. "Seal Fisheries." In *Marine Products of Commerce*, edited by D.K. Tressler and J. McW. Lemon, 816–32. New York 1951.

Thornton, Patricia. "Some Preliminary Comments on the Extent and Consequences of Out-Migration from the Atlantic Region, 1870–1920." In *Merchant Shipping and Economic Development in Atlantic Canada*, edited by Lewis R. Fischer and Eric W. Sager, 185–218. St. John's 1982.

Townsend, Charles Haskins. "The Fur Seal of the California Islands." *Zoologica* 9 (1931): 443–50.

– "The Fur Seal of the Galapagos Islands." *Zoologica* 18 (1934): 43–56.

– "The Northern Elephant Seal." *Zoologica* 1 (1912): 159–73.

– "The Northern Elephant Seal and the Guadalupe Fur Seal." *Natural History* 24 (1924): 567–78.

Tucker, Albert H. "Walrus Hunting in the Arctic." *Pacific Motor Boat* 17 (1925).

Tupper, Charles Hibbert. "Crocodile Tears and Fur Seals." *National Review* 28 (1896): 87–93.

Van Aarde, R.J. "Fluctuations in the Population of Southern Elephant Seals 'Mirounga leonina' at Kerguelen Island." *South African Journal of Zoology* 15 (1980): 99–106.

– "Harem Structure of the Southern Elephant Seal 'Mirounga leonina' at Kerguelen Island." *Revue Ecologique (Terre Vie)* 34 (1980): 31–44.

–, and M. Pacal. "Marking Southern Elephant Seals on Isles Kerguelen." *Polar Record* 20 (1980): 62–5.

Van Nostrand, Jeanne. "The Seals are about Gone ..." *American Heritage* 14 (1963): 10–18, 78–80.

Vaz-Verreira, R. "Factors Affecting Numbers of Sea Lions and Fur Seals on the Uruguayan Islands." In *Biology of the Seal ... Guelph Conference, 1972*, edited by K. Ronald and A.W. Mansfield, 257–62. Charlottenlund slot, Denmark 1975.

Volwiler, Albert T. "Harrison, Blaine, and American Foreign Policy, 1889–1893." *Proceedings of the American Philosophical Society* 79 (1938): 637–48.

Wadel, Cato. *Now, Whose Fault Is That? The Struggle for Self-Esteem in the Face of Chronic Unemployment*. St. John's: Memorial University of Newfoundland, Institute of Social and Economic Research 1973.

Waite, Edgar R. "Vertebrata of the Subantarctic Islands of New Zealand." In *The Subantarctic Islands of New Zealand*, edited by Charles Chilton, 542–600. Wellington 1909.

Walker, Philip L., and Steven Craig. "Archeological Evidence Concerning the Prehistoric Occurence of Sea Mammals at Point Bennett, San Miguel Island." *California Fish & Game* 65 (1979): 50–5.

Wartzok, Douglas, and G. Carleton Ray. *The Hauling-Out Behavior of the Pacific Walrus*. Washington, D.C.: U.S. Marine Mammal Commission 1980.

Wheeler, Mary E. "Empires in Conflict and Cooperation: The 'Bostonians' and the Russian-American Company." *Pacific Historical Review* 40 (1972): 419–41.

Whittow, G. Cause. "Tropical Seals." *Sea Frontiers* 17 (1971): 285–7.

"Wigs and Clapmatches." *Historical Footnotes: Bulletin of the Stonington Historical Society* 6 (1969): 9–10.

Wilke, Ford. *Pelagic Fur Seal Research Off Japan in 1950*. Wildlife leaflet 338. Washington, D.C.: U.S. Dept. of Interior, Fish and Wildlife Service 1952.

– , and Karl W. Kenyon. "Migration and Food of the Northern Fur Seal." *Transactions of the 19th North American Wildlife Conference* (1954): 430–40.

Willett, G. "Elephant Seal in Southeastern Alaska." *Journal of Mammalogy* 24 (1943): 500.

Williams, William. "Reminiscences of the Bering Sea Arbitration." *American Journal of International Law* 37 (1943): 562–84.

Williamson, Tony. "Inuit, the Innocent Victims." *Decks Awash* (Extension Service, Memorial University of Newfoundland) 7, no. 1 (1978): 60–1.

Winter, Sir M.G. "Trip to the Seal Fishery, March 1907." *Newfoundland Quarterly* 75 (1979): 21–7.

Woodhouse, Charles D., Jr, Robert K. Cowan, and Larry R. Wilcoxon. *A Summary of Knowledge of the Sea Otter "Enhydra lutris, L." in California and an Appraisal of the Completeness of Biological Understanding of the Species*. Washington: U.S. Marine Mammal Commission 1977.

Woodward, Arthur. "Isaac Sparks – Sea Otter Hunter." *Historical Society of Southern California Quarterly* 20 (1938): 43–59.

– "Sea Otter Hunting on the Pacific Coast." *Historical Society of Southern California Quarterly* 20 (1938): 119–34.

PHOTOGRAPHIC CREDITS

The photographs appear by permission of the following:
California Fish and Game Department Library (G.R. Chute photos) 48, 49
Coast Guard Museum, N.W. 46
J.S. Cumpston, *Macquarie Island* 38, 39, 40, 41, 42
Maritime Museum of British Columbia 30
National Maritime Museum, San Francisco 26, 31, 44, 45
Newfoundland Provincial Archives 5, 6, 8, 10, 11, 12, 13, 14, 16, 17, 18, 19, 20, 21
Newfoundland Provincial Reference Service 15, 22, 23
Oregon Historical Society 28
Provincial Archives of British Columbia 29, 32
Public Archives of Canada 1, 2, 3, 4, 7, 9, 24, 25, 27
Seattle Historical Society (F.H. Nowell photo) 37
University of Washington, Seattle 34, 35, 36, 43, 47
Vancouver Maritime Museum 33

Index

Names of vessels are listed under "Vessel" according to the era and area in which they operated

Index